Hormones, Signals and Target Cells in Plant Development

The term target cell, first conceived in animal biology, is generally taken to mean a cell that has a pre-determined competence to respond in a defined way to a specific hormone signal. In this volume, the authors present the theory that every plant cell is a target cell for one or more of the plant hormones or other regulatory signals. The different types of cells seen in a mature plant arise as a result of selective responses of meristematic cells to chemical inputs. In this context, the authors provide an overview of plant signals as well as evidence for both short- and long-distance cell-to-cell communication. An interpretation of the target cell concept at the biochemical and molecular levels is then presented using a wide range of examples. This volume will serve as a valuable reference for those working in the field of plant developmental biology.

Daphne J. Osborne is a Visiting and Honorary Professor with the Oxford Research Unit of the Open University, Oxford, United Kingdom, where her research focuses on the hormonal control of physiological and biochemical processes in plant differentiation and development.

Michael T. McManus is Associate Professor in Plant Biology at the Institute of Molecular BioSciences, Massey University, Palmerston North, New Zealand. His research is concerned with the control of biochemical pathways in plants, including the biosynthesis of hormones.

T0297310

Developmental and Cell Biology Series
SERIES EDITORS
Jonathan B. L. Bard, *Department of Anatomy, Edinburgh University*
Peter W. Barlow, *School of Biological Sciences, University of Bristol*
David L. Kirk, *Department of Biology, Washington University*

The aim of the series is to present relatively short critical accounts of areas of developmental and cell biology, where sufficient information has accumulated to allow a considered distillation of the subject. The fine structure of cells, embryology, morphology, physiology, genetics, biochemistry, and biophysics are subjects within the scope of the series. The books are intended to interest and instruct advanced undergraduates and graduate students, and to make an important contribution to teaching cell and developmental biology. At the same time, they should be of value to biologists who, while not working directly in the area of a particular volume's subject matter, wish to keep abreast of developments relevant to their particular interests.

RECENT BOOKS IN THE SERIES

Hormones, Signals and Target Cells in Plant Development

DAPHNE J. OSBORNE
Open University, Oxford, United Kingdom

MICHAEL T. McMANUS
Massey University, Palmerston North, New Zealand

CAMBRIDGE
UNIVERSITY PRESS

CAMBRIDGE UNIVERSITY PRESS
Cambridge, New York, Melbourne, Madrid, Cape Town,
Singapore, São Paulo, Delhi, Tokyo, Mexico City

Cambridge University Press
32 Avenue of the Americas, New York, NY 10013-2473, USA

www.cambridge.org
Information on this title: www.cambridge.org/9780521177450

First published 2005
First paperback edition 2011

A catalog record for this publication is available from the British Library

Library of Congress Cataloging in Publication data
Osborne, Daphne J.
Hormones, signals and target cells in plant development / Daphne J. Osborne,
Michael T. McManus.
    p.   cm. – (Developmental and cell biology series ; 41)
Includes bibliographical references and index.
ISBN 0-521-33076-9 (hardback)
1. Plant hormones.   2. Plant cellular signal transduction.   3. Plants – Development.
I. McManus, Michael T.   II. Title.   III. Series.
QK898.H67O83   2004
571.7´42 – dc22   2004019647

ISBN 978-0-521-33076-3 Hardback
ISBN 978-0-521-17745-0 Paperback

Contents

viii CONTENTS

Preface

This volume presents a conceptual approach to plant cell differentiation that differs in a number of respects from those already present in the literature. We seek to show how every cell has an individual competence to respond to the signal inputs that may impinge upon it and how every cell then has an individual qualitative and quantitative response. Central to this target cell concept is the premise that each cell is selective and can therefore discriminate amongst the many incoming signals to which it is exposed by an ability to perceive them and to respond to them.

Because each cell occupies its individual position within the plant body, the intensity or diversity of the signal inputs that it receives are not themselves identical. Hence, each cell is a unique individual and displays a unique target status even though it may also possess considerable commonality with its neighbours. We define this target status of a cell as the selectivity of its response to a signal and the intensity of that response.

The target cell concept arose originally from notions that were current amongst insect and mammalian scientists stating that a regulatory chemical produced in one organ would be perceived and activated upon by the cells in a distant organ – a specificity that operated between two distinct cell types. As the evidence for specificity of response to hormonal inputs increased during the twentieth century, developmental biologists saw this ability of cells to discriminate amongst the multitude of chemical signals to which they were exposed as a marker of the cell's ability to discriminate between them. For a passing hormone, the cell that responded was a target cell.

For the purposes of this volume, we will consider those signals that are endogenously produced or transmuted to chemical signals within the plant. We define such signals as the agents of cell-to-cell communication. This does not imply, however, that we consider electrical signals, gravity signals or differential light inputs to be unimportant, but the scope of this book focuses on the molecular communicators that can be isolated from plants as chemical entities with growth regulatory properties. It is these signal molecules that provide the messages that coordinate the processes of differentiation. We shall not, therefore, consider in any detail the formative influence of all the many external cues that a plant encounters directly from the environment.

Our aim is to present the reader with our interpretation of how the unique target status of each cell is expressed at the biochemical and molecular levels and how this forms the basis for specificity in signal-directed responses. Some of the examples that we discuss here are selected from research studies fitting most easily into the target cell concept. Other examples are those that we have re-interpreted in this light. The target cell concept, as we present it here, has arisen from the many years of our own research. However, it is also our purpose to stimulate debate on the validity of this concept when interpreting other studies of plant cell differentiation and development.

In terms of the structure of the volume, we begin by defining the concept of target cells (Chapter 1), then introduce the repertoire of signals that operate in plants (Chapter 2), and present evidence for both short- and long-distance cell-to-cell signalling (Chapter 3). The concept of the higher plant body in which the majority of cells retain a flexible differentiation status, while some functionally specialised cell types attain a state of terminal commitment, is introduced in Chapter 4. We submit though, that every cell, irrespective of its differentiation status, is a target cell and so we examine the target status of both flexible cell types and terminally committed cell types in Chapters 5 and 6 respectively. Finally, our understanding of the molecular mechanism of hormone action has the potential to be advanced by the identification of receptors and proteinaceous regulators of these signals. In Chapters 7 and 8, we review this literature and speculate on its current and future impact on the target cell concept. In Chapter 9 we consider implications of signal cross-talk.

We hope that this volume will become a useful reference to those working in the field of plant developmental biology.

We wish to acknowledge the assistance of Mrs. Vivian Reynolds, Mrs. Cynthia Cresswell, Ms. Rae Gendall, and Glenda Shaw during the preparation of this volume.

1

Introduction

Although it was evident from Darwin's studies of tropisms in plants that informational signals passed from one part of the plant to another, the proof that it was a chemical substance that passed awaited the famous *Avena* coleoptile experiments of Frits Went (1928). These showed that a molecule (later identified as indole-3-acetic acid [IAA]) was the active agent that was water soluble and would pass across an agar barrier placed between one tissue and another – in his earliest experiments this was between the coleoptile tip (producing IAA) and the IAA-regulated elongating region of the coleoptile below. A tremendous amount of work, both in studying the physiology of this response to IAA, and in identifying the many analogues to IAA, sought the molecular structures required to provide an active molecule. It was from this highly intensive period of plant physiology study that the agricultural revolution of herbicides, defoliants and growth regulators of the 1940s and 1950s was originally generated.

But it was the insect physiologists with their identities of hormone-producing glands and hormone-responding tissues remote from the glands who developed the concepts of target tissues, signalling molecules and receptor sites. Perhaps the most spectacular to record, as an example of the approaches followed later by plant scientists, is the work in the 1930s and 1940s concerning the processes of moulting of larval epidermal skins and of metamorphoses to the adult state (Karlson, 1956).

The prothoracic gland produces the steroid hormone ecdysone (the moulting or juvenile hormone) that is transported in the haemolymph to the insect epidermis. Along the way, ecdysone binds to specific proteins and then at the site of moult, induces enhanced transcription and modification of coordinated gene activity. Importantly, ecdysone-induced puffing at specific sites in polytene chromosomes could be demonstrated in salivary glands and related to each developmental stage of differentiation. The insect field was set for determining the

1

cascade events of signals, receptors and the hormonal regulation of gene expression in specific target cells. The plant field followed fast; but in the absence of specific glands or polytene chromosomes and the absence of distinguishably specific target cells the now recognised parallels between plant and animal signalling were more difficult to explore and took longer to resolve.

The concept of target cells in plants arose originally from the knowledge that cells of coleoptiles and etiolated shoots would enlarge and extend in response to auxin and that ethylene would arrest such elongation growth. Those of us who worked in the field of abscission became equally aware that the cells that made up an abscission zone enlarged prior to their separation and did so with alacrity if exposed to ethylene, but if given auxin these same cells would neither grow nor separate. The insect developmental biologists were aware that only certain cells would perceive and respond to particular hormonal signals in a particular way. It became inevitable to learn that although most plant cells might look the same, they were all as individual and distinct from one another as those making up the highly responsive organs of mammals and insects. As the cuticular cells of a larva were targets for the moulting hormone ecdysone, and the mammalian liver for insulin, so the cortical cells of the plant shoot were targets for auxin and ethylene. The concept for target cells in the developmental biology of plants was born.

A first substantive evidence for different target types in cortical cells arose from observations of the differences in their growth responses to auxin and ethylene. Whereas the immature cells of young dicotyledonous shoots such as those of the pea *Pisum sativum* will elongate in the presence of the auxin passing downwards from the meristem, the addition of ethylene to those shoots will cause them to arrest elongation growth and instead to expand laterally, with the cell volume remaining essentially unchanged and determined by the availability of auxin from the meristem (Osborne, 1976). Cells of abscission zones, however, such as those in the leaves of the bean *Phaseolus vulgaris,* behave in a quite different way: their expansion is enhanced by ethylene but not by auxin. The existence of a third type of cortical cell is found in the stems and petioles of many species of flooded or aquatic habitats. *Ranunculas sceleratus* or the water fern *Regnellidium diphyllum,* for example, possesses cortical cells that will expand and extend with either auxin or ethylene (Figure 1.1). In 1976, these three distinguishable cell types were designated as Type 1, Type 2 and Type 3 (Osborne, 1976, 1977a, b) with respect to their responses to auxin and ethylene.

Once we understood that cells that looked similar to the eye had quite specific responses to hormonal signals, it became evident that strict regulatory controls operated to maintain these coordinated patterns in the cell society. Clearly, not every cell differentiated along the same developmental pathway or had the competence to respond in a similar way to the same hormonal signals. More examples of target cells other than those with a highly specific perception and response to auxin and ethylene signals have now engaged the physiologists and molecular biologists, none less so than the terminally differentiated cells of the aleurone tissue in graminaceous seeds. In aleurone cells, the competence to respond to a gibberellin/abscisic acid control determined their final response and cell fate.

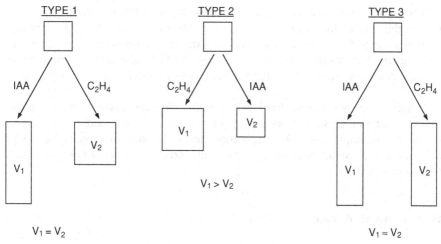

Figure 1.1. Overview of Type 1, Type 2 and Type 3 target cells in terms of their responses *in planta* to auxin and ethylene. V = volume expansion

The terms target cell, signal, cellular competence and tissue sensitivity are now used widely when describing the development of eukaryotes, but differences have emerged in terms of the definitive definitions of their meaning. Thus before embarking on our dissection and discussion of the target cell concept in higher plants, it is pertinent to begin with a series of definitions that are used in this volume.

What is a target cell?

The term target cell was first conceived in animal biology and is generally taken to mean a cell that has a pre-determined competence to respond in a defined way to a specific hormonal signal. Only the specific hormone (or a chemically related analogue) can evoke that particular response in the cell. In plant biology, the definition is essentially the same. For the purposes of this book, we take the view that every plant cell is a target cell for one or more of the plant hormones or other regulatory signals.

Cells, simple and complex tissues

The body of higher plants arises from the mitotic activity of apical meristematic regions, and thence through cell elongation and differentiation the primary plant body is formed. In the gymnosperms and dicotyledonous angiosperms, and certain monocotyledonous species, secondary growth occurs through additional cells arising from meristems remote from the apex, principally those of the vascular and cork cambium tissues.

The higher plant body is considered commonly to comprise three systems of tissues: the dermal, the vascular and the ground tissues (Esau, 1965). The primary dermal tissue is the epidermis, with periderm tissue forming in secondary growth. The phloem and xylem are the principal vascular tissues, and the ground tissue systems comprise all the remaining tissues, with parenchyma the primary cell type.

Within these tissues, a further layer of complexity emerges with the organisation of cell types. In some tissues, only a single cell type is found and these are referred to as simple tissues. In others, for example the stele, many cell types comprise the tissue, including those that are dead, and these are commonly referred to as complex tissues.

How are signals defined?

Within the plant body, cells must, perforce, communicate by chemical or physical means. These intertissue messages are the chemical signals that evoke specific biochemical and molecular events in each responsive target cell. From the first of these signalling molecules to be discovered, namely auxin in 1928, a wide spectrum of such molecules has now been identified and while the chemistry of these compounds is quite diverse they all exhibit certain shared characteristics. Although there are major sites of synthesis within the plant, signals all move readily between cells (either long or short distances) and evoke responses both at the site of synthesis or at sites that are remote. Classically, for a cell to recognise a signal, it must have a receptor for that signal. All cells must, therefore, possess an array of receptor systems, but so far comparatively few of these have been precisely characterised. However, the identification and characterisation of these receptors is currently a very active research area in plant biology and as such represents areas of high relevance to the target cell concept. Accordingly, full chapters are devoted to the identification and characterisation of receptors and their significance in the mechanisms for the relief of gene repression.

The examples of intertissue signal molecules considered in this volume include both old and new. The five major classes of plant hormones (auxin, ethylene, abscisic acid, gibberellin, cytokinins) are well established, but many more endogenously produced molecules with equally diverse structures are now known to induce a range of specific responses in specific tissues. These include

- steroid-like molecules, e.g., the brassinolides,
- low molecular weight compounds such as jasmonic acid and salicylic acid,
- oligosaccharins, including free N-glycans, and
- peptides, such as systemin.

It is not our intention in this volume to devote space to considering what constitutes a plant hormone and whether, for example, brassinolide should be included in that group. Rather, our focus is to consider the role of such molecules as signals that convey information to cells in tissues and organs – that is, to the cells that are their recipient targets.

Intracellular signal transduction

Although the question of the mechanisms by which a cell may perceive an external intertissue signal from which it can then interpret a directive in biochemical terms is still open to debate, it is not the central focus of this volume. However, the considerable progress in defining how plant hormone and other signal receptors are evaluated, particularly with respect to the degree to which such characterisation further defines the target cell concept, means that discussion is inevitably included. Signal transduction in the cytosol and nuclei will not be described here in detail, but the authors are aware that the more recent dissection of such biochemical events is of direct relevance to the target cell response. The emerging ideas of hormonal cross-talk are constantly reiterated as each new signal transduction pathway is elucidated. The aim of this volume, therefore, is to interpret such descriptions within the framework of the target cell concept.

Cellular competence

We define cellular competence in both qualitative and quantitative terms. A cell must possess the ability to perceive, transduce and respond to a signal, otherwise it is not a target for that particular input. In plants, unlike animals, all the living cells are exposed to the same hormones, though the signal transduction pathway is not necessarily the same for each target cell. Whereas a positive response to the signal is evidence of a cell's specific target state, lack of response cannot be taken as evidence of a non-target condition. For in every signal input there must be a threshold level that a cell can interpret; this must exceed the presence of non-specific inhibitors and homeostatic controls exerted by specific repressor controls of the target cell response (see Chapter 5). Furthermore, a cell may perceive, transduce and respond to a particular signal in a different way from its neighbouring cell. This we take as clear evidence of their individual and different target states.

Cell and tissue 'sensitivity' to signal inputs

For the purposes of this volume, we define cell and tissue sensitivity in terms of the concentration of inputs necessary to evoke a measurable response. Thus a highly sensitive target tissue requires a low concentration of signal input to evoke a predetermined response – i.e., display its tissue competence. It should be noted that any one cell may be a target for more than one signal; therefore the sensitivity of the tissue for each signal may vary, and must be defined in terms of a specific signal. The identification of definitive receptors for some of the plant hormones – for example, the ETR-like family of proteins as ethylene receptors – has afforded the possibility of quantifying sensitivity in terms of receptor abundance. Nevertheless, researchers who have attempted such exercises are still bound to equate such values with the extent of the physiological response (Klee, 2002).

2

Hormones and Signals: Identification and Description of Signalling Molecules

As a first step in developing the target cell concept for plants it is important that the major players in the known informational and signal repertoire are set out briefly at the start of the discussions. There are five major players: auxin, gibberellins, cytokinins, ethylene (and its precursor 1-aminocyclopropane-1-carboxylic acid, ACC), and abscisic acid. The first hormone to be discovered and isolated, auxin, is the best understood, the most important and without doubt the most remarkable. As well, the more recent signal molecules to be discovered are described in greater detail for some of them link more closely to molecules in the animal kingdom.

Auxin

Indole-3-acetic acid (IAA) is the most abundant naturally occurring auxin, with indole-3-butyric acid (IBA) and 4-chloroindole-3-acetic acid (4-Cl-IAA) also occurring naturally (Figure 2.1). IAA was discovered in 1928 by Frits Went (Went, 1928) in the search for the chemical substance that was transported from the apex of the oat coleoptile and caused the cells below to elongate. In higher plants, several pathways of synthesis are possible. IAA is an indole derivative, and both *in vivo* and *in vitro* evidence indicates routes of synthesis from the aromatic amino acid, tryptophan, although more recent genetic and biochemical experiments have suggested that tryptophan-independent pathways may also operate to yield the final product (Bartel, 1997).

Auxin biosynthesis

Tryptophan was proposed originally as the precursor of IAA due to structural similarities of the two molecules and when a clearly defined conversion was identified

6

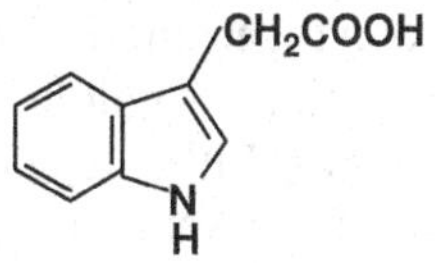

Figure 2.1. Structures of naturally-occurring auxins: indole-3-acetic acid (A), indole-3-butyric acid (B), and 4-chloroindole-3-acetic acid (C).

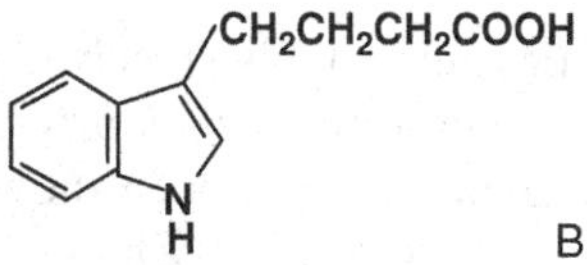

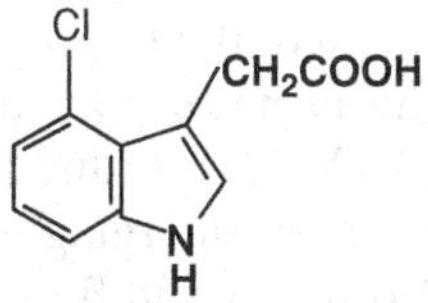

in plant-associated microbes. Further, labelling studies *in vivo* in many plant species have shown that tryptophan can be metabolised to IAA (Normanly et al., 1995). However, the efficiency of this conversion to IAA was either not tested or shown to be very low, a concern when set against the background observation that tryptophan is readily converted to IAA non-enzymatically.

To resolve the role of tryptophan as the precursor of IAA, Wright et al. (1991) used the *orange pericarp* mutant of maize, a tryptophan auxotroph. This mutant arises from lesions in two unlinked loci of the tryptophan synthase B gene, and the total IAA produced in aseptically grown plants was found to be 50-fold greater than in normal maize seedlings. When aseptically grown mutant plants were labelled with [15N]-anthranilate, a tryptophan precursor, IAA was more enriched than tryptophan, leading the authors to conclude that IAA can be produced *de novo* without tryptophan as an intermediate. Further studies with other mutants have localised indole, another intermediate on the tryptophan pathway from anthranilate, as an IAA precursor, although some workers still challenge the validity of the tryptophan-dependent pathway (Muller and Weiler, 2000).

Nonetheless, there is increasing evidence to support tryptophan as the IAA precursor, although no pathway arising from tryptophan has been definitively established. Currently, three routes are now considered: the indole-3-pyruvate pathway, the tryptamine pathway and the indole-3-acetonitrile (IAN) pathway, although the IAN pathway appears to be restricted primarily to the Brassicaceae (Normanly et al., 1995). The conversion of tryptophan to indole-3-acetamide and then to IAA is most likely attributable to plant-associated microbes.

The recent characterisation of two cytochrome P450s, designated CYP79B2 and CYP79B3, that catalyse the formation of indole-3-acetaldoxime has created further interest in the IAN pathway (Hull et al., 2000; Zhao et al., 2002b). Indole-3-acetaldoxime can be converted to IAN and then to IAA, or to indoleglucosinilates creating a metabolic branch point. Nevertheless, the identification of a nitrilase gene that can convert IAN to IAA (Bartling et al., 1992) does support this route of IAA biosynthesis, although it appears to be restricted to only certain plant families.

The tryptamine biosynthetic route has also received support recently with the identification of the *YUCCA* gene from *Arabidopsis*, the product of which is a flavin monooxygenase-like enzyme that has been shown to catalyse the conversion of tryptamine to *N*-hydroxyl-tryptamine (Zhao et al., 2001).

Such studies suggest that more than one pathway for IAA formation exists in plants, and further that more than one pathway may operate in the same tissues. An emerging consensus appears to be that tissues that produce transient, high levels of IAA utilise a version of the tryptophan-dependent pathway, while a tryptophan-independent pathway may supply the lower levels of IAA required for the maintenance of growth (Normanly and Bartel, 1999; Sztein et al., 2002). For readable accounts of the issues emerging in auxin biosynthesis, the reader is referred to the reviews of Normanly and Bartel (1999) and Bartel et al. (2001).

Auxin conjugation

Of perhaps further significance is the extent of the conjugation of IAA, and its role in the regulation of IAA homeostasis in cells and tissues. The auxin-like biological activity of IAA-conjugates has been known for many years and these compounds have been proposed as slow release forms of free IAA (Hangarter and Good, 1981) to support the 'IAA homeostatic model' (Cohen and Bandurski, 1982) (Figure 2.2). The regulation of such slow release forms in any particular cell indicates how significant the enzymes and their genes can be in the developmental context of target cells. It is now widely appreciated that the formation of IAA-aspartate represents an irreversible conversion of IAA that marks the conjugate for eventual degradation (Monteiro et al., 1988). However, enzyme activities that can hydrolyse IAA conjugates back to free IAA occur in a number of species (Ludwig-Muller et al., 1996) and the cloning of genes coding for these hydrolases have shown that their expression is tightly developmentally regulated (Bartel and Fink, 1995; Davies et al., 1999; LeClere et al., 2002) demonstrating the tissue-specific location of IAA conjugates that have been identified in plants (Kowalczyk and Sandberg, 2001). For the target cell concept, the developmental regulation of free IAA release from conjugates provides a myriad of control points by which a competent auxin-responding cell can perceive differences in the levels of the hormone released.

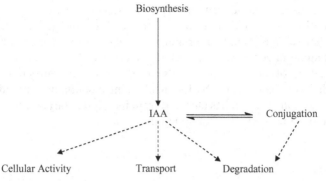

Figure 2.2. Diagrammatic representation of the 'IAA homeostatic model' in which the many regulators of IAA concentration in plant cells are indicated.

Sites of auxin biosynthesis and transport

In a careful study of the sites of auxin biosynthesis in *Arabidopsis*, the youngest leaves were found to contain the highest endogenous levels of IAA and also the highest capacity to synthesise the hormone (Ljung et al., 2001). A correlation of high IAA content and high rates of cell division has also been shown in developing tobacco leaves.

The importance of auxin in both cell-to-cell and long-distance signalling cannot be overestimated and the role of young shoots as sites of highest auxin concentration, cell division and highest rates of synthesis is critical. Because auxin can be stored in cells and tissues in biologically inactive forms such as amino acid conjugates and sugar esters, and then released again as the active auxin molecule (as in the gravity-stimulated nodes of grasses, or in germinating seeds), many cells have the potential to become sites of free IAA formation. In meristem parenchyma and the living non-vascular tissues of cortex or pith, auxin movement is from cell to cell, by a metabolic system unique to this molecule and its synthetic analogues. Remarkable amongst hormonal substances, auxin molecules are transported from the apical to basal end of each cell with respect to the shoot apical meristem; this polarity is continuous throughout the plant, progressing to the root apex where auxin efflux has been recorded. In tissue with a vascular supply, the transport of auxin is additionally served by the connection to both xylem and phloem. In immature tissue, below an apical meristem, the basipetal polarity of auxin movement is high; but as cells reach maximum size and mature, this differential between downward (polar) and upward (diffusion) movement becomes barely detectable as determined by auxin transport assays (Goldsmith, 1977). The major transport pathways then reside in the vascular tissues, which are predominantly long distance. Both long- and short-distance transport pathways therefore operate and coordinate an auxin-signalling mechanism between all parts of the plant and throughout the plant's developmental lifetime (see Chapter 3 for a description of auxin transport).

Whereas auxin is a major hormone in promoting cell expansion and elongation growth, additions of auxin do not cause mature cells to enlarge further. This does not mean that auxin is no longer a regulator of cell performance, but rather that the role of auxin in gene expression changes as the cell grows older. Instead of regulating events of cell size, auxin levels then determine the entry into flowering, fruit ripening, senescence and abscission, all being essentially terminal stages of cell differentiation. The cell has changed from its original target status, although the signal molecule, auxin, has remained the same.

Gibberellins

This group of hormones was isolated and characterised following a study of inter-species signals. A fungal disease of rice, common in Japan, was the cause of ab-normally high growth of the plant associated with yellowing leaves and wilting. In 1926, Eishii Kurosawa added a culture filtrate of the fungus to normal rice plants and demonstrated that a substance present in this filtrate led to the enhanced cell elongation that resulted in extra tall plants (Kurosawa, 1926). This, and the even later isolation of GA_A and GA_B by Yabata and Sumiki (1938), was published in Japanese and remained untranslated until after World War II. The knowledge that auxin analogues could be used as selective herbicides led to a crescendo of research in the West to discover more natural growth-regulating compounds in plants. The Japanese literature then revealed the potential of the family of gibberellin-related substances, and their presence in higher plants as well as in fungi. Although present in all plants, gibberellins are unlike auxin in not being transported in a polar way, nor are they involved in phototropic or gravity-induced curvatures of plant parts. Of special significance is their ability to induce flowering in long-day plants held under non-inductive short-day conditions (see Zeevaart, 1976). This signal to flower is not, however, necessarily applicable to short-day plants held under non-inductive daylengths, so gibberellins cannot properly be considered as flowering hormones. Also, gibberellins can direct the apices of plants with unisexual flowers (for example, Cucurbits) to the production of pre-dominantly stamenate expression, while auxin treatment favours the formation of female flowers. Here then, was the second major discovery of a family of signalling molecules. But whereas auxin seemed essential and universal to the growth of plants, as Frits Went wrote "Ohne Wuchsstoff, kein Wachstum" (Went, 1928), gibberellins showed much greater specificity – the cell growth enhancement response, for example, was greatest in genetic dwarfs and almost absent in the nor-mal wild-type (Phinney, 1956). For gibberellins, therefore, the target status of the cell was recognised early on as genetically as well as developmentally determined.

Gibberellin biosynthesis

Structurally, gibberellins are all sesquiterpenes, and they exist in plant tis-sues as the C_{20}- or C_{19}-GAs (Figure 2.3). These structures are derived from

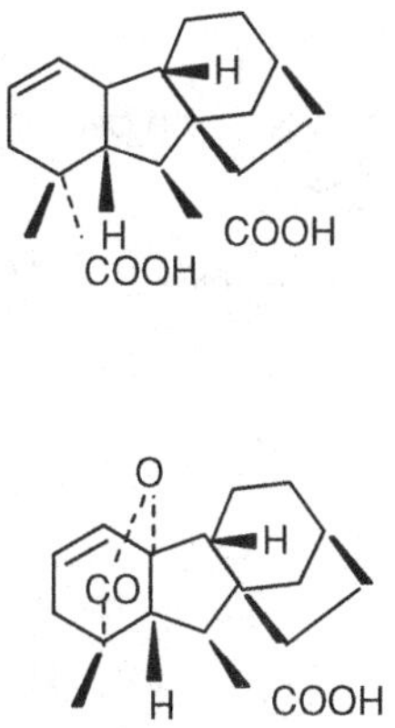

Figure 2.3. Core structures of the C_{20}(A) and C_{19}(B) gibberellins that occur in plant cells.

geranylgeranyl diphosphate via *ent*-kaurene, the latter synthesis occurring in developing chloroplasts or leucoplasts (Figure 2.4A). The pathway of the conversion of *ent*-kaurene to GA_{12}-aldehyde is common to all higher plants studied thus far, after which at least two parallel pathways operate to produce the bioactive gibberellins, GA_1, GA_3, GA_4 and GA_7 (Figure 2.4B), and modifications thereof that have been found to be species and family specific. The total number of structures identified so far in plants, fungi and bacteria is 126, and comprises biosynthetic precursors, the bioactive molecules themselves, and their catabolites. For an in-depth review of the biosynthesis and chemistry of the gibberellins and the central role of GA_{20}-oxidase in their interconversion, see the reviews of Sponsel (1995) and MacMillan (1997).

In common with IAA, gibberellins exist in most floral and vegetative tissues at very low levels (0.1–100.0 ng g^{-1} FW), suggesting that the biosynthetic enzymes are also in low abundance. However, the genomic revolution coupled with increasingly sophisticated techniques to identify and clone genes has led to the identification of the functions of many genes in the GA biosynthetic and catabolic pathways. While these are relatively recent discoveries, it is already clear that these enzymes are coded by multigene families with complex patterns of differential expression in plant tissues. Further description of the gene codings for the biosynthetic enzymes and the many environmental inputs that can influence the regulation of their expression is covered by Hedden and Phillips (2000) and Olszewski et al. (2002).

Cytokinins

The discovery of this group of compounds came from a number of lines of research but the most significant in the context of this book was the absolute need, established by Van Overbeek and co-workers in 1941, for a signal substance released from the endosperm that was essential for maintaining the life of a growing embryo. He showed that young embryos of *Datura* seeds removed prematurely

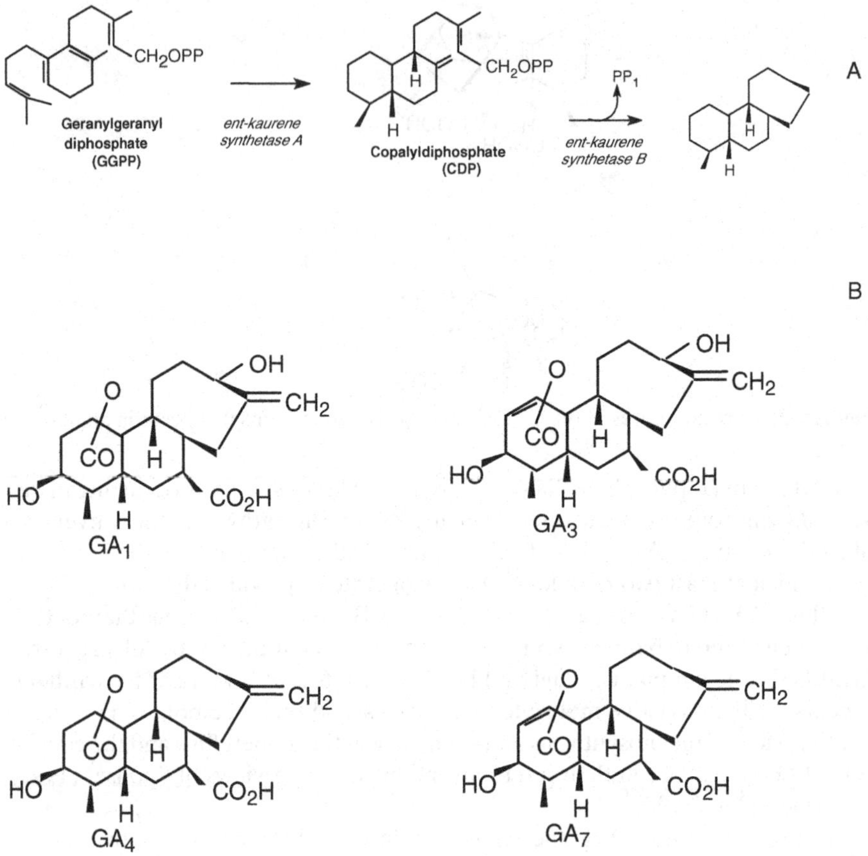

Geranylgeranyl diphosphate (GGPP)

ent-kaurene synthetase A

Copalyldiphosphate (CDP)

ent-kaurene synthetase B

A

B

GA$_1$

GA$_3$

GA$_4$

GA$_7$

Figure 2.4. A. Biosynthesis of *ent*-kaurene from geranylgeranyl diphosphate. B. Structures of four gibberellic acids with biological activity in plant cells.

from the maternal fruit would not survive and grow on a culture medium of nutrients and salts. If supplied with extracts of *Datura* seed tissues, however, the embryos continued cell division and achieved successful maturation. Some signal substance from the endosperm was essential for continued survival and growth. To obtain a large supply (200 mL, at least), the liquid endosperm (the 'milk') of coconuts was tested and found to be an equally successful substitute for the *Datura's* own endosperm signal. The eventual isolation of a synthetic compound (kinetin) from DNA and characterisation of this essential informational molecule for the maintenance of cell division took many years with the successes of Shantz and Steward (1952) and Skoog and Miller (1957). Finally, a natural product, zeatin, was isolated from maize by Letham (1963) in New Zealand (Figure 2.5). Almost all living cell types subsequently tested were induced to renew cell division when exposed to a cytokinin.

As with many other development-directing molecules, the cytokinins isolated from plants, or from bacteria or old commercial preparations of DNA, were found

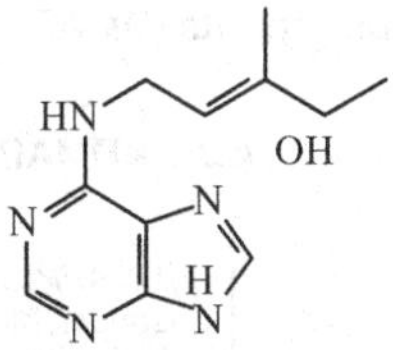

Figure 2.5. Structure of zeatin, the first naturally occurring cytokinin identified in plant cells.

to control many more events than the one that led to their original detection. Also, as with all the other directive molecules, they appear to be present in all living tissues and they act in concert with other signals also present in the cell. So it is that cytokinins have clear interactions with auxin in the control of cell division, and in root and shoot meristem initiation in callus cultures. These molecules can also enhance ethylene production and gibberellin-induced flowering. In those plant species where infections of *Agrobacterium tumefaciens* can occur on wounding, tumours and fasciated apices arise through an overproduction of cytokinin and auxin, which is mediated by the insertion and subsequent expression of bacterial cytokinin and auxin biosynthesis genes (Akiyoshi et al., 1984; Barry et al., 1984). But these *in vivo* events are restricted to certain species, indicating that the host cell too plays its selective role as the recipient target.

All the natural cytokinins are *N*-substituted adenine derivatives that generally contain an isoprenoid-derived side chain, and a wide diversity of structures have now been elucidated in plants (McGaw and Burch, 1995). Initially, a proposed pathway of synthesis in higher plants was based on the catalysis performed by the integrated *Agrobacterium* isopentenyltransferase in the crown gall tissue in which isopentenyladenosine 5′-monophosphate is produced by the addition of dimethylalyll diphosphate to the N6 position of adenosinemonophosphate (AMP). This yields the isopentenyl ribotide from which ribosides and free bases are thence derived, all with different degrees of activity and stability (Binns, 1994). However, no evidence for a plant homologue of the isopentenyltransferase from *Agrobacterium* was forthcoming until the *in silico* examination of the genome of *Arabidopsis* revealed a small multigene family of enzymes that are structurally related to bacterial adenylate isopentenyltransferase and tRNA isopentenyltransferase (Takei et al., 2001; Kakimoto, 2001). At least two members of the isopentenyltransferase gene family of *Arabidopsis, AtIPT1* and *AtIPT4*, have been expressed in *E. coli* and the enzymes shown to synthesise isopentenyladenosine 5′-monophosphate from dimethylalyll diphosphate and ADP or ATP, in preference to AMP (Figure 2.6). The discovery of these plant enzymes now affords the opportunity for detailed investigation of the biosynthesis of cytokinins in higher plants and in the regulation of the genes encoding the biosynthetic enzymes (for reviews, see Sakakibara and Takei, 2002; Haberer and Kieber, 2002). While roots, particularly root tips, are proposed to be the major sites of synthesis from whence the signal is transported to the shoots, few will doubt that all cells probably have the ability to synthesise some level of cytokinin. In a comprehensive survey, Miyawaki et al. (2004) showed, using promoter:GUS fusions,

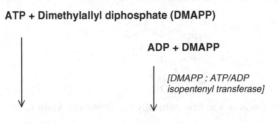

Figure 2.6. Proposed isopentenyladenine formation from the reaction of dimethylallyl diphosphate (DMAPP) with ATP or ADP in plant cells.

that the multigene family of ATP/ADP isopentenyltransferases of *Arabidopsis* (*AtIPT1, AtIPT3, AtIPT4, AtIPT5, AtIPT6, AtIPT7* and *AtIPT8*) displayed differential regulation of expression in the tissues examined, thus confirming that tissues expressing the *AtIPT* genes are widely distributed throughout the plant, including the root tips, leaf axils, ovules, endosperm tissue, developing inflorescences and fruit abscission zones. Further, they showed that cytokinin could down-regulate the expression of different members of the gene family, while the expression of two members, *AtIPT5* and *AtIPT7* was up-regulated by auxin.

The less active zeatin-riboside appears to be the long-distance transport form in the xylem and the *O*-glycosylated molecule is the candidate for a storage non-active derivative. As with other sequestered signal molecules, the extent of conversion to an active form offers a fast generating system that is, in the short term, independent of the rate of synthesis.

Abscisic acid (ABA)

The isolation of ABA was the culmination of several separate chemical searches, with ABA becoming the first naturally occurring cell-growth inhibitor to be characterised. It was shown to interfere with auxin-induced cell elongation of *Avena* coleoptiles (Ohkuma et al., 1963), to arrest bud growth in birch, *Betula pubescens*, and in sycamore, *Acer pseudoplatanus*, and as 'dormin', it was intimately linked to the onset of bud dormancy (Eagles and Wareing, 1963; Robinson et al., 1963). Levels of this inhibitor rose in concentration in senescing and shedding cotton bolls, hence the association with abscission and its naming as an abscission-accelerating signal, abscisin II (Ohkuma et al., 1963). Subsequently, Cornforth et al. (1965), using mass spectrometry and infra-red spectroscopy, determined that dormin and abscisin II were the same molecule, and the name abscisic acid was adopted soon after (Addicott et al., 1968).

ABA is a sesquiterpene consisting of three isoprene units with, originally, two proposed pathways of synthesis, either from cyclization of a C_{15} precursor from mevalonic acid (MVA) or as a cleavage product from carotenoids (Figure 2.7A). It is now proposed that ABA is not generated from MVA in the cytosol, but

A

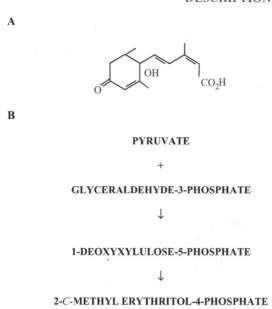

B

PYRUVATE

+

GLYCERALDEHYDE-3-PHOSPHATE

↓

1-DEOXYXYLULOSE-5-PHOSPHATE

↓

2-*C*-METHYL ERYTHRITOL-4-PHOSPHATE

+   CYTIDINE TRIPHOSPHATE

↓

4-DIPHOSPHOCYTIDYL-2-*C*-METHYL ERYTHRITOL

↓

↓

ISOPENTENYL DIPHOSPHATE

Figure 2.7. A. Structure of the naturally occurring and active (*S*)-ABA enantomer. B. Biosynthesis of isopentenyl diphosphate *via* methyl erythritol in chloroplasts.

instead is synthesised in the chloroplast of cells where isopentenyl diphosphate (IPP) is generated from pyruvate and glyceraldehyde-3-phosphate via the methyl erythrose phosphate (MEP) pathway (Figure 2.7B). IPP is proposed also to be synthesised from MVA in the cytosol of cells where it may be imported into chloroplasts for subsequent steroid biosynthesis. Eight IPP molecules form β-carotene which is then oxidised to violaxanthin and then to 9'-*cis*-neoxanthin. The C_{15} ABA intermediate, xanthoxal, is then cleaved from 9'-*cis*-neoxanthin, converted to xanthoxic acid and then, through a series of yet to be determined steps, to ABA (Figure 2.8) (reviewed in Milborrow, 2001).

Despite the different physiological leads that provided the impetus to its discovery, the signal role for ABA is now seen as the indicator of the state of a cell's water deficit (Wilkinson and Davies, 2002) and a regulator of the multiple physiological and biochemical changes associated with seed maturation, dehydration and tissue desiccation (Kermode, 1997). Excised wheat leaves, for example, were long ago shown to increase their ABA content forty-fold within the hour

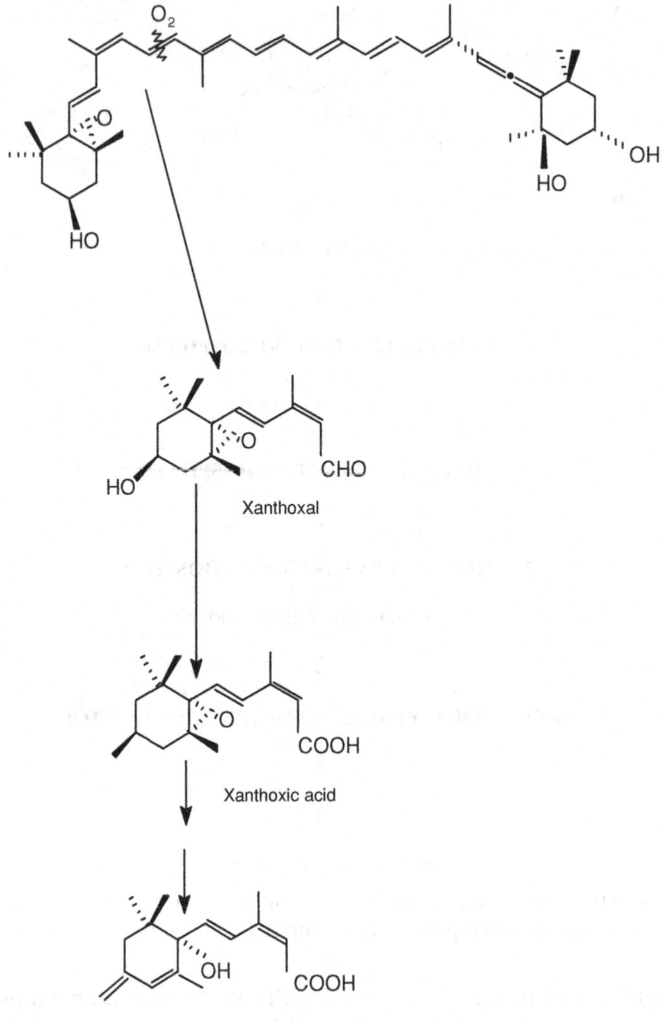

Figure 2.8. The proposed pathway of the biosynthesis of ABA from carotene in chloroplasts.

if the leaves were wilted in warm air (Wright and Hiron, 1969); in leaves of *Valerianella locusta*, ABA has been shown to regulate the closure (turgor) of stomatal guard cells (Hartung, 1983). In the wilty mutant (*flacca*) of tomato, in which stomatal closure does not occur in response to a water deficit, earlier work showed that young shoots of the mutant had a lower level of ABA when compared with wild-type plants (Tal and Nevo, 1973), but the stomata were responsive to applied ABA (Imber and Tal, 1970). It was shown later that an inability to synthesise ABA was the cause of the failure to close stomata and hence to control water loss in the mutant (Parry et al., 1988). Thus in common with that of other plant hormones, the regulation of ABA biosynthesis is responsive to many and varied environmental or developmental cues (reviewed in Milborrow, 2001).

Two particular functions are important in molecular biological terms. Firstly, abscisic acid can act as a repressor in a number of induction processes in specific target cells such as the gibberellin induction of α-amylase in the aleurones of graminaceous seeds, secondly, it can act as an inhibitor of the replicative (but not the repair) synthesis of DNA (Elder and Osborne, 1993).

Finally, the internal level of abscisic acid, as with most other hormones, also depends upon the rate at which it is released from conjugated or other sequestered forms (Milborrow, 2001).

Ethylene

The existence of perhaps the most researched and exciting signal molecule for plant physiologists today was first suspected as long ago as 1864, when Girardin (1864) reported that leaves of trees near to faulty gas mains shed their leaves prematurely. In 1911, Molisch noted that abnormal hypocotyl growth of seedlings occurred when traces of smoke or illuminating gas were present in the air (see translation, Molisch, 1938), and Californian citrus growers discovered that fumes from kerosene anti-frost pots could ripen green lemon fruits (Sievers and True, 1912). Here was an environmental signal with profound effects upon plants. Neljubov (1901) showed that the signal in illuminating gas was the two carbon volatile olefin, ethylene; but it was not until 1934 that Gane demonstrated that the active signal was a natural plant product that reached very high levels in ripe and ripening fruit. Indeed, values as high as 3,000 μL L^{-1} were reported from the cavities of apple and melon fruit (Burg and Burg, 1962a). The volatile nature of this signal explained why a ripe fruit, when enclosed with an unripe one, would induce the latter to ripen. In fact, even today ethylene is referred to as the ripening hormone. There is no doubt that ethylene can be a volatile environmental signal but within the plant, the molecule is present in solution at less than 20–30 μL L^{-1} so the steep diffusion gradient between water and air favours the continual loss of ethylene synthesised within the cell first to the intercellular spaces and then to the external air. As far as we know, no ethylene conjugates have been found, so the levels in the plant are eventually dependent upon the efficiency of biosynthesis.

For higher plants, we now know that biosynthesis proceeds from methionine via S-adenosylmethionine to 1-aminocyclopropane-1-carboxylic acid (ACC) with the formation of the latter being determined by one of the many ACC synthases produced by plants (Yang and Hoffman, 1984). Conversion of ACC to ethylene is controlled by another multigene family of enzymes, ACC oxidases (Figure 2.9). The expression of the genes for these different synthases and oxidases occurs in response to stimuli associated with ripening, tissue wounding or the state of cell development, particularly senescence. The activity of the synthase can be induced by auxin or other regulatory molecules within the cell; ACC oxidases are subject

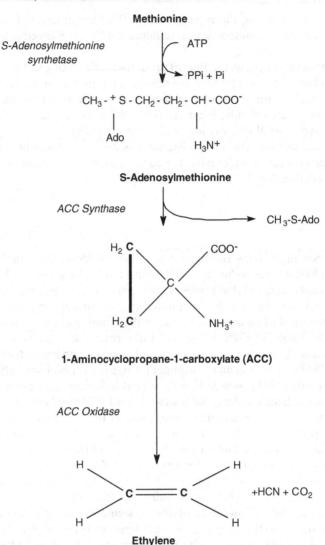

Figure 2.9. The ACC-dependent pathway of ethylene biosynthesis in higher plants. The recycling of 5'-methylthioribose *via* the Yang cycle is not shown nor are the possible conjugation routes of ACC to form 1-(malonylamino)cyclopropane-1-carboxylic acid (MACC), or 1-(gamma-L-glutamylamino)cyclopropane-1-carboxylic acid (GACC).

to autoinduction by ethylene (Kende, 1993). Whereas all plants (except for *Potamogeton pectinatus*; Jackson et al., 1997) are known to produce ethylene not all plants follow the same pathway for ethylene biosynthesis. Lower plants (liverworts, mosses, ferns and lycopods) do not produce their ethylene from methionine nor is ACC an intermediate, so an alternative pathway exists (Osborne et al., 1996). In evolutionary terms, this is very significant and it remains to be seen whether any cells in higher plants retain this early primitive route for ethylene biosynthesis, or can turn it on if the ACC-dependent pathway becomes blocked.

1-Aminocyclopropane-1-carboxylic acid (ACC)

The discovery of ACC is an interesting story. Although ethylene is the oldest identifiable plant growth regulator, the pathway of its biosynthesis in higher plants has not been easy to unravel. In apple tissue, methionine was identified as the precursor with carbons 3 and 4 providing those of the final ethylene molecule (Leiberman et al., 1966). The rest of the pathway took a further 10 years to resolve. First, S-adenosylmethionine (SAM) was shown to be a next intermediate (Adams and Yang, 1977) followed quickly by the identification of 1-aminocyclopropane-1-carboxylic acid (ACC) as the immediate precursor derived from SAM (Adams and Yang, 1979). The twist in the story is that ACC was first isolated in 1957 from both ripe cider apples and perry pears by Burroughs (1957) and also from ripe cowberries (Vahatalo and Virtanen, 1957). The ACC was related to the ripe condition and Burroughs (at the Long Ashton Cider Institute) found that ACC levels increased with ripening in concert with the fruit climacteric. It took the next twenty years for ACC to be connected directly with the biosynthesis of ethylene. The original investigation of this amino acid must be seen in context with the intense interest at the time in protein synthesis and non-protein amino acids. The information, therefore, awaited rediscovery on the resurgence of research into the missing ethylene intermediate. We now see ACC as the mobile precursor in the plant for ethylene and one of the most important regulators and coordinators of target tissue and target cell cross-talk.

The movement of ethylene molecules in the plant is by simple physical diffusion in aqueous solution or via the intercellular air spaces, so the signal range is not great. However, the movement of the immediate ethylene precursor, ACC, is open to both local and long-distance transport. As a highly water-soluble molecule, it is readily transported in the xylem and, in common with cytokinins, represents a communicating signal molecule from the root to the shoot. Since conversion of ACC to ethylene is an oxygen-requiring process, ACC assumes particular significance when roots are subjected to anoxic stress or disrupted water balance. The ACC that then accumulates is transported to the aerobic aerial parts of the plant there to be converted to ethylene. Hence, the visible response of many plants to root flooding is a visible epinasty of the stems and leaves above (reviewed in Jackson, 2002).

Brassinosteroids

Currently, at least 40 brassinosteroids have been isolated as naturally occurring compounds in higher plants. Categorised as C_{27}, C_{28} or C_{29} plant steroids (which refers to the number of carbon atoms in the side chain), each group arises from the corresponding sterol carrying the same carbon side chain. These plant

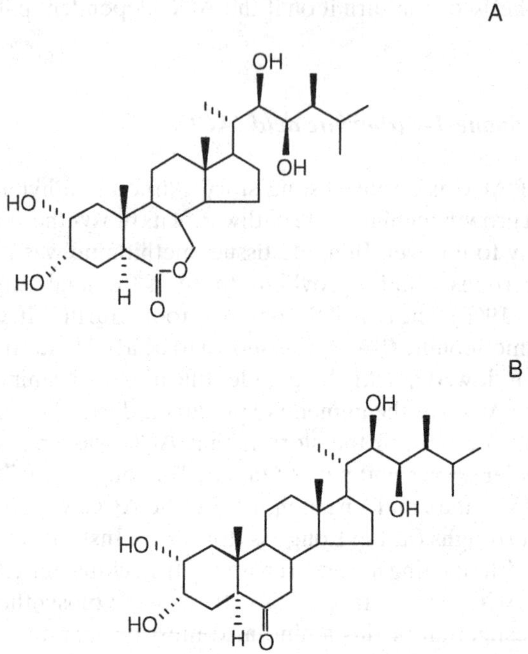

Figure 2.10. Structure of the biologically active brassinosteroid, brassinolide (A), and its immediate precursor, castasterone (B).

sterols all arise from a 2,3-epoxysqualine precursor which cyclizes to produce cycloartenol.

The most abundant and biologically active brassinosteroid is the C_{28} molecule, brassinolide, derived from campestrol (Figure 2.10). A more complete description of the structures and biosynthesis of the brassinosteroids in plants is provided in Fujioka and Yokota (2003), and the use of the early mutants in characterising the biosynthetic pathway is presented by Li and Chory (1999).

First isolated from rape pollen in 1979 (Grove et al., 1979), brassinolide has been shown to regulate a number of developmental processes in higher plants including pollen tube growth, stem elongation, the inhibition of root growth, leaf bending and unrolling, proton pump activation and the induction of xylogenesis in cultured mesophyll cells of *Zinnia*. A synergism of brassinosteroids with auxin in cell growth has been demonstrated as an ability to increase the sensitivity of plant tissue to auxin, or to stimulate auxin biosynthesis (reviewed in Mandava, 1988). These observations suggested originally that brassinosteroids may only act in concert with auxin. For example, both hormones have been shown to induce the transcription of the xyloglucan endotransglycosylase (XET)-encoding *TCH4* gene of *Arabidopsis* linked to cell wall loosening and growth (Xu et al., 1995). In support of a direct link between brassinosteroids and XET, a gene that is specifically up-regulated by applied brassinosteroid, *BRU1*, was cloned using differential display techniques. The gene is post-transcriptionally activated by brassinolide within 2 hours after treatment; other hormones are ineffective (Zureck and Clouse, 1994). Interestingly, the *BRU1* sequence has homology with XET genes from various plant sources, suggesting that brassinosteroid-induced elongation may be mediated via XET-induced cell wall modifications (Oh et al., 1998).

However, Schlagnhaufer and Arteca (1991) have implied a possible role for brassinosteroids in promoting the production of ACC in tomato leaf discs which suggests a regulatory link between ethylene and the steroid compounds. Further, experiments with the auxin-insensitive mutant *axr1* (auxin-resistant) of *Arabidopsis* show that application of the brassinosteroid, 24-epibrassinolide, will inhibit root elongation in the mutant, but application of 2,4-dichlorophenoxyacetic acid will not (Clouse et al., 1993), while applied brassinosteroids will elongate the tomato auxin-insensitive mutant *dgt* (diageotropica) without inducing the auxin-dependent *SAUR* gene. Although it is likely that brassinosteroids are modulators of the major hormones and other signal molecules, a significant school of thought has suggested that they exert their effects solely through regulating the endogenous level of auxin (reviewed in Mandava, 1988). Counter to this suggestion is the evidence that brassinosteroids have specific biological activities that are not influenced by auxin (Yokota, 1997) since brassinosteroid-induced elongation of soybean hypocotyls was shown not to be accompanied by the induction of the same genes that are up-regulated during auxin-induced elongation (Clouse et al., 1992). More recently, molecular evidence, gained using a range of brassinosteroid mutants, confirms that these compounds can directly influence plant development. These mutant studies can be broadly divided into two types – those in which the mutant phenotype can be rescued by the addition of exogenous brassinosteroid, most commonly brassinolide (the biosynthetic mutants), and those in which added brassinolide does not rescue the phenotype (mutations in brassinosteroid signalling). Some of the brassinosteroid mutants are discussed here, but it is the identification of the brassinolide receptor and the intracellular signalling components that regulate response to this signalling group and thus the target cell response that is of most relevance (see Chapter 8).

Some of the earliest mutants described included two dwarf mutants of wheat, *det2* (de-etiolation; Li et al., 1996) and *cpd* (constitutive photomorphogenesis and dwarfism; Szekeres et al., 1996) that display abnormal phenotypes in the light; both can be rescued with applied brassinolide. Subsequent studies have shown that the lesions occur in enzymes in the biosynthetic pathway of brassinolide. For *det2*, evidence suggests that this is a true homologue of the mammalian steroid 5α since *det2* can catalyse the biosynthesis of steroids including testosterone and progesterone when expressed in mammalian cells (Li et al., 1997). In pea mutants, dwarfism, long considered to be regulated solely by gibberellin, can also be relieved by brassinosteroids (Nomura et al., 1997). Two of these, *lka* and *lkb*, contain normal endogenous gibberellin levels but the *lkb* mutant is deficient in brassinosteroids and is rescued by adding brassinolide or brassinosteroid precursors suggesting that a synthesising lesion occurs in the steroid biosynthetic pathway (Nomura et al., 1997).

Feldmann and colleagues have characterised the *dwf4* and *dwf5* series of biosynthetic mutants. Again, the dwarf phenotype of *Arabidopsis* could be rescued by applied brassinolide, but not with the other hormones tested. Subsequent analysis has determined that the *dwf4* lesion disrupts a cytochrome P450 that catalyses multiple 22α-hydroxylation steps in the biosynthetic pathway (Choe et al., 1998) and the *dwf5* mutants are disrupted in a sterol Δ^7 reductase step (Choe et al., 2000).

For the second class of mutants (concerned with brassinosteroid signalling) one of the earliest characterised was the brassinosteroid-insensitive mutant, *bri1* (Clouse et al., 1996). This single mutant, isolated from a population of 70,000 homozygous M_2 ethyl methyl sulphonate (EMS) mutants of *Arabidopsis*, displayed a phenotype in which roots elongated in response to added $10^{-6}/10^{-7}M$ 24-epibrassinolide (EBR), whereas this concentration of brassinosteroid inhibited wild-type root growth by 75 percent. The mature mutant plants displayed severe growth abnormalities including extreme dwarfism (plants were less than 10 percent of wild-type in terms of height at a similar developmental stage), the rosette leaves were stunted and dark green with a thickened, curled appearance and with petioles that failed to elongate. This phenotype could not be rescued by added brassinosteroid. In further characterisation studies using a root inhibition assay, added brassinosteroid (24-epibrassinolide) did not inhibit root growth in the mutant, but did do so in the wild-type. However, the other hormones tested, 2,4-D, IAA, BAP, kinetin ABA and ethylene did inhibit root elongation to the same degree in the *bri1* mutant as in the wild-type (Clouse et al., 1996). Added GA_3 had no effect on root elongation in the wild-type or the *bri1* mutant. These experiments suggested that the mutation was concerned with brassinosteroid signalling and not biosynthesis, and subsequent characterisation of *bri1* has shown that it functions as a brassinolide receptor. Further, many of the downstream elements that both negatively and positively regulate brassinolide responses have now been identified; discussion of these elements is included in Chapter 8.

These mutant-based studies, with others, have continued to define brassinosteroid biosynthesis but further, these have confirmed that endogenously produced brassinosteroids do regulate developmental processes in plants. Moreover, in *Arabidopsis*, brassinosteroids have been shown to be synthesised in all organs tested, but are most actively synthesised in young, actively developing tissues, and less so in mature tissues, consistent with their action as a growth promoter (Shimada et al., 2003). Whether the brassinosteroids should be elevated to join the classical plant hormones remains to be determined. However, we consider the evidence is such that these compounds should be included as plant signalling molecules with a bridge to the target steroid signalling found in animals and insects.

Jasmonates

Jasmonic acid (JA) was first isolated from the fungus *Lasiodiplodia theobromae* (Aldridge et al., 1971), and later as the methyl ester and senescence-promoting

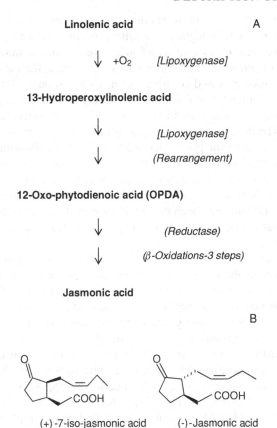

Linolenic acid   A

↓ +O₂   *[Lipoxygenase]*

13-Hydroperoxylinolenic acid

↓   *[Lipoxygenase]*

↓   *(Rearrangement)*

12-Oxo-phytodienoic acid (OPDA)

↓   *(Reductase)*

↓   *(β-Oxidations-3 steps)*

Jasmonic acid

B

(+)-7-iso-jasmonic acid   (-)-Jasmonic acid

Figure 2.11. A. Biosynthesis of jasmonic acid from linolenic acid. B. Structure of (−)-jasmonic acid and its stereoisomer (+)-7-iso-jasmonic acid.

substance from wormwood (*Artemisia absinthium*) by Ueda and Kato (1980). The essential (and fragrant) oils of *Jasminium grandiforum* L. and *Rosmarius officinalis* L. are also sources of the volatile ester (Demole et al., 1962; Crabalona, 1967). The commercial interest in these essential oils ensured the early elucidation of the biosynthesis and structure of methyljasmonate (MeJA), but it was not until Ueda and Kato (1980) showed that application of the free acid promoted leaf senescence in oat (*Avena sativa*) that these compounds became of interest to plant physiologists. Chemically, jasmonates are cyclopentanone derivatives of linolenic acid via the octadecanoid pathway (Figure 2.11A). The first product is (+)-7-isojasmonic acid, which is rapidly converted to its more stable stereoisomer, (−)-jasmonic acid (Figure 2.11B) (Vick and Zimmerman, 1984), although both substances are physiologically active (Wasternack and Parthier, 1997).

In common with other hormones, (−)-jasmonic acid exists as a number of forms within all plant tissues. It has been found as the free acid, as the methyl ester or as conjugates with certain amino acids, as glucose esters or as hydroxylated derivatives (Wasternack and Parthier, 1997). Biological activity has been attributed to all of these chemical forms. Jasmonic acid is ubiquitous amongst higher plants, with highest concentrations in shoot and root apices and developing leaves and fruits, and with lower levels found in mature roots and leaves (Creelman and Mullet, 1995). In terms of function, the molecule has two major roles as a mediator and coordinator of plant responses. For the first, applied jasmonate will hasten and enhance flowering, tuberization, fruit ripening, synthesis of storage compounds and senescence. For its second role, the molecule is synthesised in response to wounding, pathogen attack, mechanical damage and drought stress, including the possibility of JA acting as a long-distance signalling molecule (Ryan and Moura, 2002; Stratmann, 2003). As a stress-induced signal, its induction occurs where membrane lipids are disturbed under the same conditions that cause rises in ethylene formation, so the accumulation of jasmonate, in concert with ethylene, has been proposed to activate an array of genes which constitute the plant's response to environmental cues (Turner et al., 2002; Devoto and Turner, 2003).

The use of mutants of *Arabidopsis* has confirmed that the jasmonates act as growth regulators in their own right. Three mutants, *jar1, coi1*, and *jin1* have been shown to be insensitive to applied jasmonate but not to the other hormones tested, thus confirming that the jasmonates act as a primary signal in plant growth and development, although these compounds very commonly also work in concert with other growth regulators. The *jar1* mutant was isolated from a population of EMS mutants of *Arabidopsis* that did not show MeJA-induced inhibition of root growth (Staswick et al., 1992), and subsequent characterisation has shown that the mutant affords resistance to the opportunistic soil fungus, *Pythium irregulare* (Staswick et al., 1998). Interestingly, *jar1* mutants are fertile although it is known that JA is required for male fertility in plants, suggesting that the lesion is not required for all JA-mediated responses. The lesion has been characterised and shown to occur in the acyl adenylate firefly luciferase family, a large group of enzymes that activate –COOH groups for subsequent modification (Staswick et al., 2002). The *jar1* enzyme is specific for JA, but other members of the family have been shown to interact with salicylic acid (SA) and IAA.

Mutants have also been important in determining that JA and the JA precursor, 12-oxo-phytodienic acid (OPDA), both show biological activity. Sanders et al. (2000) showed that the *delayed dehiscence1* (*dde1*) mutant of *Arabidopsis* was male sterile, with the lesion occurring in the JA biosynthetic enzyme, 12-oxo-phytodienoate reductase (OPR3), the substrate of which is OPDA. However, a normal phenotype could be restored if JA was applied to the plants, suggesting that JA is required for stamen and pollen development. In another study, Stintzi and Browse (2000) determined that the *opr3* mutant, in common with the *dde1* mutant, accumulates OPDA (and not JA) in response to wounding (the *dde1* and *opr3* are both mutant alleles of OPR3), but that these mutants still have a competent defence against insect (the dipteran *Bradysia impatiens*) and fungal (*Alternaria brassicicola*) pests (Stintzi et al. 2001). Although application of JA to these mutant plants could induce known JA-induced defence genes, OPDA accumulation alone was sufficient.

Salicylic acid

The biosynthesis of salicylate is via decarboxylation of *trans*-cinnamate (a product of the phenylpropanoid pathway) to form benzoic acid which then undergoes 2-hydroxylation to form salicylic acid (Figure 2.12) (Raskin, 1992). *In vivo*, salicylic acid (SA) has been identified in methyl or glucose ester form or as glucose or amino-acid conjugates. The conjugated and esterified forms of SA are of particular interest since they, in common with those of other hormone substances, are inactive storage forms and the regulation of release to the free acid represents an important and immediately operable mechanism to control the extent of a biological response.

Until relatively recently, research involving salicylates was sporadic and a number of apparently non-related facts indicated a possible signal role. In earlier studies, applied salicylic acid was shown to alter floral induction (Khurana and Maheshwari, 1978; Kaihara et al., 1981) and mineral uptake (Harper and Balke, 1981). Ethylene production in pear cell cultures was lowered when salicylic acid was added to the medium (Leslie and Romani, 1988), and accumulation of the wound-inducible transcript encoding ACC synthase was reduced in tomato fruits (Li et al., 1992). Since early work demonstrated that salicylate inhibited ethylene production in intact plants (Leslie and Romani, 1986), researchers presumed that it was via this interaction that the acid eventually mediated certain aspects of development. Perhaps, in common with other plant hormones, salicylate acted antagonistically to ethylene; but many of these implied effects were only speculative, particularly since the endogenous concentrations of salicylate did not always support this correlation (reviewed by Raskin, 1995). More recent evidence now more firmly supports a role for salicylate in coordinating certain responses in higher plants to plant pathogens (Gaffney et al., 1993), and exposure to ozone (Sharma et al., 1996) and UV-B (Surplus et al., 1998). This activity as a signalling compound in these responses has implications for the target cell concept and is discussed further in Chapter 3. In contrast, there is less evidence for the importance of salicylate as a regulator of natural plant development. Two exceptions are (i) the role of the acid in the control of thermogenesis in the spadix of the *Arum* lily (Raskin et al., 1987) and (ii) its role as a signalling factor regulating gene expression during leaf senescence (Morris et al., 2000). In the warming of the mature spadix, the cyanide-insensitive non-phosphorylating electron transport pathway operates in which energy is dissipated as heat, rather than coupled to ATP production. The process is particularly pronounced in floral parts of the aroid lily *Sauromatum guttatum* Schott (the voodoo lily), where the increased temperature aids the volatility of certain secondary compounds that act as attractants to pollinators. The trigger of the process is a sudden rise in the endogenous concentration of salicylate (Raskin et al., 1987).

In leaf senescence, salicylate appears to play an important signalling role. While cytokinins and ethylene have been shown to regulate the timing of senescence in opposing ways (Gan and Amasino, 1995; John et al., 1995), other yet-to-be-identified age-related factors are seen as the inducers (Hensel et al., 1993; Grbic

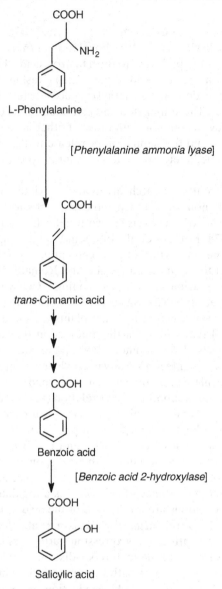

Figure 2.12. Biosynthesis of salicylic acid in plants.

and Bleecker, 1995). Using mutants of *Arabidopsis* with altered SA signalling (*npr1* and *pad4*) or transgenic plants overexpressing a salicylate hydroxylase gene (*nahG*) so that SA would not accumulate in these plants, Morris et al. (2000) examined the induction and expression of specific senescence-associated genes in these genetic backgrounds and concluded that salicylate can be classified as one of the primary regulators of the process.

*Methylsalicylate and methyljasmonate as
signalling volatiles*

Interplant communication is open to signalling by volatile molecules. Ethylene is the most obvious candidate but any plant volatile has this potential. Methyljasmonate, for example, from *Artemisia tridentata* was shown originally to induce the synthesis of proteinase inhibitors in the leaves of neighbouring tomato plants, a response that is similar to wounding (Farmer and Ryan, 1990). Indeed, in response to wounding, ethylene and jasmonate have been shown to act together to enhance the transcription of proteinase inhibitor (*pin*) genes (O'Donnell et al., 1996). In further evidence of biological activity of MeJA, Seo et al. (2001) generated transgenic plants of *Arabidopsis* that overexpressed the MeJA biosynthetic gene, *S*-adenosyl-$_L$-methionine:jasmonic acid carboxyl methyltransferase (*JMT*). The transgenic plants displayed a three-fold increase in MeJA content, and critically, JA-responsive genes (e.g., *VSP* and *PDF1.2*) were observed to be constitutively expressed, and the plants shown to be resistant to the floral pathogen, *Botrytis cinerea*.

The critical issue with volatiles as plant signals is the threshold concentration to evoke a response. For ethylene, this can be less than 1 ppm; for MeJA values of 100 µM have been reported; and for MeSA, values of 10 µM (Ding and Wang, 2003) and 2.5 ug L^{-1} (Shulaev et al., 1997). For a more detailed description of the role of volatiles in plant communication, particularly methylsalicylate, the reader is referred to Chapter 3.

Nitric Oxide

Nitric oxide (NO) has only recently become established as an important signalling compound in animal cells, but through comparative research between animals and plants, it can now be included in a description of plant signals (Leshem and Haramaty, 1996). Full information on its role in plants is yet to be gathered, so a detailed assessment of an involvement with specific target cells is a little premature. However, what is known about NO in plants is summarised briefly here and the reader is referred to Durner and Klessig (1999), Wendehemme et al. (2001) and Neill et al. (2003) for more detailed accounts.

In plants, NO can be generated from several potential sources both enzymatic and non-enzymatic (Figure 2.13); evidence suggests that it may be synthesised apoplastically in a non-enzymic reaction utilising nitrite (Bethke et al., 2004). It has been shown to be synthesised enzymically from NO_2^- by an NAD(P)H-dependent nitrate reductase. In addition, and in common with animal cells, there is some evidence that NO can be synthesised by the action of a mammalian-type NO synthase (NOS) that catalyses the conversion of L-arginine to L-citrulline and NO. Although this enzyme is yet to be purified from plants, activity has been demonstrated in a range of tissues, and antibodies raised to mammalian NOS have detected the presence of NOS-like proteins in plant extracts. In maize roots, an element of developmental regulation has been discerned with NOS-like

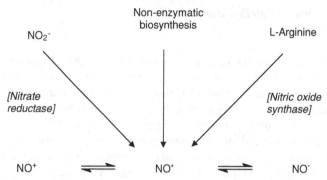

Figure 2.13. Proposed formation pathways for nitric oxide (NO•) in higher plants.

protein localised to the cytosol of cells in the division zone and in the nucleus of cells in the elongation zone (summarised in Neill et al., 2003).

Researchers propose that biologically, NO functions in the coordination of plant defence reactions since there is now good evidence that the signal can induce the expression of key pathogen-associated plant genes (Delledonne et al., 1998; Durner et al., 1998). This induction occurs in a salicylate-dependent or salicylate-independent pathway. Further evidence for the role of NO in plant defence arises from the speculation of similar interactions of NO and aconitase in plants and animals. In animal cells, NO binding to aconitase inactivates the enzyme and converts it to an mRNA-binding protein that can regulate the turnover of mRNA encoding the transferrin receptor and ferritin (so regulating the iron content of cells). In plants, NO can inhibit aconitase and since the protein has some homology to mammalian aconitases, it seems that conversion to an mRNA-binding protein is possible. The utilisation of this mechanism to regulate iron levels in cells provides scope to envisage how an environment can be created that is lethal to the cell (as part of the hypersensitive response) or is lethal to the pathogen.

In terms of the regulation of developmental processes, NO has been shown to inhibit ethylene biosynthesis in pea leaves, bananas and strawberries and so can influence tissue senescence and fruit ripening (Leshem and Pinchasov, 2000). In addition, it has been found that ABA induces rapid NO synthesis in epidermal tissues and NO enhances ABA-induced stomatal closure (Schroeder et al., 2001). The demonstration that NO can reduce ethylene biosynthesis and interact with other hormones provides for many potential modulating roles for NO during the development of higher plants (see Neill et al., 2003).

Oligosaccharins

Historically, consideration of oligosaccharins began with Albersheim and his colleagues who studied the mechanisms of phytoalexin synthesis in the hypersensitive responses of pathogen-infected plants (Albersheim and Valent, 1978).

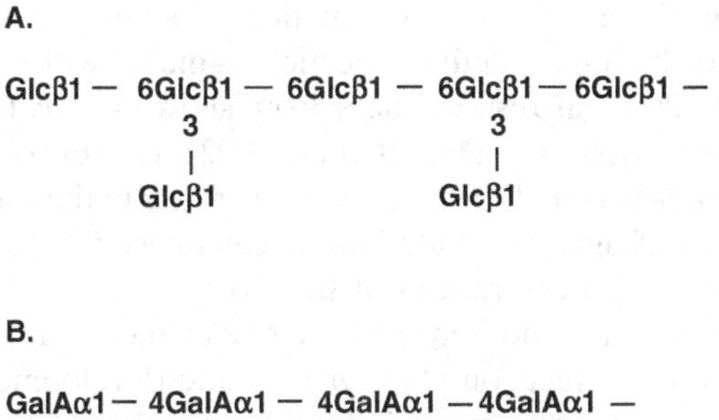

A.

Glcβ1 — 6Glcβ1 — 6Glcβ1 — 6Glcβ1 — 6Glcβ1 —
        3   3
        | |
    Glcβ1   Glcβ1

B.

GalAα1 — 4GalAα1 — 4GalAα1 — 4GalAα1 —

Figure 2.14. Diagrammatic representation of (A) the structure of an active heptasaccharide β-glucoside alditol, a phytoalexin elicitor from fungal cell walls, and (B) a pectin-derived oligosaccharide (the oligogalacturonides) from plant cells.

The signal substances that induced phytoalexin accumulation in response to pathogens were called 'elicitors'. These molecules were heat-stable and pH-stable and of much higher molecular weight (5–20 kDa) than those of the major plant hormones, and they were all neutral glucan polysaccharides of 3-, 6-, and 3,6-linked glucosyl residues in the β-configuration. Branching appeared necessary since unbranched β-3-linked glucans (e.g., laminarin) had little or no activity. Of the very many fragments isolated from fungal cell walls, only one was active as a phytoalexin elicitor, a heptasaccharide β-glucoside alditol (Figure 2.14A). This molecule was effective at 10^{-9} to 10^{-10} M, a lower concentration than those of optimal hormone responses. Many lower molecular weight fragments of plant cell walls (12–14 degrees of polymerisation, DP) also appeared to function as elicitors of phytoalexin formation, both in whole plants and in tissue cultures. These were produced by endo-polygalacturonase activity in either plant or pathogen cell walls. Since the cell wall is not a static entity but is constantly under modification with time during growth and maturation, it is not surprising that the wall could be the source of liberation of very many different small fragments of glycan oligosaccharides. Such changes represent part of the target transformations that every cell undergoes during the progress of differentiation from meristematic initial until cell death. What is more surprising is the remarkable capacity and high activity of these oligosaccharins to change the direction and type of cell growth. Additionally, flowering could be inhibited but growth enhanced in cultures of duckweed (Gollin et al., 1984) and thin layers of tobacco cells could be induced to root or flower depending upon the concentration and length of the oligomers supplied (Tran Thanh Van et al., 1985; Eberhard et al., 1989). It was from studies such as these that the concept arose that oligosaccharides could act as signalling molecules in plant developmental processes.

Oligogalacturonides (OGAs)

Structurally, these compounds are homopolymers of α-1,4-linked D-galacturonic acid released from homogalacturonan (the major constituent of the cell wall pectin matrix), via the action of polygalacturonases and pectate lyase (Figure 2.14B). In terms of biological activity, pectic fragments with a degree of polymerisation of 10 to 20 sugar residues have been shown to elicit a range of plant defence responses (reviewed in Darvill et al., 1992). However, it was the observation by Tran Thanh Van et al. (1985), originally made in thin-layer leaf cultures of tobacco, that these fragments could initiate cell divisions and the induction of flowering primordia at concentrations of 10^{-8} to 10^{-9} M, that led to the role of OGAs as signal molecules and hence to their relevance to the target cell concept. In terms of other effects on plant growth and development, OGAs were shown originally to antagonize auxin-induced elongation of pea stems (Branca et al., 1988), although more recent work with seedlings of cucumber (*Cucumis sativa* L.) indicates that this antagonism occurs by an indirect mechanism and not simply by inhibition of auxin action (Spiro et al., 2002). OGAs will stimulate ethylene production in a variety of tissues, including pear cell suspension cultures (Campbell and Labavitch, 1991a) and tomato fruit pericarp discs (Campbell and Labavitch, 1991b), and OGAs in the size range of 4–6 DP can induce ACC oxidase expression in leaves of tomato within 1 hour of treatment (Simpson et al., 1998).

Xyloglucan derivatives

Xyloglucan is a heterogeneous cell wall hemicellulosic polysaccharide with a structural role in plant cells, typically comprising up to 20 percent of the primary walls of dicotyledonous plants. Oligosaccharins of mixed sugar composition are derived from these structural xyloglucans by hydrolysis and/or by natural endotransglycosylation (Figure 2.15) (McDougall and Fry, 1991). Although very low in concentration *in vivo*, some have specific sugar sequences that have been shown to regulate cell growth and plant development at remarkably low concentrations (Aldington and Fry, 1993). The best characterised of these is a nanosaccharide termed XXFG which, structurally, is composed of Glc_4-Xyl_3-Gal-Fuc (Figure 2.15A) and was originally shown by York et al. (1984) to inhibit auxin (2,4-D)-stimulated growth of pea stems at nanomolar concentrations. A second xyloglucan, termed XLLG which structurally is Glc_4-Xyl_3-Gal_2 (Figure 2.15B), will stimulate elongation of pea stem segments (McDougall and Fry, 1990) at equally low levels, although in the absence of auxin. It is not clear why two such similar molecules should behave in opposing ways unless the perception by the target cells can discriminate between the two. A more recent study has suggested a mechanism as to how xyloglucan-derived oligosaccharins might enhance cell elongation (Takeda et al., 2002). Using XXXG, which structurally is Glc_4Xyl_3 (Figure 2.15C), Takeda et al. showed that this xyloglucan structure will cause solubilization of xyloglucan from the cell wall but maintain the microfibrils in a transverse orientation, so leaving the cells free and competent to expand under the influence of other plant growth substances.

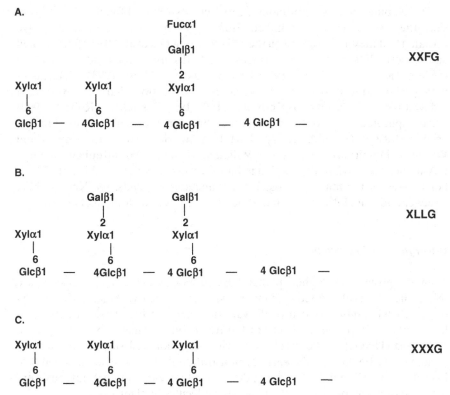

Figure 2.15. Active mixed xylose-containing oligosaccharins derived from cell walls by enzymic activity.

Arabinogalactan proteins

Complexes of sugars and proteins are common secondary products in plants. The arabininogalactan proteins (AGPs) are proteoglycans that occur bound to cell membranes through a glycosyl-phosphatidyl-inositol (GPI) anchor (Youl et al., 1998) in the cell wall. Structurally, these compounds contain less than 10 percent protein and more than 90 percent carbohydrate. The protein moiety is rich in hydroxyproline, so AGPs are classified as hydroxyproline-rich glycoproteins. The carbohydrates attached to the protein core consist largely of arabinose and galactose linked by *O*-glycosylation to the OH group of hydroxyproline or serine residues, although AGPs with glucosamine and *N*-acetyl-$_D$-glucosamine have been reported (Van Hengel et al., 2001). The detailed carbohydrate structure is yet to be determined but the general structure is of a β-1-3-galactan chain that is *O*-linked to the protein with branches of galactose, arabinose and glucuronic acid linked at carbon-6. The carbohydrate side chains can be large, consisting of more than 50 residues; the total molecular mass of AGPs has been calculated at 60 to 300 kDa (see Majewska-Sawka and Nothnagel, 2000).

These compounds are not believed to have a structural function in the wall since they are not covalently linked. Studies with monoclonal antibodies have revealed that they are localised on the cell surface (Knox et al., 1989, 1991; Pennell and Roberts, 1990); and in cell suspensions, AGPs are secreted into the culture medium (Komalavilas et al., 1991; Kreuger and van Holst, 1993). In terms of biological function, it is clear from the earliest investigations that AGP expression is developmentally regulated (Knox et al., 1989, 1991; Pennell and Roberts, 1990) and is implicated in the control of somatic embryogenesis in cell cultures (Kreuger and van Holst, 1993), guidance of pollen tube growth in stylar tissues (Cheung and Wu, 1999), inhibition of root growth (Willats and Knox, 1996) and enhancement of GA-induced α-amylase production in barley aleurone cells (Suzuki et al., 2002). For reviews of the many proposed AGP functions in plants, see Knox (1995), Kreuger and van Holst (1996) and Majewska-Sawka and Nothnagel (2000).

Unconjugated N-glycans

N-linked glycans in higher plants have a pentasaccharide core structure [$Man_3GlcNAc_2$] substituted with mannosyl, xylosyl, fucosyl, N-acetylglucosamyl or galactosyl residues that provide variations around four basic structures in higher plants. With the discovery of the free or unconjugated N-linked glycan structures (UNGs) in the medium of white campion cell suspension cultures (Priem et al., 1990b) and the early demonstration that these structures had biological activity (Priem et al., 1990a), the concept developed that N-glycans were yet another group of saccharide signalling molecules in plants.

The structures of a series of UNGs isolated from tomato and white campion have now been elucidated and two of these with biological activity are shown as Figure 2.16. Many of the UNGs so isolated have been determined as N-linked glycans originally linked to glycoproteins, and one structure, $Man_3AraGlcNAc_2Fuc$ has been described from plant tissues for the first time (Priem et al., 1994) (Figure 2.16). It is suggested that UNGs are synthesised as glycan moieties on plant glycoproteins and then released as free N-glycans through the action of specific de-N-glycosylation enzymes (Berger et al., 1996).

The mixed oligosaccharide $Man_3XylFucGlcNAc_2$ is biologically active in the nanomolar range either alone or synergistically with the synthetic auxin 2,4-dichlorophenoxyacetic acid (2,4-D), and has been shown to increase the elongation rate of segments of flax hypocotyls. Effects are, in common with the other oligosaccharins, highly dependent upon concentration, and at higher concentrations, the same glycan inhibits the promotory effects of 2,4-D. This glycan and another, the oligomannose $Man_5GlcNAc$, were both found to promote ripening in tomato fruits. Again, at higher concentrations, ripening was delayed (Priem and Gross, 1992). This is not an unusual response to a regulatory compound for most become inhibitory above an optimal concentration for enhancement. What is unclear is how the two different concentration responses are brought about. For example, galactose, when added alone to a medium with stem segments reduces the promotory effect of auxin, and free galactose and auxin can, depending

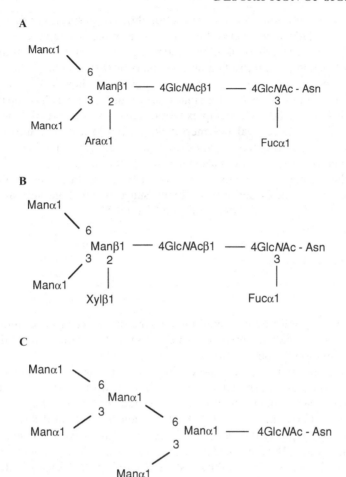

Figure 2.16. Diagrammatic representation of unconjugated *N*-glycans (UNGs) structures identified in plant cells with known biological activity (A, Man$_3$AraFucGlcNAc$_2$; B, Man$_3$XylFucGlcNAc$_2$), and a novel unconjugated *N*-glycan structure (C, Man$_5$GlcNAc).

on the concentration of the sugar, attenuate the biological effect of the *N*-glycan (for a review, see Priem et al., 1994). Whether these responses simply reflect a change in the uptake of auxin or limitations or excesses in glycan building blocks in polysaccharide synthesis or are showing that the glycans are true signal molecules remains to be resolved. Whatever their mode of action, there is no doubt that they afford a means of temporal control in cell growth.

Lignans as signalling molecules

Although the lignification of plant cell walls can readily be distinguished anatomically by appropriate stains (e.g., safranin) or chemically by phloroglucinol, we now know that lignin content and composition differ from species to species and tissue to tissue, and a high level of difference exists in their metabolic formation. Of special interest in this book is the knowledge that cell wall lignin and the lignans (either non-polymerised lignin precursors or their oligomeric polymers) could be signal molecules. All wall lignin is composed of phenolic polymers with multiple inter-subunit linkages. More than one biosynthetic pathway probably exists providing the considerable complexity, particularly when cross-linking bonds are formed with other cell wall polymers including polysaccharides and proteins. What is important in the present context of target cells is not just the synthesis of lignans and their products, but also the signal potential of lignin-derived fragments during degradative events. None has yet been identified chemically as a signal molecule but the evidence is now highly suggestive that one such molecule is central to the induction of abscission (see Chapter 5).

Peptides as signals in plants

Systemin

Although well established in animal biology, the discovery of lower molecular mass proteins (normally referred to as peptides) that contain intercellular signalling information has been slower to emerge in higher plants. The first peptide, systemin, was isolated as a wound-inducible factor following insect damage to tomato leaves. Earlier work had shown that OGAs (with a DP of ∼20 uronide residues) could act as the proteinase inhibitor inducing factor (PIIF), but these compounds are not mobile. Pearce et al. (1991) then used standard chromatography approaches to purify the mobile signal from tomato leaves and determined that systemin was an 18 amino acid peptide and a synthetic version of the peptide could induce both proteinase inhibitors I and II. Using [^{14}C]Ala-labelled systemin, researchers showed that the peptide moved from the wound site through the phloem to other plant parts. McGurl et al. (1992) designed oligonucleotide probes to residues 12–18 of systemin, screened a tomato leaf cDNA library and then rescreened the positives with a probe from residue 1–6. Only one clone was isolated and sequenced and the ORF* determined. This coded for a polypeptide of 200 amino acids with systemin occupying residues 179–196 (Figure 2.17A). The full-length prosystemin protein was predominantly hydrophilic with few hydrophobic residues and no identifiable N-terminal leader sequence; furthermore, no processing sites bordering the systemin peptide were common to mammalian proteolytic processing sites. Homologues to the prosystemin protein were found only in potato, nightshade and pepper, but not in tobacco. McGurl et al. (1992) determined that on wounding older leaves of tobacco, prosystemin mRNA was detectable in unwounded upper leaves. Further, the induction of prosystemin mRNA occurred earlier (at 3–4 hours) than the observed induction of PIs (at 8 hours), and tomato plants expressing an anti-sense prosystemin showed an attenuated induction of inhibitor I or II following wounding. Like all signalling

---

* ORF = open reading frame

1   M G T P S Y D I K N K G D D M Q E E P K V K L H H

26   E K G G D E K E K I I E K E T P S Q D I N N K D T

51   I S S Y V L R D D T Q E I P K M E H E E G G Y V K

76   E K I V E K E T I S Q Y I I K I E G D D D A Q E K

101 L K V E Y E E E E Y E K E K I V E K E T P S Q D I

126 N N K G D D A Q E K P K V E H E E G D D K E T P S

151 Q D I I K M E G E G A L E I T K V V C E K I I V R

176 E D L <u>A V Q S K P P S K R D P P K M Q T D</u> N N K L

A

Plant Origin		Sequence
Tomato		
(*Lycopersicon esculentum*)		A V Q S K P P S K R D P P K M Q T D
Potato		
(*Solanum tuberosum*)	I	A V <u>H</u> S <u>T</u> P P S K R D P P K M Q T D
	II	A <u>A</u> <u>H</u> S <u>T</u> P P S K R D P P K M Q T D
Black night-shade		
(*Solanum nigrescens*)		A V <u>R</u> S <u>T</u> P P <u>P</u> K R D P P K M Q T D
Red pepper		
(*Capsicum annuum*)		A V <u>H</u> S <u>T</u> P P S K R <u>P</u> P P K M Q T D

B

Figure 2.17. A. Amino-acid sequence of the prosystemin protein from tomato, with the systemin peptide sequence underlined. B. Comparison of the amino-acid sequence of systemin from tomato with systemin-like peptides from other species of the Solanaceae.

molecules, the effectiveness of systemin can be altered by the presence of other regulatory substances such as jasmonic acid (Scheer and Ryan, 1999); the extent of this induction was used to isolate a 160 kDa high-affinity receptor binding site located (in tomato) on the plasma membrane (Scheer and Ryan, 1999) (for further discussion of systemin as a signalling molecule and its perception, see Chapter 3).

Systemin-like peptides

Since the discovery of systemin in tomato, two other systemin-like 18 amino acid peptides have now been characterised in tobacco. While homologues of tomato systemin had been identified in other solanaceous species such as potato and nightshade (Figure 2.17B), it had not been identified in tobacco even though tobacco, in common with many plant species, displays a systemic wound response with the induction of trypsin inhibitors which are members of the potato proteinase inhibitor II family. To identify systemin-like peptides, Pearce et al. (2001a) used a sensitive assay in which added systemin could rapidly (over 10–15 minutes)

$$\downarrow$$

```
1     M R V L F L I Y L I L S P F G A E A R T L L E N H
                      TOBACCO SYSTEMIN I
26    E G L N V G S G Y G R G A N L P P P S P A S S P P

51    S K E V S N S V S P T R T D E K T S E N T E L V M

76    T T I A Q G E N I N Q L F S F P T S A D N Y Y Q L

101   A S F K K L F I S Y L L P V S Y V W N L I G S S S
                                    TOBACCO
126   F D H D L V D I F D S K S D E R Y W N R K P L S P
      SYSTEMIN II
151   P S P K P A D G Q R P L H S Y
```

Figure 2.18. Amino-acid sequence of the tobacco systemin I and tobacco systemin II precursor protein. A predicted N-terminal signal peptide sequence is underlined, and the tobacco systemin I (TOB SYS I) and tobacco systemin II (TOB SYS II) sequences are double underlined.

increase the pH (alkanilise) of medium-supporting tomato cell suspension cultures, with changes of 0.4–1.0 commonly observed. This rapid alkalisation assay utilising tobacco mesophyll cells in culture was used to identify and then purify two peptides designated tobacco systemin I and II. Each could induce the synthesis of trypsin inhibitors in tobacco leaf tissue (although whether they are mobile signals is yet to be established), and their biosynthesis was up-regulated by MeJA (leaf tissue treated with MeJA was used for the purification of tobacco I and II). The 18 amino acids of both tobacco systemin I and II showed no homology to each other or to systemin itself. Unlike systemin, each peptide was glycosylated with pentose sugars, and removal of these sugars severely reduced activity (i.e., the induction of trypsin inhibitors). Using oligonucleotide primers designed to the Tob I sequence, Pearce et al. (2001a) isolated a cDNA using RT-PCR and determined that both Tob I and II were derived from the same 165 amino acid pre-protein, with tobacco systemin I arising from near the N-terminal (residues 36–53) and tobacco systemin II from the C-terminal end (residues 144–161) (Figure 2.18). The treatment of leaves with MeJA induces an up-regulation of the pro-protein in leaves, and both tobacco systemin I and II can induce the activity of a 48 kDa mitogen-activated protein (MAP) kinase in tobacco mesophyll cell cultures.

RALF peptides

During the purification of tobacco systemin I and II, Pearce et al. (2001b) used the rapid alkalinisation assay with tobacco cells to purify another peptide which they termed the rapid alkalinisation factor (RALF). This peptide, of about 5 kDa, was more potent in terms of timing (being more rapid than Tob I and II) and in terms of magnitude of the pH change. These RALF peptides did not induce proteinase inhibitors in tobacco but could induce the activity of the 48 kDa MAP kinase when added to tobacco cell cultures; they have also been shown to be potent inhibitors of root growth of tomato seedlings. A tobacco EST* was then identified that

* EST = expressed sequence tag

↓

1   M G V P S G L I L C V L I G A F F I S M A A A G D

26   S G A Y D W V M P A R S G G G C K G S I G E C I A

51   E E E E F E L D S E S N R R I L A T K K Y I S Y G

76   A L Q K N S V P C S R R G A S Y Y N C K P G A Q A

101 N P Y S R G C S A I T R C R S

Figure 2.19. Amino acid sequence of the RALF precursor protein. A predicted N-terminal signal peptide sequence is underlined, and the RALF polypeptide sequence is double underlined.

contained the exact sequence of the RALF peptide and was used to isolate a cDNA from a tobacco leaf library. The cDNA encoded a pre-pro-protein of 115 amino acids, with a putative signal peptide that was cleaved between residues 23 and 24 (Figure 2.19). The RALF peptide was derived from the C-terminal portion of the pro-protein and a putative protease dibasic motif (Arg-Arg) was identified two residues upstream from the RALF peptide. Unlike tomato systemin, RALF peptides have been shown to be homologous in many species and the RALF cDNA is present in libraries made from roots, shoots, leaves and flowers. Whereas systemin and tobacco systemin I and II exert a general response in the plant, the RALF peptide may operate more specifically in roots.

Another signalling peptide that clearly directs a response in specific target cells has been isolated from nodulating roots of legumes (van de Sande et al., 1996). Here nodule formation requires the root pericycle cells to divide, but this is initiated only where cortex is adjacent to the protoxylem of the root vascular system. Bacterial infection of epidermal root hair cells by *Rhyzobia* elicits cell cycle activation only in these particular pericycle cells and induces only in those cells the associated activation of the gene that codes directly for the specific ENOD40 peptide required before the new cell divisions start (Compaan et al., 2001). Exactly how the peptide functions is unclear but, as with systemin, evidence suggests that the ENOD40-peptide-induced response can be modified by the presence of other hormones such as cytokinins (Mathesius et al., 2000) and ACC oxidase/ethylene (Heidstra et al., 1997).

CLAVATA peptides

A second well-characterised peptide hormone shown to act directly as a signal on specific target tissues is CLAVATA3, which interacts with its putative receptor CLAVATA1 in association with another closely related protein CLAVATA2. The identification of the CLAVATA3 peptide arose from the characterisation of the *clv3* mutants, *clv3-1* (by EMS) and *clv3-2* (by δ-irradiation) (Clark et al. 1995), the phenotypes of which display an enlarged central (stem cell) zone of the apical meristem, suggesting that the regulatory signals for maintaining mersistem size have become interrupted. Clark et al. (1995) showed that the phenotypes of the *clv3* mutants were similar to the recently characterised *clv1* mutants (Leyser and Furner, 1992; Clark et al., 1993), and the *clv1/clv3* double mutants were epistatic suggesting a direct molecular interaction. Fletcher et al. (1999) cloned the *CLV3* gene using two tagged mutant alleles: a weak allele, *clv3-3*, generated by T-DNA tagging (Feldman and Marks, 1987) and a strong allele, *clv3-7*, tagged by the insertion of the maize transposable element, *En-1* (Wisman et al., 1998). The *CLV3* gene encodes a protein of 96 amino acids that shows no appreciable homology to other proteins in the database (Fletcher et al., 1999). It includes an 18 amino acid residue N-terminal signal sequence that suggests that the protein is targeted to the secretory pathway, but since there is no ER retention sequence, the protein is presumed to be released directly into the extracellular space. Further, Fletcher et al. (1999) carried out cell localisation studies using *in situ* hybridisation and determined that *CLV3* is expressed predominantly in the L1 and L2 layers and in a few underlying cells of the L3 layer, while *CLV1* is expressed more deeply into the L3 layer, with no *CLV3* mRNA detectable in the L1 cells. The putative interaction between CLV3 and CLV1, in association with CLV2, is described in more detail in Chapter 3.

Phytosulfokines

As shown with the MeJA interaction with systemin, peptides function in conjunction with other plant hormones. The recently identified phytosulfokines (PSKs) are another such example. Matsubayashi and Sakagami (1996), working with mesophyll cell cultures of *Asparagus officinalis*, identified what they referred to as the mitogen from 'conditioned medium'. The concept of conditioned medium is well recognised as one in which a higher density of cells has been dividing previously, and so when cells that are grown at a sufficiently low density such that they will not divide, even in the presence of auxin and cytokinins, are transferred to conditioned medium, cell division is initiated. Matsubayashi and Sakagami (1996) isolated a sulfated pentapeptide, designated phytosulfokine-α [Tyr(SO_3H)-Ile-Tyr(SO_3H)-Thr-Glu], and a sulfated tetrapeptide, designated phytosulfokine-β [Tyr(SO_3H)-Ile-Tyr(SO_3H)-Thr], as the mitogenic components of their conditioned media, and they determined that synthetic PSK-α and PSK-β could induce cell division of cell cultures when cells were at a low density (4×10^4 cells/mL). If these peptides were not sulfated, then no mitogenic activity was recovered. Matsubayashi et al. (1999) showed, using mesophyll cultures of *Zinnia*, that cytokinin and auxin as well as the critical cell density are required to induce the differentiation of mesophyll cells into tracheary elements. If the cell density becomes too dispersed then auxin and cytokinin alone are not sufficient to induce cell division. However, if PSK-α (10 nM) is added, in the presence of auxin and cytokinin, then cell division is re-established.

Yang et al. (1999) using 15-mer oligonucleotide probes based on the sequence of the PSK-α pentapeptide, isolated a 725 bp cDNA from a rice cell culture library. The cDNA encoded an 89 amino acid preprosulfokine protein, with a molecular mass of 9.8 kDa and with no homologues in other plant species (although PSK-α

↓
1 M V N P G R T A R A L C L L C L A L L L L G Q D T

26 H S R K L L L Q E K H S H G V G N G T T T T Q E P

51 S R E N G G S T G S N N N G Q L Q F D S A K W E E

76 F H T D Y I Y T Q D V K N P

Figure 2.20. Amino-acid sequence of the PSK-α precursor protein. A predicted N-terminal signal peptide sequence is underlined, and the PSK-α peptide sequence is double underlined.

peptides had been isolated from cell cultures of several plant species). The prepro-sulfokine protein had a 22 amino acid residue N-terminal sequence, and a single PSK-α peptide was derived from the C-terminal portion of the protein (residues 80–84) with putative protease cleavage sites identified at the N- (D/Y) and C-terminal (D/V) of the PSK-α peptide (Figure 2.20). If the preprosulfokine gene is expressed in rice, then PSK-α is secreted into the medium. Although PSK-α is always observed in the medium of cell cultures, Yang et al. (1999) demonstrated, using RT-PCR, that preprosulfokine mRNA can be expressed in shoot and root apices as well as in leaves, which suggests that these peptide substances may afford another means of regulating cell proliferation *in vivo*. Although the sequences of PSK-α and PSK-β are conserved amongst the plant kingdom, the preprosul-fokine genes are not. Yang et al. (2001) used the sequence of PSK-α to identify four preprosulfokine genes from EST libraries of *Arabidopsis*. Two, designated *AtPSK2* and *AtPSK3*, were shown to be differentially expressed in vegetative tis-sue, with *AtPSK2* expressed predominantly in roots while *AtPSK3* is expressed in roots and stems and shoot apices, again suggesting a role for these peptides in plant development *in vivo*. However, Yang et al. (2001) also demonstrated that transformation of *Arabidopsis* with either *AtPSK2* or *AtPSK3* increased the rate of callus formation, but no change in mitogenic activity was observed when plants were transformed with *AtPSK2* and *AtPSK3* in the anti-sense orientation suggesting that redundancy within the gene family also operates.

  As happened with some of the other peptide hormones, a receptor for the phytosulfokine peptide has been identified. Matsubayashi and Sakagami (2000) photo-labelled 120 and 160 kDa peptides in the plasma membrane of rice suspension using the photoactivable ^{125}I-labelled PSK-α analogue $[N^C$-(4-azidosalicyl)Lys^5]-PSK-α. They found that labelled peptides could be competed with unlabelled PSK-α, but not with inactive analogues of the phytosulfokine peptide. In further work, Matsubayashi et al., (2002), using an affinity column comprising [Lys^5]PSK-Sepharose, purified a major protein of 120 kDa and a mi-nor protein of 150 kDa from microsomal membrane preparations of a specific cell line of carrot. They sequenced the major 120 kDa protein in its entirety, some of the 150 kDa protein (which was shown to be identical to the 120 kDa), and used parts of the 120 kDa sequence to design oligonucleotide primers; they also cloned the gene from RNA isolated from the same carrot cell lines using RT-PCR. The isolated cDNA had an ORF comprising 1021 amino acids which generated a protein of 112 kDa, with a molecular mass of the mature protein (less the signal peptide) of 109.5 kDa which agrees with the calculated molecular mass of the purified 120 kDa glycosylated protein (as identified by affinity column chromatography). The translated protein had a high degree of homology with other members of the receptor-like kinase (RLK) family in plants, with 21 extracellular leucine-rich repeats (LRRs), a transmembrane domain and a cytoplasmic kinase domain. The RLK family, which comprises serine/threonine kinases, is the largest recognisable class of transmembrane sensors identified (with 340 members) in the genome of *Arabidopsis*. The occurrence of the extracellular LRRs in the phytosulfokine receptor indicates that it is one of 174 members comprising the LRR transmembrane kinases, termed the RLK-LRRs (The Arabidopsis Genome Initiative, 2000). This receptor protein has a 36 amino acid island within the 18th LRR, a variable domain amongst the RLKs that is the putative ligand binding site. Finally, Matsubayashi et al., (2002) expressed the phytosulfokine RLK gene in the sense orientation in carrot cells and demonstrated an increase in callus formation (although no root or shoot formation occurred), but when the gene was expressed in the anti-sense orientation, then a substantial inhibition of callus growth was observed.

S-locus cysteine-rich proteins (SCRs)

To conclude our brief survey of plant peptides, another that should be included, albeit briefly, is the family of intercellular signalling molecules, the S-locus cysteine-rich proteins (SCR) or the S-locus protein, SP-11. These SCR proteins form one recognition component of the self-incompatibility (SI) system of pollen and stigma operating in the Cruciferae, which is controlled by a multi-allelic dominant locus, the S-locus. (This is discussed in greater detail in Chapter 4.) The SCR/SP-11 peptides, of 74–83 amino acid residues (47–60 after removal of the secretory N-terminal sequence), are secreted from the developing microspores in the tapetum to reside in the pollen coat exine layer. At pollination, the SCR translocates into the cell walls of the stigma epidermal cells and activates the SI process. The putative SCR receptor has been identified as the S-receptor kinase (SRK) which is a membrane spanning Ser/Thr kinase which, in common with other RLK protein in plants, has extracellular leucine-rich regions (LRRs), a transmembrane domain and a cytoplasmic Ser/Thr kinase domain (Stein et al., 1991; Goring and Rothstein, 1992). The SRK is expressed specifically in the stigma, and the occurrence of this protein is critical to the SI interaction. However, an associated protein, the S-locus glycoprotein (SLG), has been shown to enhance the interaction. The SLG protein is a secreted stigma glycoprotein that is homologous to the SRK extracellular domain, but it does not have the cytoplasmic kinase activity (Takasaki et al., 2000; Takayama et al., 2001).

From these studies in plants, it is evident that the functional peptides can arise from cleavage of precursor molecules or by direct transcription and that the products can evoke responses in specific cells. The possibility now arises that different target cells will have different capacities to cleave precursors. Since the isolation of the first peptide, more have been identified (Lindsey et al., 2002; Ryan et al., 2002) and it is certain that many more remain to be discovered (e.g., Hanada et al., 2003; Yamazaki et al., 2003).

Major and satellite signals

Whereas there is no doubt of the powerful informational directives a plant cell receives from the five major hormones, there is also no doubt that the absolute intensity of the response can be modulated by the presence of other signal molecules. Just as pH, temperature, ionic concentrations, osmolarity and many other physical factors can all determine the quantitative nature of a response, so may chemically induced molecular changes in ion channels, plasmodesmatal connections, membrane domains and transcriptional/translational/post-translational events shift cell performance from optimal to sub- or supra-optimal. The brassinosteroids, jasmonates, oligosaccharins and peptides described here, while not the major players in plant growth control, can all exert at different times and in different ways a differentiation- and performance-directed influence in both the short and long terms. They are substances that can modify the effectiveness of hormones and the capabilities and potential of the cell. At any one time, each cell is exposed to a multi-array of external and internal signals. In all instances, it is the cell itself that must discriminate and sort the inputs; the response that ensues then becomes the outcome of the precise limitations placed upon that cell by its own target state.

We should perhaps add that all the signalling molecules we have considered are essentially plant produced and plant orientated, but our consideration is not exhaustive. For example, polyamines are common to both eukaryotes and prokaryotes and as low molecular weight compounds they can regulate and influence metabolic events (Kumar et al., 1997). Sugars (sucrose or glucose) levels in a cell operate a complex signalling network with ethylene and abscisic acid to inhibit or promote specific growth responses (Leon and Sheen, 2003). Such substances are universal signal molecules. Perhaps as every cell is a target cell, so, given the appropriate circumstances, every molecule can become, at least temporarily, a signal molecule.

3

Cell-to-Cell Signalling: Short and Long Distance

Despite enormous biological diversity, uniformity within a species remains remarkably constant. Trees can be instantly recognised by their shape or form, and leaf and flower structures are a basis of identification and classification. Roots conform to each species expectation, and the natural mutant that does not conform to the species type is a rarity.

This means that the society of cells that make up the plant body is under very strict control. No one branch can outgrow its neighbours and distort the overall shape of the tree. The buds that are terminal on any branch keep pace with, but do not outstrip those on the neighbouring branches. The phenomenon of apical dominance, used by crop growers from time immemorial as the basis to reshape by pruning or to improve yields, is intertissue signalling at its most evident. The consistency with which lateral buds will grow when the terminal bud is removed is central to plant culture and pruning systems throughout the world. Not until Frits Went demonstrated the presence of a chemical substance (indole-3-acetic acid) in *Avena* coleoptile tips and then showed that it would replace the terminal bud in inhibiting the growth of laterals was the first intertissue signal molecule properly established. With the knowledge that the auxin molecule is transported in a polar fashion from cell to cell as it passes from terminal bud to tissues below, it is not difficult to understand how a terminal bud can control the growth of fellow buds below over distances that are relatively short. What is more difficult to understand is the mechanism by which all the terminal buds of a tree keep pace with each other, thus enabling trees to maintain recognisable shapes. This requires a much more complex and long-distance mechanism of intercommunicating tissue signalling and response.

Origins of cell-to-cell signalling

The evolution of epidermis and vascular differentiation will have brought long-distance information exchange and short-distance molecular signalling to the primitive land plant. Lignins became a critical component of the walls of mechanical support cells, and those of higher plants are present as cross-linked polymers of p-hydroxyphenyl, guaiacyl and syringyl units in varying proportions (Lewis and Yamamoto, 1990; Fukuda, 1996). The importance of lignins and wall thickenings in the present context of target cells is the potential opportunities they offer for differences in cell-to-cell communication. The ability to synthesise the wall-stiffening and wall-supported lignin-like polymers goes hand in hand with the progressive colonisation of the land. There could be a link here between the unusual feature that although all lower plants studied so far produce ethylene, they do so by an alternative non-ACC, non-methionine pathway (Osborne et al., 1996) and exhibit a variety of wall thickenings that differ from those of higher plants.

Mosses or liverworts in general show almost no lignified tissue, though elongated cells, presumed to be adapted to transport activities, are present in the central regions of thallose and foliose species. In the variable group of chemical constituents that make up these secondary cell walls, there is still controversy as to whether higher plant lignins are present at all in these lower orders. ^{13}C-NMR spectra have indicated an absence of p-hydroxyphenyl, coniferyl and sinapyl units and instead the presence of 1,3,5-hydroxypolyphenols, which could be derived from an alternate branch of the phenylpropanoid pathway (Wilson et al., 1989). Phenylpropanoids are potential precursors for ethylene, and the switch between lower plant ethylene biosynthesis to that of the higher plant methionine-mediated pathway may be linked to the rerouting of the lignin-like pathway found in bryophytes and pteridophytes. The accepted regulation of secondary wall rigidification expressed in the tracheary elements of higher plants and in the transdifferentiation of parenchyma cells to tracheary cells in various cell cultures, e.g., *Zinnia* (Fukuda, 1994), would appear to differ from those of lower plants. The signalling potential of these complex substances, specifically when the lignin-containing cells are associated with differentiating or senescing cells, may eventually be seen to be of major importance. We have one example to date. Abscission (even in the presence of ethylene) can be blocked if the vascular tissue is removed from excised abscission zones of bean (explants) and cell separation is only re-initiated when the vascular tissue is replaced, indicating a clear signal function of stelar tissue and perhaps also for specific degradation products derived from the accompanying lignin polymers (Thompson and Osborne, 1994). We know that it takes about 24 hours following excision of an explant for the abscission-inducing signal from stelar tissue to be generated and then passed to the abscission zone cells; there is no inducing stele signal present from freshly cut tissue. The lag period before the abscission-inducing signal is released is closely linked in temporal terms to the senescence changes of the pulvinus and its vascular cells.

Traces of sucrose have been shown to increase the production of ethylene and stimulate xylogenesis in lettuce pith explants (Warren-Wilson et al., 1994), and so it is interesting to speculate that the transportation of inductive concentrations of sucrose along the conductive tissues of the phloem in higher plants and along the elongated cell pathways carrying metabolites in primitive plants could have been one of the causal signals to the evolutionary development of lignin-like elements in the plant body.

Short-distance signalling

Informational flows from cell to cell are not a question of simple molecular diffusion without constraints. The presence of the cytoskeleton, membranes and cell walls can provide impedimentia that offer restraints to certain molecules and free passage to others. All information between cells is therefore monitored and in a sense censored, so no matter how close the cells are, the smallest differences in their target states provide a means for determining the extent and kinds of signal transfers that are made between them.

In a uniform environment, the free-floating single cell is open to similar signals on all sides, but as soon as polarity is developed, such that the internal components become differentially distributed or the cell becomes part of a cluster, signal discrimination between the cell parts and neighbour cells is established. The organization of a callus tissue into a somatic embryo requires these events to take place. The maintenance of a meristem throughout the lifetime of a plant requires precise and regimental control of a series of ligand-receptor signal transduction pathways and gene controls from which an apex can escape only by mutation, damage or an external introduction of foreign signalling substances from either pathogens, predators or man.

In making the presumption that a cell, providing it is alive, is always in a target state we can question how long any one target status may be sustained during the lifetime of a cell. Perhaps the fastest changing conditions reside beside the meristem where division is followed swiftly by enlargement and differentiation hence the potential for an altered target cell expression is therefore greatest.

Meristems

Early in meristem differentiation, cells dividing from either the root or shoot apex take on a differentiation state that is dictated by their position with respect to that apex and information transmitted from their immediate neighbours. If this community association remains unchallenged then a normal pattern of development progresses with epidermis, cortex and vascular cells forming in concert as the cells extend. But this society of cells, although stable, is at first highly receptive of either new information or any change in the intensity of the initial informational cues. In this respect, those cells close to the meristem are vulnerable to redirection and readily acquire new types of target status from their immediate surroundings. With maturity, however, the ability to assume a new identity is lessened and long-distance cues assume greater force. For the root meristem of *Arabidopsis*, it has become clear that removal (killing) of a cell by laser microbeam treatment permits a neighbour cell to take on the target status of the eliminated cell. For example, by using two β-glucuronidase promoter fusions specific for vascular and root cap cells, researchers demonstrated that if the several cells of the quiescent centre are killed, those of the vascular initials that abut onto the quiescent centre cease to express vascular cell markers and begin to express the root cap marker of the nearest neighbour cell type (van der Berg et al., 1995). This tells us two things: that meristematic cells are highly flexible with respect to their potential target states, and that they can respond to short-distance signalling within the range of only a few cells. Additionally it is clear that each cell is also a source of signals and it is by the co-operational interaction of target state and signal recognition that the society of the meristem develops into its species-specific form.

Evidence that target state and signalling occur at the earliest stage of primordia cell development comes from another means of removing short-distance signal sources. For this, tissue-specific promoters have been used to express genes for enzymes that control the formation of cytotoxic endproducts within the transformed cell. For example, *Diptheria* toxin A (DTA), when regulated by the *APETALA* 3 promoter and expressed in tobacco flowers, leads to localised nuclear DNA fragmentation and cell death. The petal and stamen primordia are killed by DTA but this event occurs only after the initiation of the sepal primordia. Despite elimination of petals and stamens, the sepals still develop normally indicating either an absence of petal/stamen interaction with sepals or that the earlier induction of sepals dissociates them from a closed petal/stamen signal regulation (Day et al., 1995).

In seeking to understand how cells communicate in the signal/target situation, we must not overlook some of the most notable significant pioneering studies that were made on lower plants and which are still of significance today. In ferns, cleavage of a single apical initial has long been believed to be the source of almost all the cell lineages that follow (Bierhorst, 1977), and for *Equisetum*, the presence of this single apical cell seems certain (Golub and Wetmore, 1948). These single cells are therefore the source of all the ensuing cells that make up the body of the shoot, and their early cleavage products are hence producing progeny with the maximum flexibility possible. With the development of meristems of increasing size there emerged the more complex higher plant apex with multiple cell initials. Then, the total flexibility of each progenitor cell was reduced to that of the origin of a specific stem line and specific cell lineages. Functional interpretation of the different layers (L1, L2, and L3) as proposed by Satina et al. (1940) and extended by Poethig (1987, 1989) provided for different developmental programmes with different target states: the L1 cells being initials for epidermal tissues, L2 for cortical tissue, and L3 for the internal provascular origins. That the history of these origins is inherited and retained as part of the target information of a cell, even late in its differentiation progress, is seen from the performance of chimearic layers of abscission zones formed by tissue grafting. It is clear, however, that each layer still influences the performance of its neighbour and that informational molecules pass between them (see Chapters 5 and 6).

The epidermal (L1) line of cells rarely undergoes redifferentiation *in vivo* to another cell state. This may be due in part to their stable positioning as the external barrier of the plant corpus with the outside environment. But there are occasions when the epidermis does convert to a different and internal status reminiscent of L2 lineage. This can occur in the floral apex when carpels adjoin and fuse in the absence of any union of the cytoplasmic contents of the adhering cell but with the close appression of the cell walls (Cusick, 1966). These kinds of examples in which it is the cell wall that provides the informational signal to cell change have been demonstrated by other studies. In the gynoecium fusions of *Catharanthus roseus*, for example, Verbeke and Walker (1985) showed that the appressed epidermal cells that lose their cuticle assume an isodiametric shape, vacuolate and start to divide in the short time-space of 9 hours; an impermeable barrier inserted between the carpels prevents these changes though a membrane that permits passage of water-soluble molecules does not. Furthermore, transfer of a membrane containing the signal molecules onto the outside epidermis of a carpel induces differentiation changes similar to those that occur in the normal epidermal fusions of the carpels (Siegel and Verbeke, 1989). The question now arises, is it a part of the wall itself and its attendant cuticle that holds the intact signal molecule or is the signal molecule(s) a breakdown product of the wall? Whichever, it is clear that the cell wall is the source of short-distance informational signals that can programme the performance of neighbour cells and determine their future target cell status. Moreover, very strict specificity exists in the molecular informational transfer, for whereas carpel-to-carpel epidermal contact results in redifferentiation, a carpel cannot self-induce against a barrier and requires another carpel epidermis in order to do so. The change of target state that ensues requires the combined recognition of the two cuticular surfaces and the joint production of the inductive signal. Carpel contact with another epidermis type, for example, will not induce dedifferentiation (Verbeke, 1992). The nature of the informational molecules derived from the cell wall is now a focus of considerable study and it is becoming clear that the developmental pathway that a cell pursues is as highly influenced by its nearest neighbour signal as it is from the longer-distance influence of the major hormonal cues. The likely diffusable water-soluble molecules that could provide this short-distance communication must be products of the cell wall – i.e., peptides or oligosaccharides. Others that do not diffuse are those that are retained as part of the cell wall cuticular waxes, structural proteins, pectins and mixed sugar polysaccharides. The epicuticular waxes that are secreted outside the cell walls of *Arabidopsis,* for example, give important informational cues to stomatal development. The *hic* mutants of wax synthesis pathways (the *HIC* gene encodes a putative 3-ketoacyl CoA synthase) have been shown to alter both the wax composition of the epidermal cuticle and the pattern of stomatal distribution (Holroyd et al., 2002). Of particular interest related to the mechanisms of communication in these examples of new cell-to-cell contacts is the evidence that the localised domains of modified plasmodesmata permit the passage of proteins, transcripts or other molecules between them (Van Der Schoot et al., 1995).

Because the embryo of a higher plant and the formation of the first meristems are difficult to study *in situ*, much effort has been directed to understanding simple model systems such as the free-swimming zygote of *Fucus* or somatic embryos formed in tissue culture. The first division of the *Fucus* zygote is asymmetric, initiating a polarity between the two cells and demarking the rhizoidal and apical ends of the thallus. Laser microsurgery to free protoplasts of *Fucus* embryos at this two-cell stage shows that there is dedifferentiation of the protoplast and temporary loss of polarity until new walls are synthesised. Then the thallus and rhizoidal cell resume their thallus or rhizoidal type growth and polarity. However, if contact between the two protoplasts with the parent cell wall is precluded, neither rhizoid nor thallus protoplast will properly regenerate (Berger et al., 1994). Cell wall contact with the protoplast is required to maintain the stable polarity of both cell types and their in-line successors, but the nature of the directing molecules is still unknown although here too, glycans and peptides remain likely candidates.

The fundamental requirement for a zygote to develop even to the earliest changes of embryo cell specification is the establishment of asymmetry. Within the ovule the input from adjoining maternal tissues and gravity are all directional forces that can supply such asymmetry. Again, the *Fucus* zygote affords an amenable experimental system, which even if not supplying the correct answers for the higher plant zygote, offers instead the opportunity to follow mechanisms that lead to comparable differentiation patterns.

Chemical gradients and morphogenic fields, much studied by early developmentalists, are critical to the induction of polarity in the multicellular system but in a free-swimming Fucoid zygote they must be initiated first by a physical asymmetry. Long ago it was established that a differential level of light was sufficient to bring this about (Jaffe, 1958; Quatrano, 1978). The unlit side of the zygote becomes an entry region for ionic currents, and a Ca^{2+} flux is generated within, from the dark to the light side, with a current gradient of 2 pico-amps of Ca^{2+} between the two (Brownlee and Wood, 1986). The position of the inward current demarcates a domain of cytosolic vesicle accumulation and the side of wall weakness from whence the rhizoid eventually protrudes. Polarity of the cell is established. It has been shown that vesicles carry novel cell surface polypeptides, including a β-1,3-exoglucanase, that can modify the wall at their sites of accumulation (Belanger et al., 2003). It may also be assumed that the cell region which accommodates the emerging rhizoid (the new rhizoidal cell) has cytoplasmic surface and wall molecular markers that distinguish its target status from its non-rhizoidal, thallus neighbour. Indeed, polypeptides secreted during asymmetric cell growth have revealed signal sequence similarities between many *Fucus* proteins and those of cell surface signal proteins in other eukaryotes, although the *Fucus* polypeptides also encode unique sequences (Belanger et al., 2003). In this example it is the short-distance generation of electrical polarity within the cell that first determines the cell's subsequent developmental fate. As soon as there are two cells adjoining, an additional dimension of signalling is introduced, that of chemical neighbour-to-neighbour informational cues.

The multicellular primordia of higher plant meristems will always be receiving multiple signals, electrically driven ionic currents and fluxes, and direct cytosolic intercommunication via plasmodesmatal connections between adjoining cells. There is good evidence that even in small groups of 20–40 cells in tissue cultures of carrot (in a Murashige and Skoog medium containing auxin and a cytokinin) a through current of approximately 2 μamps^{-2} is generated. Where the current enters, cotyledons eventually initiate and where the current exits the root will arise (Gorst et al., 1987). With the addition of auxin (in these experiments, 2,4-D) cell clusters enlarged but failed to differentiate and would eventually fragment. Only when the 2,4-D was removed from a cluster would somatic embryo formation resume. What is of considerable interest is that throughout the period of 2,4-D treatment the through currents that were first generated when the cluster was small continued to be maintained and were therefore inherently conserved as part of the initially induced polarity. Other studies with tobacco cells suggest that 2,4-D is less effective than IAA in maintaining the stability of direction or intensity of transdifferentiation currents (Goldsworthy and Mina, 1991) and they offer this as a reason for the failure of 2,4-D-treated cultures to differentiate a properly organised plant growth.

In vivo, certain mutations cause a failure of cotyledon and meristem apex formation so that the root and hypocotyl appear topless. The *TOPLESS (TPL)* gene in *Arabidopsis* is one that is expressed very early in embryogenesis. Another, the *SHOOT MERISTEMLESS (STM)* gene is essential for only the embryonic meristem formation so that the homozygous recessive embryo will germinate with root, hypocotyl, cotyledon but no apex. The small bump derived from the dividing L1, L2 and L3 tunica and corpus organisation fails to arise in the *stm* or the *tpl* mutants. Whereas the transcript of *STM* is detectable in one or two cells by the late globular stage of embryo formation and subsequently between the cotyledons in all later stages, the cotyledon and leaf primordial development are apparently independent of this particular gene control.

This genetic control of shoot apex formation occurs also in tissue cultures. Callus from roots of *stm* mutants are unable to form apical meristems *in vitro*, whereas callus from wild-type will do so. Apical cells, therefore, possess the capacity for self-renewal that is lacking in the *stm* mutants (Barton and Poethig, 1993).

Up-regulating cytokinins will induce adventitious shoots *in planta* (in the close community of cells that constitute a meristem) by overexpression of the isopentenyltransferase *(ipt)* gene and will also give rise to a "shooty" type of development in transformed cells in culture (Smigocki and Owens, 1988). Whether cytokinins are central to the positional expression of genes for apical meristem initials in *Arabidopsis* is not entirely clear but another indicator that meristems are indeed cytokinin target cells comes from the altered meristem programme *(amp)* mutant which has an abnormally enlarged apex. The *amp*1 mutant does have accompanying high cytokinin levels.

Within the apical dome itself, neighbour cell layers can exhibit an independent target cell status. In maize, for example, *KNOTTED1* gene transcripts are detected in the L2 layer but not in L1, whereas the *KNOTTED1* protein is always present in both cell layers (Jackson et al., 1994). How is the protein transported, therefore, between cells in these two layers? Evidence is now good that there is selective trafficking of proteins such as the *KNOTTED* homeodomain protein and its mRNA across plasmodesmata of certain cell types. Certainly in the close community of cells that constitute a meristem such short-distance selective signal transmission and perception is possible (Lucas et al., 1995).

Genetic evidence, based on the isolation of membrane-associated proteins from mutant *Arabidopsis* plants that exhibit alterations in genes controlling cell numbers in the meristematic apex, gives reason to believe that signalling complexes of proteins under a multiple phosphorylation control of kinases and phosphatases form a basis for intercellular communication. Receptor lipoprotein complexes determine the entry of molecular signals at a plasma membrane and, in turn, their transfer across cytoplasmic interfaces, through MAP kinase-regulated cascades that reach to gene sites in the nucleus. Once bound at the cell surface, a signal molecule encounters a series of discriminatory events on the cytosolic side of the plasma membrane, so that initially small differences between neighbour cells can be amplified or attenuated. Thus each cell will possess an individual target identity, and however small the difference may be between them, the many modifications in conformational states and activity states that develop during a downstream cascade can result in a significant impact upon the transcription output from the nuclear chromatin-located genes that are the final destination of external signals.

Controlling meristem size

Studies of the CLAVATA (CLV) family of proteins (CLV1, CLV2 and CLV3) and a large family of similar proteins have led to some understanding of the biochemical nature of cell-to-cell interaction and signalling in the meristem. In *Arabidopsis,* a series of *clv* mutants has been generated with phenotypes that display an accumulation of undifferentiated stem (central zone) cells, so providing giant meristems in which the normal controls that regulate meristem size do not operate (Leyser and Furner, 1992; Clark et al., 1993). Using the *clv1-1* mutant (Koornneef et al., 1983), Clark et al. (1997) cloned the *CLV1* gene and determined that the CLV1 protein was a member of the RLK-LRR protein family comprising an identifiable N-terminal signal peptide, 21 complete extracellular LRRs, a transmembrane domain and a serine/threonine kinase domain. Williams et al. (1997) demonstrated that the CLV1 protein did undergo autophosphorylation at the Ser/Thr kinase domain and that CLV1 can transphosphorylate another CLV1 protein with an inactive (mutated) kinase domain.

Mutant studies on the second member of the CLV protein family (*clv2* mutants) determined that these plants showed a similar, albeit weaker, phenotype to the *clv1* and *clv3* mutants, suggesting that the CVL2 protein may also interact with CLV1 and CLV3. Jeong et al. (1999) used a T-DNA tagged mutant of *Arabidopsis, clv2-5,* to isolate the *CLV2* gene and determined that the CLV2 protein comprises 720 amino acids with homology to CLV1, but with a truncated cytoplasmic region and no kinase domain. Of further interest from this study is the finding, using anti-CLV1 antibodies, that the CLV1 protein does not accumulate in the *clv2-3* mutant background. This observation, together with an earlier finding that full-length *CLV1* mRNA can be detected in the *clv2-1* mutant background (Kayes and Clark, 1998), suggests that CLV2 is necessary to stabilise the CLV1 protein.

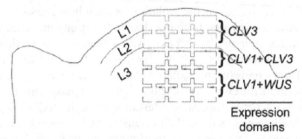

Figure 3.1. Diagrammatic representation of the distribution of CLV3, CLV1, and WUS proteins within the L1, L2, and L3 cell layers of the shoot apical meristem of *Arabidopsis* (modified from Rojo et al., 2002).

The third member of the CLAVATA family, CLV3, has been characterised in some detail by Fletcher et al. (1999), and details of the cloning of *CLV3* and the translated CLV3 protein are provided in Chapter 2. The genetic evidence indicates that *clv1clv3* double-mutant analysis shows that these genes are mutually epistatic, suggesting that there is some molecular interaction of the protein products.

From such genetic evidence, the possibility exists that CLV3 acts as a ligand for CLV1/CLV2, although there is, as yet, no substantiated biochemical evidence for such a relationship. Nevertheless, the work of Rojo et al. (2002) has shown that CLV3 must be localized in the apoplastic space (as predicted from the occurrence of an N-terminal signal sequence) to activate the CLAVATA signalling pathway. Using reporter gene (green fluorescent protein, and GUS):CLV3 translational fusions, Rojo et al. (2002) showed that a full-length CLV3 protein is sufficient to target the protein into the apoplastic space. If the N-terminus is removed, then the translational fusion is not targeted to the apoplast. To determine the significance of secretion, the CLV3 protein was directed through the secretory system to the vacuole using translational fusions. These constructs were transformed into a *clv3-2* genetic background and were shown not to be able to restore the wild-type phenotype, suggesting that secretion into the apoplast is necessary for activation of the meristem control pathway.

The regulatory CLAVATA genes have important implications also for the function of other genes. For example they interact in turn with another gene *WUSCHEL (WUS)* which operates in a feed-back loop to control CLV-mediated signal transduction pathways. In *Arabidopsis*, WUS encodes a homeodomain protein that is localized in the nucleus and is restricted to a small group of cells that lies beneath those of the CLV3 expression domain in the L3 region of the apical meristem (Figure 3.1). Two further genes (FAS1 and FAS2), which encode components for chromatin assembly, are essential for maintaining WUS transcription (Fletcher, 2002). The feed-back loop between *WUS* and *CLAVATA* expression is therefore subtle and intricate so that the maintenance of a controlled meristem performance that retains the integrity of expression of the progenitor stem cells requires and offers very fine tuning by each target cell within the complex. In one model, Lenhard and Laux (2003) have proposed that CLV3 is a mobile signal that diffuses laterally from the stem cells and represses WUS expression. However, in the underlying central organizing region, CLV3 is already bound to CLV1, so the availability to repress the expression of *WUS* must be reduced as the cells extend.

Recognition by membrane surface receptors is not the only way that signal molecules may enter a cell. Plasmodesmatal pores that traverse both cell walls and plasma membranes provide a major traffic route and conduits for direct continuity of the cytosol between adjoining cells (reviewed in Lucas, 1995).

Although the plasmodesmata are of highly complex structure at maturity, their origins lie at the earliest stages of cell formation as part of the cytoplasmic connections that remain between daughter cells at cytokinesis. Initially, these connections of approximately 1–1.5 nm in diameter remain as highways for small or large molecules, viral particles, nucleic acids and proteins. They provide an unassisted passage from cell to cell, limited only by the exclusion size of the pore and the modifying effects of the array of protein particles aligned against the outer surface of the plasma membrane that lines the pore and the endoplasmic reticulum that fills the pore channel. In the meristem, the closely adhering cells show simple unbranched pores, with a generally low exclusion size, which protects meristems from major invasions by viral particles that normally exceed 2.0–2.5 nm. However, since dyes and dextrans appear to have free access, small signal molecules presumably also move freely by this path implying that a high state of cytosolic control must operate to maintain the secure target state. Once cell expansion commences, however, secondary plasmodesmata develop, the simple plasmodesmata can become branched and the molecular size exclusion or permissive traffic control is altered (Ghoshroy et al., 1997). Signal exchange between neighbour cells therefore changes as the target status of the cell changes, essentially from that of primarily short-distance information transfer to that of an increasing accommodation for the long-distance signal information from and to other differentiated tissues.

To have a direct effect upon the transcriptional events of gene expression, the transport-aided signal must not only cross cytoplasmic barriers but also reach and enter the nucleus. This means that for traffic both into and out of the nucleus, control is at least twofold: by nuclear membrane discrimination and by nuclear pore regulation (Dingwall, 1991). In this sense, both plasma membrane and the nuclear membrane provide optional conveyancing systems. Nuclear targeting and trafficking in plants for hormonal or other signal molecules is still relatively unclear, but accessory proteins, nuclear localisation sequences and phosphorylation levels must play their part.

Other short-distance signals

Contact pressures as signals. Touch or contact signalling is essentially an epidermal-based event, but cell expansion growth itself can also evoke internal pressures between the different layers of cells of an expanding primordium.

A natural "buckling" of the cells occurs during expansion growth as each cell exerts pressure upon its neighbour during the downward displacement from the shoot apical meristem. Mechanical buckling is seen as a "minimal energy phenomenon" by which folds and ridges originate as more cells are produced from the meristem apex. The directional stresses so induced, feed back upon the cytoskeletal behaviour of the cells involved. In this way, topographical cell buckling or wall stretching is coupled to internal cellular change and altered cell boundary conditions (see Green, 1999, for detailed discussion). The ongoing cell–cell interactions that proceed from growth open up continuing opportunities for changes in microtubule orientations and in the plasma membrane operation of plasmodesmatal trafficking and signal perceptions and transmissions at the plasma membrane. Even though cells may be neighbours, it is evident that those in the meristem, where short-distance information sharing plays a major part in L1, L2 and L3 organization, neither receive nor transmit identical signals, either in the form of hormones or other organic or inorganic molecules or ions or by external or internal contact pressures. Each cell in the community expresses an ever-changing, dynamic target identity and in concert, a continually changing response in gene expression.

Another important aspect of contact signal transmission is the sensitivity of plants to touching, the thigmostimulus. We already know that moving a dish of germinated pea plants and the rattling together of the etiolated pea stems can evolve enough contact-induced "wound" ethylene to modify the lipid synthesis 3 hours later (Irvine and Osborne, 1973). Thus the implications for the status of lipoproteins in plasma membranes and their receptor sites is clear. Plant tendril epidermis cells, for example, are highly perceptive to a contact touch. Peptides containing the sequence arginine-glycine-asparagine (RGD) when applied to the external walls of *Characean* cells can interfere with a wide variety of responses including gravity sensing (Wayne et al., 1992). Proteins (integrins) that bind this RGD sequence are located in the plasma membrane of both fungi and higher plants (Laval et al., 1999). Strands that link the plasma membrane to the cell wall even during plasmolysis of the cell are now accredited with specific properties. First described by Hecht (1912) these are now shown to contain actin microfilaments and microtubules (Lang-Pauluzzi and Gunning, 2000). Walled cells exposed to RGD-containing peptides lose these Hechtian strands with a loss of the signal transmission pathway between the cell wall and the plasma membrane (Kiba et al., 1998). The integrin proteins and the Hechtian strands are now considered to be linked to microtubules in the cortical cytoplasm which in turn can regulate ATPase activity and cell growth (Nick, 1999). A model for the pressure response across a membrane proposed by Jaffe at al. (2002) indicates how the thigmo-input would be transduced to integrin-like receptors in the plasma membrane via Hechtian strands.

ACC as a short-distance signal. One of the better characterised systems in which there is evidence that ACC, the ethylene precursor, can act as a short-distance signal is in the coordination of post-pollination ethylene production in flowers of the orchid, *Phalaenopsis* spp. In a detailed study, Bui and O'Neill (1998)

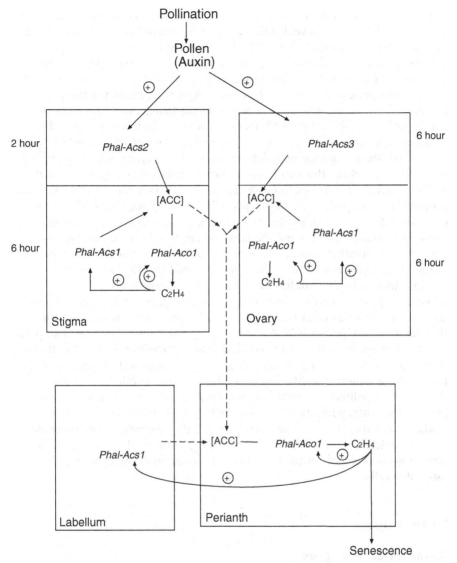

Figure 3.2. Diagrammatic representation of ACC as a short-distance signal in response to pollination-induced changes in different floral tissues of the orchid (modified from Bui and O'Neill, 1998). ACS = ACC synthase; ACO = ACC oxidase.

examined the differential expression of ACC synthase genes in different floral tissues post-pollination. In this system, the exact nature of the primary pollination signal was unknown, but the ethylene produced acted as a secondary pollination signal. Post-pollination, ACC synthase activity was observed first in the stigma, then in the ovary and in the labellum but not in the perianth tissues, although in a previous study all these tissues were shown to produce ethylene in response to pollination (O'Neill et al., 1993). Examination of the spectrum of ACC synthase genes revealed that *Phal-ACS2* was induced first in the stigma after 2 hours and its expression peaked at 24 hours; expression of this gene was not observed in the labellum or perianth tissues. At 6 hours post-pollination, expression of a second ACC synthase gene, *Phal-ACS3*, was observed in the ovary, with a peak of expression at 24 hours. Again, expression of this gene was confined to the ovary, it was not observed in the labellum and perianth tissues and, in common with *Phal-ACS2*, its expression was induced by auxin. Earlier investigations with orchid flowers had identified another ACC synthase gene, *Phal-ACS1*, the expression of which was shown to be up-regulated by ethylene (O'Neill et al., 1993). Bui and O'Neill (1998) showed that expression of *Phal-ACS1* occurred in the stigma, but with a peak of expression after *Phal-ACS2*. They proposed that the pollination signal (possibly auxin) up-regulated the expression of *Phal-ACS2* in the stigma, the ACC produced was converted to ethylene by basal ACC oxidase activity and the ethylene produced turned on expression of *Phal-ACS1*. The ethylene-induced *Phal-ACS1* was also shown to be expressed in the ovary and the labellum where it is induced by ethylene, but not in the perianth tissues. Earlier studies by O'Neill and colleagues (Nadeau et al. 1993; O'Neill et al., 1993) had shown that ACC oxidase gene expression occurred in the stigma at 6 hours (that then peaked at 24 hours), later in the labellum and then finally in the perianth 24–48 hours post-pollination to coincide with petal and sepal senescence. Bui and O'Neill, therefore, proposed that pollination induces ACC synthase gene expression and activity first in the gynoecium and then in the labellum. The ACC produced is converted to ethylene which acts as the secondary pollination signal (and induces *Phal-ACS1*), but ACC is also transported to the outer perianth tissues (that lack ACC synthase activity) where ACC oxidase converts it to ethylene, thereby regulating the senescence programme of these different target cell types (Figure 3.2). Thus in these tissues, at least, there is good evidence for the short-distance transport of ACC between closely associated cells.

Inter-wall signals

Saccharide-derived signals

Oligogalacturonides. The earliest observation that applied oligogalacturonides could regulate plant growth and development came from experiments with tobacco thin-layer explants (Tran Thanh Van et al., 1985). The explants consisted of three to four cell types (epidermis, collenchyma, chlorenchyma and parenchyma) arranged in four to ten cell layers and were usually excised from basal internodes of primary floral branches. Concentrations of a synthetic auxin (IBA) and cytokinins and the pH of the medium can all be manipulated to induce flower formation, vegetative bud growth, and root or callus formation (Mohnen et al., 1990). When Tran Thanh Van and colleagues added products of endo-α-1,4-polygalacturonase-treated sycamore cell walls to such thin layers, the differentiation of the primordia was altered to express new developmental patterns. Subsequent studies involved the purification and characterisation of the active fragments released by the action of polygalacturonase. Eberhard et al. (1989) showed that fragments released from sycamore cell walls inhibited root formation and induced a polar-directed tissue enlargement and flower formation. Marfa et al. (1991) used the induction of flowering in these layers as an assay to purify active compounds and established that oligogalacturonides of 12–14 DP (degrees of polymerisation) were optimal in terms of activity. Bellincampi et al. (1993) found that oligogalacturanans inhibited the auxin (IBA)-induced formation of roots both on thin layers and on tobacco leaf explants. Also, inhibition could be reversed by the addition of yet higher concentrations of auxin. Further examination of this interaction with auxin found that in leaf explants, auxin is needed for at least 4 days to induce roots, but OGAs must be applied at days 1 or 2 to be effective. The inhibitory effects of the OGAs are, however, reversible so that auxin can again induce root formation in tobacco leaf explants when the applied OGAs are withdrawn. These effects have been shown to be optimal at 10-14 DP. To determine the nature of the interaction of these cells with auxin and OGAs, Bellincampi et al. (1996) studied explants excised from *rolB*-transformed tobacco plants with high rooting expression; OGAs were shown to inhibit *rolB*-driven root morphogenesis. Furthermore, the use of *rolB*-β-glucuronidase gene fusions demonstrated that OGAs inhibit at the level of the auxin-induced transcription of *rolB* genes.

Further definition of the interaction between OGAs and auxin has been investigated by Spiro et al. (2002) using intact cucumber (*Cucumis sativa* L.) seedlings. These workers showed that IAA inhibited root growth, but OGA had no effect. When IAA and OGA were added together, OGA initially did not retard IAA-induced growth inhibition, but it did result in the earlier re-establishment of growth when compared with auxin-treated tissues without added OGA. Of most relevance to the target cell concept is that OGA induced its maximal response (as determined by extracellular alkalinization) in the basal region of the root, whereas IAA-induced alkalinization occurred in the growing apical portion. These results suggest that OGAs do not operate by simply inhibiting IAA effects directly, and some support for this concept has been provided by Mauro et al. (2002) who determined that in tobacco seedlings OGA inhibits the induction of the late (i.e., cycloheximide-sensitive) auxin-induced genes, but not in the earlier (primary) auxin-induced genes (i.e., those whose induction is not sensitive to cycloheximide) (see Chapter 7 for a description of auxin-induced genes).

In terms of a regulation by oligogalacturonides *in vivo*, these molecules have not been found as soluble factors *in planta* during normal growth or as products in cell culture media, but they are released from cell walls by the action of polygalacturonases (PGs) produced by fungal pathogens and hence these products are critical to plant defence responses (Darvill et al., 1992). The PGs produced by pathogen attack are regulated by plant cell-wall-localised polygalacturonase-inhibitor proteins (PGIPs) which are widespread amongst plant species (Yao et al., 1995). The complexity in the regulation of cell wall enzymes controlling OGA release from pectins has implications for inter-wall signals during every stage of plant development as well as in responses to pathogens. The PG enzymes belong to multigene families that are precisely developmentally regulated (Hadfield and Bennett, 1998), so it is easy to envisage how their differential expression can confer a significant degree of control on oligogalacturonide signalling and the subsequent perception and response of each target cell *in planta*.

Oligogalacturonides (with an effective range of 9–16 DP; optimal at 12–14 DP) can cause K+ efflux, alkalinization of the incubation medium, acidification of the cytoplasm and an influx of calcium in cultured tobacco cells (Mathieu et al., 1991). These effects are likely to be involved in signal transductions induced by specific OGAs *in planta*, although the precise significance of each has yet to be determined.

How OGAs might be recognised by plant tissues as part of a response has been hinted at with the identification of a 34 kDa plasma membrane bound protein, originally purified from potato or tomato leaves, the phosphorylation of which *in vitro* is enhanced by applied oligogalacturonides (Farmer et al., 1989), with a minimum length calculated at 14–15 residues (Farmer et al., 1991). This protein has been identified in plasma membranes of tomato roots, hypocotyls and stems as well as leaf membrane preparations of soybean leaves (Reymond et al., 1995). Of particular interest with regard to this putative receptor is the molecular size of the DP recognition. Reymond et al. (1995), using purified oligogalacturonide fragments, determined that a DP range of 13–26 residues was sufficient to induce phosphorylation of the putative receptor. This size range was larger than the usual range shown for applied OGAs to induce a defence response in plant tissues (i.e., 10–15 DP). To explain this, it was postulated that the accessibility of the applied OGAs through the cell wall may permit only lower molecular mass structures to pass through to the cell membrane, whereas the phosphorylation experiments were carried out on isolated and purified membranes. It is tempting to speculate, therefore, that part of the PG hydrolysis that produces OGAs may also induce temporal changes in the structure of the cell wall thereby permitting OGAs of critical sizes to pass for receptor binding and endocytotic internalisation that is both tissue-type-specific and developmentally determined.

Arabinogalactan proteins. The assessment of arabinogalactan proteins (AGPs) as signals during plant growth and development has involved two intimately associated experimental approaches. The first uses specific monoclonal antibodies that recognise AGPs as epitopes on the cell surface, the expression of which is developmentally regulated (reviewed by Knox, 1995). The second applies these molecules to cell cultures and observes the developmental changes that are evoked (reviewed by Kreuger and van Holst, 1996). Although the extent of the relationship between cell-wall-bound AGPs and soluble compounds in short-distance cell-to-cell signalling has yet to be established unequivocally, it is certain that the pattern of AGP expression at the cell surface changes during development, and further, the display of a specific AGP marks each cell type for differentiation along a certain pathway.

The primary evidence for AGPs as soluble signals is the demonstration of their presence in the medium of cell suspension cultures. Komalavilas et al. (1991) used "Paul's Scarlet" rose cells in suspension culture and isolated both membrane-bound and secreted AGPs. These differed in structure with the secreted forms slightly smaller in molecular mass when compared with the membrane-bound forms. Carrot cell suspension cultures also liberate biologically active AGPs to the medium (Kreuger and van Holst, 1993). The AGPs from carrot seeds (isolated by precipitation with β-glucosyl Yariv antigen), altered the growth of 3-month-old embryogenic cell lines. These cultures, with less than 30 percent of the dense, highly cytoplasmic appearance diagnostic of an embryogenic cell, were increased to 80 percent with the addition of carrot seed AGPs. Further, the addition of AGPs from carrot seed to a 2-year-old non-embryogenic cell line resulted in the re-induction of embryogenic potential. Kreuger and van Holst (1995) produced two monoclonal antibodies to AGPs from carrot seeds which precipitated specific AGP fractions from both tomato and carrot seeds. However, the tomato and carrot AGP fractions had differing biological activities. The first (recognised by monoclonal ZUM15) induced vacuolation of embryogenic cells in culture such that these cells then failed to produce embryos. The second (recognised by monoclonal ZUM18) increased the percentage of embryogenic cells from 40 to 80 percent. Further evidence for the signalling role of these molecules has been provided by the experiments of Van Hengel et al. (2001). Using protoplasts of carrot (*Daucus carota*), they showed that the addition of secreted AGPs from the media could promote somatic embryogenesis, and this effect was enhanced if the AGPs added were pretreated with endochitinase. Together, these results demonstrate an important precedent in the consideration of AGPs as short-distance signalling molecules. It is now clear that plant cells have the competence to produce and secrete immunologically distinct AGPs with different biological activities and of varying structures; this competence represents a ready potential for generating an unlimited array of signalling information.

The second approach used in the assessment of AGPs as signalling molecules comes from studies using monoclonal antibodies to detect AGP epitopes localised either on the plasma membrane or in the extracellular matrix. In the first such study, Knox et al. (1989) found early markers of cell position, but not of cell type, in the root apical meristem of *Daucus carota* L. Using a monoclonal antibody, JIM4, isolated from clones raised against immunized carrot cell protoplasts, they identified two discrete cell clusters in the developing pericycle, in the stele, and in mature tissues of the epidermis. A developmentally regulated pattern of expression was similarly recognised in other cell types during embryogenesis (Stacey et al., 1990).

Taken together, these studies, with others not reported here, illustrate collectively an important spatio-temporal role for AGPs in plant cell differentiation, as well as the opportunities they afford to mark particular target cell types. The question that arises from such cell localisation studies is whether these changes in AGP epitope expression occur as a result of cell differentiation or the changes in expression direct such processes. The observation that applied AGPs can influence cell differentiation, albeit in cell culture, would support a role for these compounds as developmental cues *in vivo*. Suzuki et al. (2002) showed that the incubation of barley aleurone protoplasts with Yariv antigen inhibited the induction of α-amylase activity by GA. They also isolated an aleurone-specific AGP (underlining the cell-specific nature of AGP expression), further suggesting that these cell wall fragments are functional in directing a hormone-regulated process in plants. However, much further investigation is needed before AGPs can be included as truly directive short-distance signalling molecules in plants.

Xyloglucans. In terms of signalling molecules, xyloglucans share important similarities with the oligogalacturonides. Both are cleaved from larger polysaccharide precursors which are important structural components of the cell wall and, in addition to the information contained within the oligosaccharin structure, the local release of these molecules provides a control point which can influence every neighbour cell within reach.

The first demonstration of such molecules came from the search for the molecular basis of auxin-induced cell elongation. York et al. (1984) purified an endo-β-glucanase activity from the culture fluid of *Trichoderma viride* and added it to the soluble polysaccharides from suspension cultures of sycamore. Two of the structures derived were then isolated and characterised in detail as $Glc_4Xyl_3Fuc_1Gal_1$ (XXFG) and Glc_4Xyl_3 (XXXG) (for structural details, see Fry et al., 1993). Studies with 8–9-day-old pea stems showed that XXFG, but not XXXG, inhibits 2,4-D-induced stem elongation. A single concentration of auxin and a range of concentrations of XXFG were used to determine that the concentration optimum for the xyloglucan was as little as 10^{-9} to 10^{-11}M. It seemed at first that the fucose moiety was essential for anti-auxin activity (McDougall and Fry, 1989; Augur et al., 1992), but the identification of an *Arabidopsis* mutant containing L-fucose substituted by L-galactose in its active xyloglucan has shown that galactose can substitute for fucose without loss of signalling information (Zablackis et al., 1996). The mutant excepted, the activity of a cell-wall-bound α-fucosidase may be important, therefore, in the regulation of the extent of auxin-induced growth by liberated XXFG fragments (Augur et al., 1993). Growth inhibition by a xyloglucan is not restricted to auxin-induced growth since the GA_3-induced (at 10^{-5}M) elongation of 6–7-day-old etiolated pea epicotyls is also suppressed by nanomolar (10^{-9}–10^{-11}M) concentrations of XXFG (Warneck and Seitz, 1993). Over this concentration range, XXFG also inhibits epicotyl growth in the absence of GA_3, presumably through interactions with endogenous auxins. The concentration of oligosaccharins is quite critical to the response that is elicited, with higher concentrations increasing the rate of cell extension rather than inhibiting growth.

McDougall and Fry (1990), used another endo-β-glucanase digested xyloglucan from Paul's Scarlet rose cell suspension cultures to isolate and identify four distinct structures [XXXG; XLLG ($Glc_4Xyl_3Gal_2$); XXLG ($Glc_4Xyl_3Gal_1$); XXFG; Fry et al., 1993)] two of which, XXLG and XLLG, promoted growth in the pea stem segment assay and stimulated an acidic cellulase activity from *P. vulgaris*. The optimal concentration was found to be in the order of 10^{-6}M, which prompted speculation that these compounds could act as substrates for the enzyme xyloglucan endotransferase (XET) (Fry et al., 1993).

Using a fluorescently labelled xyloglucan, high resolution scanning electron microscopy, and confocal microscopy, Takeda et al. (2002) showed that added

XXXG could induce elongation in epicotyl segments of pea from which the epidermis was removed, and that added auxin (2,4-D) had a synergistic effect. They also determined that XXXG solubilised xyloglucan from the cell wall but maintained the microtubules in the transverse orientation. In this orientation, cell growth occurred by elongation along the primary (apical–basal) axis of the stem. The questions of most relevance to the target cell concept are whether these xyloglucan derivatives occur naturally, and how the generation of these signalling molecules might be regulated. As short-distance signalling molecules, their biosynthesis must be intimately linked to the target state of the cell that produces them. It is still viewed currently that they are generated from larger xyloglucan precursors rather than by direct synthesis and secretion with other cell wall structural polysaccharides. The nature of the larger precursors is still unknown and the putative endo-β-glucanase which might act on polymeric substrates is not yet identified. It has long been known that cell wall hydrolases in stem tissue are hormonally regulated (Fan and Maclachlan, 1967; Ridge and Osborne, 1969), and that the activity of enzymes that can release oligosaccharins are, in turn, stimulated by oligosaccharins (McDougall and Fry, 1990; Farkas and Maclachlan, 1988). It is possible to envisage, therefore, that the hormonal regulation of cell wall hydrolases, coupled with the developmentally regulated availability of polymeric substrates and the control of the hormonal response by the oligosaccharins released, represents a feed-back loop of sufficient complexity of interaction to provide a unique target cell status for every individual growing cell.

Evaluation of the unconjugated N-glycans as signal molecules. Consideration of UNGs as signalling molecules is intimately linked to the importance of these structures on plant glycoproteins. In animal biology, evidence clearly points to a range of functions for such sugar structures including the determination of protein conformation, as protection from proteases and as motifs for interorganellar sorting and secretion (Dwek, 1995). In plants, far fewer studies have addressed the question directly, but some evidence suggests that these molecules are important for protein function (Van Huystee and McManus, 1998). Of relevance to this volume is that these glycan molecules may provide specific information for the protein as well as signalling information when the UNG becomes separated from the parent protein. This signalling scenario has attractive parallels with the intercellular communication mediated by oligogalacturonides and xyloglucan derivatives released from cell wall polysaccharides. While not covalently bound to the wall, many plant glycoproteins are ionically linked to the wall matrix and many are secreted when cells are maintained in suspension culture. Putative de-N-glycosylation enzymes are also secreted, suggesting a similar site for enzymes and substrates (Berger et al., 1996).

An important component of the assessment of UNGs as signalling molecules is whether their generation is developmentally regulated. In one study of radish seed development and germination (Berger et al., 1996), two enzymes, an endo-N-acetyl-β-D-glucosaminidase (ENGase) and peptide-N4-(N-acetyl-β-D-glucosaminyl) asparagine amidase (PNGase), were found to have different substrate specificities. The ENGase preferentially released oligomannose glycans and PNGase released more complex glycans. The activity of both enzymes was modified by ABA and gibberellin, thus affording a further tier of developmental control. Both enzymes have been linked to the increased production of UNGs into the medium of a suspension culture of *Silene alba*, particularly under carbon-starvation conditions (Lhernould et al., 1994).

Oligosaccharins and the target cell concept. Although receptors for oligosaccharins have not been characterised, this does not exclude their existence in plants. Alternatively, it has been suggested that there may be no receptors or membrane carriers for UNGs, but rather these molecules exert their influence by disrupting lectin:sugar interactions (Priem et al., 1994).

For each class of oligosaccharin signalling molecule described here, some discussion has been included about how specificity or selectivity of these short-distance signals may be implemented. In contrast to the signalling hormones such as IAA or ethylene, which have a conserved structure in all plant parts, oligosaccharin structures vary greatly. A large number of *N*-glycan structures are known so far, and the range of sugars found in the arabinogalactans and the specificity exerted by the size (DP) of oligogalacturonides are wide and possibly plant-specific with some indication for developmental regulation. The current lack of evidence for proteins that can recognise xyloglucan derivatives or the unconjugated *N*-glycans does not exclude the natural occurrence of such receptors.

For all of these oligosaccharin signalling compounds, further selectivity may also be mediated through the very probable tight developmental regulation of the enzymes which catalyse their formation – the polygalacturonases, the endo-β-1,4-glucanhydrolases and the de-*N*-glycosylation enzymes. The regulation of these enzymes is coupled with the probability that the availability of wall-specific substrates almost certainly differs from cell to cell. Further, at each differentiation state cells show differences in wall structure providing a mechanism for further diversity of signal molecules and the regulation of their production. Taken together, the modes of generating intimate signal diversity from cell wall saccharides and glycoproteins provides a multitude of developmentally regulated neighbour-to-neighbour cell-specific intertissue signals.

This also creates the possibility of target cell competence being determined by the ability of any cell or tissue to generate its own signal molecules. While direct evidence for such mechanisms is lacking, the extent to which hormones are the ultimate regulators of specific developmental processes in the plant becomes a subject for debate. The control of such multiple signal production in target cells is a major part of the target cell concept and is addressed in more detail in Chapters 7 and 8.

Lignin-derived signals

We do not know the limits of cell wall variability which includes not just the basic pectin, cellulose, hemicellulose, and protein elements, but also those of additional molecules such as polyphenol propanoids that can be condensed into the wall matrices at specific dirigent protein directed sites. Probably all cells contain potential precursor lignans (monomers, dimers, oligomers) but not all cells show similar definitive lignin structures in their walls. Differential gene expression and the temporal and spatial activity of genes in various vascular cell types give rise to micro-heterogeneity for both lignin biosynthesis between vessels and tracheids. For example, expression of caffeoyl coenzyme A O-methyltransferase (CCOMT) appears confined to tracheids and fibres but is undetectable in vessels (Ye et al., 2001). Structural genes in the lignin pathway are mainly regulated through transcriptional control, so cell-specific expression resides in the regulatory sequences. A series of events then links lignin monomer biosynthesis with insertion and polymerisation within a pre-existing extracellular wall matrix. The down-regulation of various lignin biosynthetic genes gives transformed lines with different lignin contents and different lignin monomer compositions.

Lignin heterogeneity can exist at both cellular and subcellular levels. In *Picea*, for example, the middle-lamella and cell wall corners contain lignin of a higher p-hydroxyphenol propane content than secondary wall layers (Whiting and Goring, 1983). These differences in synthesis of lignins and the specific modes of deposition must reflect back upon the events that occur later during development and at senescence. The differentiation of xylem elements within the growing stems of angiosperms is a highly localized programme of cell death that takes place within the environment of non-senescent neighbour cells. In the cells that abut upon the abscission zones, however, all of the tissues distal to the abscission zone are involved in programmed cell death before the abscission event is achieved. Furthermore, the abscission cells, though being the immediate targets for the reception of senescence-induced signals, do not themselves become senescent. In contrast, they become highly metabolically activated and express a new array of growth and gene expressions in response to signals from their neighbours. We are currently limited in our knowledge of which tissues or cells cause the new signals to arise but we do know that if the lignified tissues of the vascular system are removed, then no signal to abscise is generated (Thompson and Osborne, 1994). The positional or directive signals that lignified cells may offer to their neighbours during the developmental stages of plant growth can also be seen as operative through a target cell communication cascade long before any degradative events of overall senescence come into play.

The higher plant lignins are all derived from monolignols produced via the shikimate pathway. As a multistep process to the polymerisation that follows, many genes have now been identified that offer a variety of possibilities for the final selective targeting to the cell wall. Genetic modifications to tobacco plants give a number of indicators of how these possibilities may be brought about in wild-type and in transgenic plants. Anti-sense inhibition of cinnamoyl CoA reductase (CCR) activity leads to a reduction of up to 50 percent of the total lignin content accompanied by a loosening of the cell wall and reduced cell wall cohesion (Piquemal et al., 1998; Ruel et al., 2001). However, a double transformant of CCR and cinnamyl alcohol dehydrogenase (CAD), although still showing the high reduction in total lignin, did not show the loosening effects on its xylem cell walls. This indicates the kinds of morphological modifications that can be induced in wall substructure and the effects that these can then have upon the subsequent wall degradation and signal products at senescence. Both the variations in the saccharides of wall composition and the irregular distribution of the lignins within the wall lead to numerous opportunities for spatially controlled wall assemblies and an almost unlimited number of lignin signal molecules that can be derived from their eventual degradation products.

Whereas we seek in molecular terms the chemical nature of the limitless numbers of signals to which a plant cell is exposed and to which it may or may not respond, the sensitivity with which other organisms can recognize these products can differ from that of the plant and at present, it lies well beyond the sensitivity of our own detection systems. A few examples suffice to illustrate the target sites in plant cell walls that are specifically recognised by micro-organisms.

Using purified isolated cell walls preparations from the *Agrobacterium rhizogenes* T-DNA-transformed cells of carrot roots, and inoculating them with a *Gigaspora* species, researchers found that appresoria form only on the cell walls of epidermal cells from the host roots, not on those of the cortical cells (Nagahashi and Douds, 1997). Non-host root wall preparations (e.g., those of beet) did not induce appressorial formation either. Since no living cells are implicated this suggests that the wall constitution alone can determine whether appresoria will form. Although the epidermis can be recognised as a check point to infection, another checkpoint lies in the ability of the pathogen to escape and then penetrate the root cortex. For example, the *li sym4-2* mutant of *Lotus japonicus* exerts a block to a successful mycorrhizal symbiosis, presumably because the hyphae cannot exit the epidermal cell on the inner periclinal wall (Bonfante et al., 2000).

The extent of hormone production by specific cells is another sensitive recognition mechanism existing between the invader and the plant target cell. Only cells of the inner cortex facing the phloem poles express ACC oxidase, as judged by *in-situ* hybridisation in pea and vetch roots. Since nodule formation is inhibited by ethylene, this may explain why nodule primordia are generally positioned opposite xylem poles and not opposite phloem poles (Heidstra et al., 1997). In the non-nodulating pea mutant *sym16,* the wild-type (readily nodulating) phenotype can be restored and the nodule primordia initiated by treatment with silver ions or by the application of the ACC synthase inhibitor, AVG (Guinel and Geil, 2002). These elegant examples of positional differentiation and target cell signalling, so difficult to demonstrate in normal plant cell-to-cell interactions, are readily displayed at the plant micro-organism–plant cell interface.

Long-distance signals

Early evidence of long-distance cell-to-cell signalling to target regions in higher plants includes the concept of a floral-inducing stimulus (Chailakhyan, 1936). Inductive photoperiods of daylength were perceived by leaves from which a stimulus was generated that travelled to the vegetative meristem. There, transition events took place in which a floral meristem was initiated in place of the previous meristematic apex. The conduit involves phloem transport. No flowering hormone has yet been isolated and a floral receptor in the vegetative apex has not been determined, but irrefutably the message passes from leaf to apex. The photo-induced leaf does not lose its photo-inductive status as a result, even to the extent of functioning as a flowering inducer when grafted onto another uninduced plant (see Salisbury, 1963).

Went (1936) had already proposed that factors other than water and nutrients travel in the xylem sap from the roots to the shoots thereby influencing and coordinating the plant's behaviour. The irrefutable evidence for the formation and transport of a flower-inducing signal laid the foundation for the long-distance transport of information by chemical substances. The flowering stimulus is now accepted as more complex than the movement of a single type of signal molecule, but there is ample evidence for the transport of many such molecules in the long-distance communication between cells.

The *pin-formed* (*pin*) mutants of *Arabidopsis* have been particularly useful in establishing the role of polar auxin transport during flowering. In the mutant, there are no flowers or abnormal flowers on the inflorescence axis. Goto et al. (1987) demonstrated that the application of polar auxin inhibitors to wild-type *Arabidopsis* induces a phenotype similar to the *pin* mutants (Okada et al., 1991). Further, expression of the *iaaH* gene from *Agrobacterium tumefaciens* in the *pin* mutant background (which increases the endogenous IAA content in these plants) does not alter the mutant phenotype which retains the non-flowering condition (Oka et al., 1999), indicating that an operational polar auxin transport pathway appears to be a pre-requisite for the formation of floral meristems.

Hormones and root-to-shoot signalling

It is not our intention to exhaustively review the literature on hormone transport pathways, but integral to our consideration of target cells is how the movement of these signalling molecules between different plant parts is established. While the development of roots from root meristems and shoots from shoot meristems is intimately controlled, the two-way traffic of root-to-shoot and shoot-to-root communication is the basis of the coordination of responses to internal and external signalling cues.

The classic view of the regulation of root-to-shoot signalling involves the role of auxin from the shoot regulating the translocation of cytokinins from the roots. Such studies have established the significance of the auxin/cytokinin interaction, but work more recently with the *ramosus* (increased branching phenotype) pea mutants has shown that other yet-to-be-identified molecules also signal as part of this hormonal interaction (Beveridge, 2000). In these investigations, the concept of a feed-back signal from the shoot regulating the production of a branching regulator from the roots has been established. Further, reciprocal grafting experiments with the *rms1* and *rms2* mutants have determined that the *rms2* mutant is impaired in the shoot-to-root feed-back signal, while the *rms1* mutant affects a signal moving from roots to shoots. However, measurements of the levels of cytokinins and auxins in these mutant backgrounds and in grafts show that signals other than auxin act as the feed-back signal from the shoot, and signals other than cytokinin act as the branching regulator from the root (Beveridge et al., 1997a,b).

A role for cytokinins as a root-to-shoot signal has been well established in the mediation of nitrate transport from roots to shoots since the early observations of Simpson et al. (1982). More recent research has established that in the nitrate-sufficient root, cytokinin biosynthesis is switched on and cytokinin translocated from the roots to the shoots where the hormone is proposed to be perceived in the leaf tissue by a His-Asp phosphorelay system with the attendant signal transduction events (Sakakibara et al., 1998, 2000). The cytokinin induces the expression of a host of nitrogen-metabolism-associated genes, while the translocated nitrate (a signal in its own right) induces the expression of genes that are more directly associated with nitrate uptake and reduction, and ammonia assimilation (Takei et al., 2002).

ACC and long-distance signalling in plants. ACC, the immediate precursor to ethylene, is known to move from cell to cell, as has been shown in the stigma-to-perianth transfer in the orchid flower (see earlier discussion in this chapter). It therefore acts as a signalling molecule in its own right, although its conversion to ethylene is still likely to be the mechanism by which the ACC is transduced in terms of developmental changes. In addition to this short-distance cell-to-cell movement, ACC can act as a longer-distance signalling molecule. A well-characterised example is the transmission of ACC from roots to shoots in response to flooding (Jackson, 2002). Flooding the roots of *Lycopersicon esculentum* and *Ricinus communis* induces ethylene production in the shoots and leaf epinastic curvature. Bradford and Yang (1980) were the first to show that ACC moving in the xylem from the roots to shoots is the signal connecting the two tissues. Subsequent work in tomato has confirmed that the lack of oxygen in the root tissues caused by flooding inhibits ACC oxidase activity but induces the expression of one member of the ACC synthase gene family (*LE-ACS7*) (Shiu et al., 1998). The ACC thus produced accumulates and is subsequently transported to the shoot. An increase in ACC oxidase activity in the shoot tissues, which is also induced soon after flooding, then converts the ACC to ethylene. The root signal that is responsible for the induction of ACC oxidase activity has yet to be identified, but if the activity of one of the members of the ACC oxidase gene family in tomato is repressed (by anti-sense technology), then less ethylene is produced and the degree of leaf epinastic curvature is reduced (English et al., 1995). It is known that flooding induces the expression of another member of the ACC synthase gene family, *LE-ACS1,* and this may also mediate the induction of ACC oxidase (Jackson, 2002). Notwithstanding that further research is required to more clearly define the nature of signals induced by flooding it is of relevance to the target cell concept that it is only the cells on the upper (adaxial) surface of the petiole that respond to ethylene and enlarge – hence causing epinasty and demonstrating a typical Type 3 target cell response.

ABA and long-distance signalling in plants. When examining the role of ABA as an informational root-to-shoot signalling molecule, studies on the translocating hormone in response to a water deficit is perhaps the best characterised example. That ABA moves from the root and is transported in the xylem to the shoot where it can exert some physiological and biochemical effects is now widely accepted, and the reader is referred to Jackson (1993) for a critique on the early evidence to support this phenomenon. However, as more research is conducted to examine the mechanism by which this signalling occurs, it is becoming clear that the mechanism of transport and its regulation is complex; see Hartung et al. (2002) for a further overview of the subject. In terms of tissue specialisation and the target cell concept, there are two aspects of this signalling system that are of specific interest: (i) the regulation of ABA biosynthesis in the root, and (ii) the supply of ABA to the target tissue of the leaf and in particular to the aperture control of the target stomatal guard cell in the epidermis.

ABA produced in the root is synthesised in the cortex and the stele, with the highest accumulation at the root tip. Much of this ABA can be lost to the rhizosphere, where ABA also accumulates. ABA can be taken up again by plant roots from an external medium, so the relative concentrations of the hormone in the soil and root tissues become an important regulator of ABA levels in root cells. It is now known that conjugates of ABA, predominantly glucose esters, are also synthesised in the root tissues, and they too can exist in the external medium from where ABA can again be taken up by plants. As a free acid or as conjugates, ABA can move into the xylem either symplastically or apoplastically before transport to the shoots. Conjugated forms are also proposed to be transported to the shoots or, as has been reported in maize, conjugated forms can first be converted to free ABA by the action of root-cortex-localised ß-glucosidases (Hartung et al., 2002).

The fate of ABA once it arrives at the shoot suggests that it is redistributed particularly into the more alkaline compartments (i.e., the cytosol) in a pH-dependent mechanism. As well as compartmentalisation, certain cell types, for example the epidermis, may act as major sites for ABA accumulation. The sequestration of ABA in the leaf mesophyll and epidermal cells is important in the context of explaining the discrepancy between measured concentrations of ABA in xylem sap and those required to induce stomatal closure. If stomata were exposed to the concentration of ABA in the xylem sap of well-watered plants, then the stomatal pores would be permanently closed. Therefore, the cells of the leaf mesophyll must play an important role in regulating the concentration of ABA to which the stomatal target cells are exposed (for a more in-depth discussion, see Wilkinson and Davies, 2002). Further, the occurrence of localised apoplastic glucosidases that can release free ABA from transported physiologically inactive conjugates may provide another pool of free ABA that is accessible to guard cells.

ABA perception in plant cells. The significance of compartmentalisation of ABA within target cells in terms of long-distance transport raises the question of the perception of the incoming ABA signal, specifically the location of possible ABA receptors. Currently, the molecular evidence for ABA receptors is not as convincing as that for ethylene or the cytokinins (see Chapters 7 and 8 for a detailed characterisation of these and other signals for which defined receptor proteins have been identified). There are, however, some good candidates for specific binding proteins of ABA, the first being the demonstration of external putative plasma-membrane-binding proteins in guard cells of *Vicia faba* (Hornberg and Weiler, 1984). These workers used photo-affinity labeling to identify proteins of 20.2 kDa (designated site A), 19.3 kDa (site B) and 14.3 kDa (site C) that were specifically cross-linked with *cis*(+)ABA. The identification of plasma-membrane receptors suggests that hormone perception is extracellular; support for this was offered by Anderson et al. (1994) who micro-injected ABA into guard cells of *Commelina communis* and determined that the hormone was ineffective at inhibiting stomatal guard cell opening. Only when applied externally was ABA effective. The guard cells remained alive post-micro-injection, and so it was concluded that the perception site for ABA-mediated inhibition of stomatal opening must be on the extracellular side of the membrane. The sole reliance on external ABA for controlling stomatal opening has been examined by Allan et al. (1994) who used caged ABA micro-injected into guard cells and then pulse-released by UV-mediated photolysis. They showed that the internal ABA could induce stomatal closure and determined that internal ABA may be effective at inducing stomatal closure but is unable to inhibit stomatal opening. This led to the suggestion that (at least) two independent ABA perception mechanisms occur at internal and external receptor sites.

In support of internal receptors, Zhang et al. (2002) purified a soluble ABA binding protein of 42 kDa from epidermal cells of *Vicia faba* using ABA-EAH-Sepharose affinity chromatography. These workers demonstrated ABA stereospecificity, determined that the binding protein has a K_D of 21 nM for ABA and, most importantly, found that monoclonal antibodies raised to the 42 kDa decreased ABA-induced phospholipase D activity in a dose-dependent manner. The induction of phospholipase activity by ABA in guard cells of *Vicia faba* has been shown previously to potentially be part of a downstream signalling cascade. Thus the further dissection of ABA binding and phospholipase D activity will become important evidence of an ensuing transduction event as a consequence of hormone binding.

Nevertheless, identification of the components of the downstream (intracellular) signalling of ABA perception, including the role of heterotrimeric G proteins (Coursol et al., 2003), irrespective of the nature of the receptor protein or its localisation, is now well advanced and readers are referred to the comprehensive review of Finkelstein et al. (2002).

Auxin (IAA) transport and shoot-to-root signalling. In addition to the establishment of a role of auxin transport during flowering, the *pin-formed (pin)* mutants of *Arabidopsis* have also been instrumental in the elucidation of the mechanism of polar auxin transport, perhaps the best characterised of all cell-to-cell and long-distance signalling systems in plants. In their earlier characterisation of the *pin1-1* and *pin1-2* mutants, Okada et al. (1991) determined that both mutant lines displayed reduced polar auxin transport. Galweiler et al. (1998) cloned the *AtPIN1* gene by insertional mutagenesis using the autonomous transposable element *en-1* from maize and determined that it coded for a 622 amino acid protein of 67 kDa with 8–12 putative transmembrane domains flanking a central region that is predominantly hydrophilic. Critically, the fact that the protein was previously localised at the basal ends of xylem parenchyma cells strongly supported a role in polar auxin transport. Galweiler et al. (1998) did not demonstrate that these proteins could transport auxin directly, but Chen et al. (1998) working with another member of the *PIN* gene family *AtPIN2* did so. Using the *agravitropic1* (*agr1*) mutant of *Arabidopsis*, (which has a phenotype that displays increased root growth when treated with auxin, a decreased response to ethylene and to auxin transport inhibitors, and an increased retention of added auxin at the root apex) they cloned the *AGR1* gene and determined that it was closely homologous with the *AtPIN1* gene. Furthermore, when the AGR1/AtPIN2 protein was expressed in yeast, they showed that the presence of the protein promoted an efflux of labelled IAA out of the transformed cells. In roots, Muller et al. (1998) determined that the AtPIN2 protein was localised to the anti- and periclinal sides of the cortical and epidermal cells. Subsequently, two further members of the *PIN* gene family, *AtPIN3* (Friml et al., 2002a) and *AtPIN4* (Friml et al., 2002b), were characterised following the asymmetric redistribution of auxin in tropic responses to light or gravity.

The PIN3 protein is localised in the membranes at the periphery of starch sheath cells within the hypocotyl (the shoot endodermis) and in the root (pericycle and columella). However, these efflux proteins can rapidly (2–5 minutes) relocate within the cell in response to a change in the direction of the gravity vector and so are proposed thereby to mediate a change in auxin flow that regulates differential growth (Friml et al., 2002a). Lateral transport of auxin within the columella cells of the root to the lateral root cap cells in response to a change in the direction of the gravity vector has been confirmed recently using a green fluorescent protein (GFP)-based reporter system in living roots of *Arabidopsis* (Ottenschlager et al., 2002).

The *AtPIN4* gene is expressed in the quiescent centre and surrounding cells in a domain below that of the *AtPIN1* gene. Interruption in the expression of *AtPIN4* results in an inability to maintain a gradient of endogenous auxin from the root tip and a failure to canalize externally applied auxin. It appears that the role of AtPIN4 is to establish a sink of auxin basal to the quiescent center, and that the reverse gradient that is subsequently established dictates the ensuing events of root patterning (Friml et al., 2002b).

Yet another member of the PIN gene family in *Arabidopsis*, *AtPIN7*, has been implicated in the establishment of an auxin polarity at the earliest stage of embryogenesis. Initially, this protein is located in endomembranes at the apical end of the basal cell from the first division of the zygote until the 32-cell stage, denoting the early polarisation of these basal suspensor cells. At the 32-cell stage, this asymmetric localisation reverses, shifting to the basal side of the cells, at the same time as PIN1 becomes localised to the provascular cells facing the basal embryo pole. Within these early forming root tips, PIN1 expression in the cells shifts towards the side of the quiescent centre cells, adjacent to the columella precursors. These reversals coincide with an apical-to-basal generation of auxin gradients. At the 2-cell stage therefore, there is already a distribution between the (basal) auxin transporting cell and the (apical) responding cell, with the first establishment of an auxin gradient. These are the very earliest target cell distinctions that direct the polar axis of a developing embryo and eventually the axiality of the adult plant (Friml et al., 2003).

While the AtPIN proteins have been shown to function as efflux proteins, the AUX1 protein has been shown to have identity to the amino acid permeases of bacteria. Since IAA is structurally related to its precursor, tryptophan, it may be a substrate for these putative influx proteins in plant cells (Bennett et al., 1996). Disruption of expression of the *AUX1* gene results in an auxin-resistant root growth phenotype and the abolition of root curvature in response to gravity. Subsequent localisation studies have revealed that the AUX1 protein may mediate two functionally distinct auxin transport pathways in the root apex of *Arabidopsis*: an acropetal transport in the (inner) protophloem cells towards the root tip and a basipetal transport in the (outer) lateral root cap cells and in those of the columella (Swarup et al., 2001).

The differential expression of the *AtPIN* genes and localisation of the AUX1 proteins reflect the nature of the requirements for a number of different pathways of auxin movement. At the simplest level, there is the mass basipetal flow from the sites of auxin biosynthesis in the developing shoot through to the root tips and a secondary acropetal flow from the root cap. A number of external stimuli such as gravity or light will also dictate the direction of root growth. Such differential growth in the two sides of the root must be mediated by the differential accumulation of both the auxin efflux and influx proteins. It is likely, therefore, that a high degree of temporal and spatial control overrides the expression of each member of the gene family as well as the extent of expression in each target cell.

For the *AtPIN* genes, some of the controls dictating this differential expression at the cellular level are beginning to be revealed. Using the vesicle trafficking inhibitor, brefeldin A, Geldner et al. (2001) showed that the critical asymmetric distribution of efflux carriers at the plasma membrane is dictated by a rapid actin-dependent cycling of vesicles between the plasma membrane and the endomembrane system. That the actin cytoskeleton is critical to the maintenance of the asymmetric distribution was shown by treatment of plant tissue with cytochalasin, which fragments the actin cytoskeleton and so disrupts the asymmetric distribution of PIN1 and thus the mechanism for polar auxin transport (Geldner et al., 2001). A protein that will bind 1-naphthyl phthalamic acid (NPA) has been proposed to link PIN1 to the actin cytoskeleton, since auxin transport inhibitors (such as NPA) also disrupt the membrane trafficking processes (Figure 3.3).

A similar vesicle-mediated asymmetric distribution of transporters has been identified in mammalian tissues and many plant researchers are attracted to the similarities with auxin transport (reviewed in Muday and Murphy, 2002). In insulin-responsive tissues, an increase in blood glucose triggers an insulin-induced, calcium-dependent, phosphoinositol/protein kinase signalling cascade that causes endomembrane vesicles containing the glucose transporter, GLUT4,

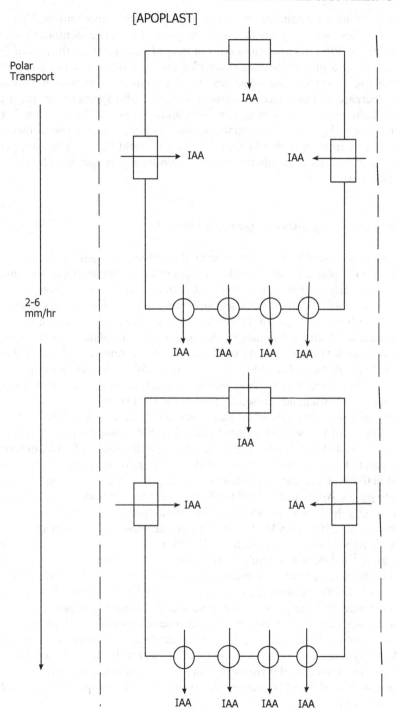

Figure 3.3. Diagrammatic representation of the localisation of PIN proteins (efflux carriers) and influx carriers that together mediate polar IAA transport. O denotes the efflux proteins; □ denotes the influx carriers.

to be distributed asymmetrically via an actin-dependent mechanism. There are now many homologues of the mammalian system that can be identified in plants, including the PINOID protein kinases of *Arabidopsis*, in which the *pinoid (pid)* mutants have a phenotype that resembles the *pin1* mutant. Thus overexpression of the *PINOID* gene produces transformants with growth defects akin to an increase in polar auxin transport and, in high-expression plants, a collapse of the primary root meristem (Benjamins et al., 2001). In animal cells, insulin is the trigger for the asymmetric distribution of the glucose transporter, and there is some speculation that auxin itself could be a primary trigger in plants of the asymmetric distribution of auxin transport proteins (Muday and Murphy, 2002).

Systemin as a long-distance signalling molecule

The systemic wound response reinforces the concept of signal molecules moving within the plant. Using the induction of proteinase inhibitors in response to wounding, early studies demonstrated that the synthesis of these proteins was induced not only locally (at the site of wounding) but also at sites remote to the wound (Green and Ryan, 1972). Several factors were later shown to induce the synthesis of proteinase inhibitors when applied to plant tissues, including a pectic fragment of 20 uronide units termed the proteinase inhibitor inducing factor (PIIF; Ryan, 1974). While these compounds can induce the synthesis of proteinase inhibitors locally, the inhibitors themselves are not mobile over longer distances, as shown in later research (Baydoun and Fry, 1985).

Systemin was first discovered and isolated by virtue of its ability to induce the synthesis of two wound-inducible proteinase inhibitor proteins (proteinase inhibitor I and II) in tomato. The 18-residue peptide was purified from tomato leaves using HPLC, sequenced, and an identical peptide synthesised which mimicked activity of the native peptide when supplied to the cut stems of young tomato plants. Approximately 40 fmol of the synthetic peptide was required per plant to give half maximal accumulation of the proteinase inhibitors I and II (Pearce et al., 1991; see Chapter 2). Perhaps the most interesting aspect of this early characterisation of systemin by Pearce et al. (1991) was the demonstration, using ^{14}C-labelled systemin, that the peptide could move in a bi-directional manner, from the site of a wounding application to the terminal leaflet through the phloem to other unwounded parts of the plant. Expression studies in tomato demonstrated that the prosystemin gene was constitutively expressed in vegetative tissues but not in roots, and it was wound-inducible both at the site of wounding and in tissues remote from the wound site – i.e., prosystemin mRNA could be systemically induced (McGurl et al., 1992). By both the constitutive and systemic induction of prosystemin, the overall internal concentration of systemin is increased together with the ability of the plant to respond to a wound by pathogens or predators.

Applied systemin has been shown to induce the accumulation of JA in tomato leaves, a product of the octadecanoid pathway (Doares et al., 1995), and to activate a 48 kDa myelin basic protein (MBP) kinase (Stratmann and Ryan, 1997). The induction of such mechanisms with a commonality of secondary messenger-based signalling has been strengthened by the identification and purification of a putative receptor (Scheer and Ryan, 1999, 2002). In preliminary purification experiments, Scheer and Ryan (1999) synthesised an iodinated, substituted but biologically active systemin molecule, ^{125}I-Tyr-2, Ala-15-systemin (involving substitutions as V_2T and $M_{15}A$) that rapidly, reversibly and saturably bound to cells of tomato in culture. Binding was increased by MeJA treatment and was shown to be restricted to a 160 kDa membrane-bound protein that could be competitively inhibited by a biologically inactive systemin analog with a $T_{17}A$ substitution. Scheer and Ryan (2002) treated 7-day-old tomato cells in suspension culture with 50 μM MeJA for 15 hours, then added ^{125}I-azido-Cys-3, Ala-15-systemin and cross-linked the ligand onto the putative systemin receptor using UV-irradiation. The labelled membrane-bound 160 kDa protein was then purified to homogeneity (8,200-fold) and the amino-acid sequence determined. An oligonucleotide probe was then synthesised based on the sequence of one of the tryptic peptides and used to probe a tomato cell suspension cDNA library. A full-length cDNA was then isolated that was 100 percent identical to the tryptic sequence identified from the purified protein. Translation of the cDNA revealed a protein (SR160) with homology to the RLK family of plant proteins, in particular the putative brassinosteroid receptor, BRI1 (described in detail in Chapter 8). The SR160 protein has 25 extracellular leucine-rich repeats (LRRs), with a 68-amino-acid island that is important for signal recognition (a 70-amino-acid island is a binding site for brassinolide to BRI1) between the 21st and 22nd LRR, a transmembrane domain and a cytoplasmic serine/threonine protein kinase. BRI1 and SR160 are highly conserved over the serine/threonine kinase domain and the transmembrane domain, but are less homologous in the extracellular LRRs possibly due to the different ligands that are recognised by each receptor.

Although applied labelled systemin can distribute through a wounded leaf of tomato within 30 minutes of application, and to the upper leaves within 5 hours (a rate comparable with that of exported sucrose in the phloem) (Narváez-Vásquez et al., 1995), a school of thought considers that a different signal (see 'X' in Figure 3.4) may be synthesised in the leaf mesophyll cell and that only after plasmodesmatal transport of this signal to the companion cell of the sieve tube can systemin be synthesised. A GUS reporter gene linked to the prosystemin promoter was used to transform tomato plants, and coupled with the use of prosystemin antibodies, histochemical analysis indicated that the prosystemin molecules were found only in the sieve tube companion cells and closely associated parenchyma cells, but not throughout the mesophyll (Jacinto et al., 1997). Tissue printing protocols applied to the petioles of methyljasmonate-treated leaves confirmed the localisation of the GUS expression to the vascular tissue. Such observations do raise the important question of the target cell role of companion cells as receptor sites and synthesis sites for many signals that are transmitted by the phloem.

Regardless of whether or not systemin synthesis resides in parenchyma mesophyll of the leaf or in the companion cells of the sieve tubes, an initial wound or

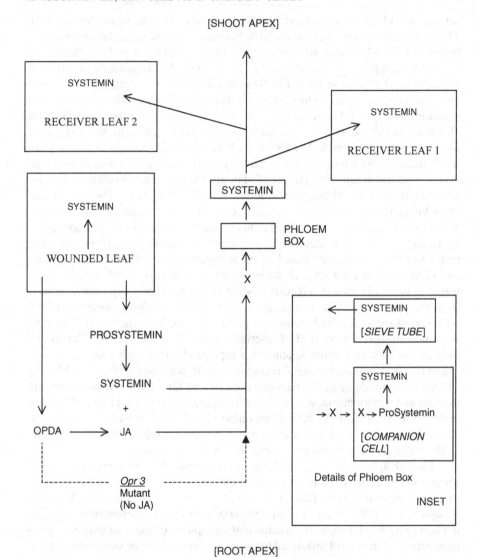

Figure 3.4. Diagrammatic representation of proposals for long-distance systemin signal transport and response following a local wounding event and increased jasmonic acid (JA) biosynthesis. A role for the 'phloem box' and the JA-independent signalling pathway via the JA-lacking *opr3* mutant of *Arabidopsis* is also shown.

photoperiodic stimulus must first pass across a series of cells (as short-distance transfer) before being unloaded to the phloem in a symplastic-regulated transfer. The reverse must take place through plasmodesmata and plasma membrane at the long-distance destination of the sink terminus; eventual sorting or exclusion of solutes or signals then being selectively discerned by the parenchyma cells surrounding the terminal phloem elements (Oparka and Santa Cruz, 2000).

Although the movement of systemin is presumed to occur through the phloem, other studies on the long-distance transmission of systemic wound signals and the induction of proteinase inhibitors at remote sites, suggest additional mechanisms of long-distance wound-induced signalling. Using the *spr1* mutant of tomato, in which activation of proteinase inhibitor gene expression by prosystemin is impaired (Howe and Ryan, 1999), Lee and Howe (2003) determined that JA accumulation in response to applied systemin was abolished in the mutant, and that the mutant accumulated only 57 percent of JA in response to wounding when compared with wild-type. Furthermore, reciprocal grafting experiments showed that the *spr1* lesion inhibits production of the wound signal rather than its perception in distant, undamaged leaves. Based on such results, the authors suggest that JA, and not systemin, may be the transmissible signal but that the role of systemin is to activate JA biosynthesis. Not all mutations of the jasmonate biosynthetic pathway, however, indicate a necessity for the induction of JA to achieve the induction of proteinase inhibitors and the systemic wound reaction (Stintzi et al., 2001). The *opr3* mutant of *Arabidopsis* is defective in the reductase that converts a JA intermediate 12-oxo-phytodienoic acid (OPDA) to the immediate precursor of JA. In the absence of JA formation, Stintzi et al. (2001) found that the mutant still shows the systemic response. Either another product of OPDA or OPDA itself could therefore possess overlapping activity with JA in certain systemic expressions. The reader is referred to the review of Stratmann (2003) for a more detailed discussion of the role of systemin and JA as long-distance wound-inducible signal molecules.

In their analysis of the *spr1* systemin mutant, Lee and Howe (2003) followed another set of wound-responsive genes that suggested the existence of a systemin/JA-independent pathway. One example may be the hydraulic dispersal model that proposes a mass flow in the xylem with molecules carried, rapidly, over long distances in both basipetal and acropetal directions (Malone, 1993, 1994). Electrical signalling has been proposed as another means for conveying a wound response over long distances (Wildon et al., 1992). Such systemic electrical signalling after wounding has been recorded in several plant species, with the electrical conductance associated with vascular tissue (Samejima and Sibaoka, 1983).

It is not our purpose in this volume to analyse the respective merits of each mechanism, but to include all of these as possibilities to account for long-distance movement of signals in plants and the potential this offers to the many target cells that eventually perceive them. However, two target cell types, the phloem and the phloem parenchyma cells, do emerge as critical to the systemic response of a wounded plant. Immunocytochemical and *in situ* hybridisations carried out at the ultrastructural level have confirmed that both the wound-induced and methyljasmonate-induced prosystemin mRNA and protein of the vascular tissue are confined exclusively to the phloem parenchyma cells, and that the protein is present both in the cytosol and in the nucleus (Narváez-Vásquez and Ryan, 2004). Although studies of these distinct cell types are difficult, it is clear that further genetic analysis will provide additional insight. In this respect, the fact that in tomato, phloem develops both externally and internally to the xylem provides special opportunities for manipulative surgery.

Plant nutrients and shoot-to-root signalling

In a discussion on root-to-shoot signalling, the role of the important nutrients (N, S, and P) must not be overlooked even though their function is not a central theme of this volume. To some extent it is obvious that nutrient concentrations convey signals in plants so their informative capacity must remain in perspective for long-distance intertissue communication (see Jackson, 2002). For example, in sulfur uptake and assimilation, the status of the glutathione content and/or the glutathione:SO_4^{2-} ratio in the shoots feeds back to the roots in the requirement to access SO_4^{2-} from the external environment (Lappartient et al., 1999; Herschbach et al., 2000). Likewise, the inorganic phosphate (Pi) levels in the shoot serve as signals to the root to regulate the activity of Pi transporters in the roots (Martin et al., 2000). For nitrogen, the amino acids in the phloem moving from the shoot to the root are proposed to regulate the activity of the high-affinity nitrate uptake system (Forde, 2000). The evidence for all three of these compounds as candidates for shoot-to-root signals is not unequivocal and the reader is referred to Forde (2002) for a more detailed consideration. Indeed, for nitrogen status, the role of auxin has been proposed as part of the signalling pathway, and it may be that auxin is also an integral part of the signalling capacity of other nutrients.

Volatiles and Signalling

Methylsalicylate

Salicylate has been conclusively shown to be involved in mediating plant responses both at the localised site of infection, the 'hyper-sensitive response', and in healthy tissues at remote sites from the point of infection, the phemenon of 'systemic acquired resistance (SAR)' (Gaffney et al., 1993; Ryals et al., 1995). In these responses, the synthesis of SA at the site of infection is mediated through the pathogen stimulation of a membrane-localised NAD(P)H oxidase, and subsequent production of the superoxide anion (O_2^-) which is then converted rapidly to H_2O_2 (via superoxide dismutase). In turn, benzoate-2-hydroxylase, the enzyme that catalyses the conversion of benzoic acid to SA, has been shown to be stimulated by H_2O_2 (Leon et al., 1995). More recently it has been proposed that the free acid may potentiate cell death in the hypersensitive response through a direct stimulation of the membrane-localised NAD(P)H reductase to increase O_2^- levels which can mediate cell death and, through the increase of H_2O_2 levels, increase the levels of SA (reviewed in Draper, 1997). In this scenario, as well as salicylate-inducing defence genes and SAR, SA may also contribute to the hypersensitive response directly by inducing cell death through its putative interaction with catalase or other protein candidates (reviewed in Barker, 2000; Kumar and Klessig, 2003). Such responses of plants and their cells and tissues to invading pathogens involve a complex series of actions comprising processes of differential recognition (avirulent versus virulent microbes), the production of signalling molecules and the induced coordinated responses. These are themes in signalling associated with SAR which are pertinent to our discussion of cell-to-cell signalling in a target cell complex.

With the discovery of SA as an inducer of defence responses in tissues remote from the point of infection, a question arises as to whether the free acid moves systemically in plants. Earlier experiments indicated that SA does not move within the plant ((Rasmussen et al., 1991; Vernooij et al., 1994), but more recent investigations using TMV-inoculated tobacco leaves showed that the volatile methyl salicylate (MeSA) can be an airborne signal (Shulaev et al., 1997) in addition to the movement of methyl salicylate and benzoic acid in the vascular system. Importantly, Shulaev et al. (1997) determined that MeSA acts by conversion back to SA, suggesting that MeSA functions as the signalling compound. In common with other signalling compounds, MeSA has been shown to act in concert with ethylene, via regulation of biosynthesis of the hormone (Ding and Wang, 2003).

An allied aspect of SA-based signalling which is relevant to its long-distance transport is the conversion to a conjugated form, SA β-glucoside. SAG is ineffective at inducing defence responses and is most probably a storage form, with the free SA being released by a cytosolic or wall-bound β-glucosidase (reviewed in Ryals et al., 1996). The nature or regulation of the enzymes that hydrolyse this and other signal molecules that are converted to sugar or peptide conjugates is still largely unknown, but it is clearly a critical part of overall signalling control. Although the availability of receptors may be one way to regulate the ability of a cell to perceive a signal, control of the equilibrium between the inactive conjugates and their free forms must be equally important in determining the signal response.

While SA is concerned principally with defence signalling in plants, there are features of SA signalling that set a precedence in our discussions on signals and their perception by target cells. The question of airborne movement is a key aspect of SA signalling. How many other related volatiles signal in this way? There are obvious parallels with the plant hormone ethylene, which, in its volatile form (rather than in solution in the cell), has major effects on plant growth. Methyljasmonate is another volatile signalling compound (outlined in Chapter 2), and there are many others to be identified (Arimura et al., 2000).

4

Population Diversity of Cell Types and Target Identification in Higher Plants

Plants are remarkable in that during evolution from the single cell to the multi-cell state they developed centres of cell division, the meristems, as the principal repositories of all genetic information. In essence, whether it be the single apical cell of a liverwort or the multicellular dome of tissue of the higher plant, the meristem holds the blueprint of the species. It is only from the divisions of their meristematic cells that the plant body can continuously enlarge and reproduce. Whereas an embryo generates polar identities and a meristematic initiation from the two opposing ends of the zygotic cell, whole plants do not have a restriction to two meristems and they develop unlimited numbers of new primordia as the body of the plant continues to increase in size and cell number.

Anatomists, biochemists, molecular biologists and developmental botanists each see the same plant from different viewpoints. The anatomist studies the structural and visible characters of cells and tissues and describes them accordingly. The biochemist homes in on the functional processes of metabolic control and enzymatic activity attributable to specific plant parts – photosynthetic activity in leaves, for example. The molecular biologist seeks the genetic control of biochemical processes and is particularly attracted to the performance of mutant plants with abnormal behaviour, using them to probe the genetic control of the normal. The developmental botanist is most concerned with the progression of events in cells and tissues which dictate how any one cell may mature to perform an explicit function in the community of its neighbours. To an extent these areas of study must overlap, so in this chapter we have selected examples of target cell types that span the interests of all these groups. Readers will immediately become aware of other examples as they consider the relevance of the information to their own special interests.

Any meristem (or cells with meristem potential) can be removed from the parent and then be grown independently to form new and clonal members of the species, so illustrating the totipotency of plant cells. But all meristems and their subtending living tissues, within the plant parent body, continue to be regulated by an informational cross-talk that passes continuously between the meristematic cells and the differentiated parts. Today's plant biologists probably accept that no event passes unnoticed by a plant, in terms of environmental perception, signal transmission or biochemical response. So no cell, or group of cells, can become independent of the sensitive and sophisticated controls that determine overall plant behaviour unless they are released completely from all internal cytoplasmic (symplastic), surface membrane and external wall matrix (apoplastic) constraints and contacts with their neighbours. Few cells are so privileged. Gametes, lenticels and the separating cells of root caps and of most shed organs are examples, but all have past developmental histories that determine their identity and all are then relatively short-lived. The embryos of shed seeds, however, are an exception. They carry within them the already differentiated meristems of the new plant and are one of the few examples of tissues that can survive the rigours of parent plant rejection *in vivo* for many years if necessary. But they do so only because these particular meristem cells have all acquired the capacity to dehydrate to water contents of 5 percent or less without harm. In this respect, all the cells of the embryo have a target commitment – the ability to withstand desiccation – while still preserving their individual positional target status after the re-imbibition of water, the re-establishment of cell-to-cell interactions and the renewed operation of signal perception and response. Bulbils or adventitious plantlets also possess meristems at shedding but they do not develop the stress resistance of what we term the orthodox seeds. Under the specialized conditions of *in vitro* tissue culture, however, a wide variety of cells isolated from different parts of plants can be induced to divide when exposed to chemical signals that act as substitutes for those perceived *in vivo*; these cells can then be induced to form new meristematic, vascular, or other specialized centres.

Armed with the knowledge that each cell in a meristem is already positionally and biochemically marked in relation to its neighbours and the evidence that its fate is already programmed by the ability to perceive and respond to those signals that dictate which genes can or cannot be expressed, we are beginning to understand how a society of plant cells operates. Central to this understanding is the concept that every cell has an identity as an individual target cell within the plant community and that each target cell also has the capacity to influence and be influenced by its neighbours.

Anatomists have long recognised that the multicellular dome of the an-giosperm shoot apex is already a community in which individual identity has been established and cell lines delineated. Much information has been gleaned from studies of chimeras, in which cells or sectors of an apex are composed of genetically different origin and are therefore visibly identifiable. Such mixed populations can arise naturally as a result of either grafting or mutation, or they can be artificially generated by induced mutations using radiation or chemical treatments. Many much prized variegated plants have been produced in this way. In his analysis of cell lines in periclinal chimeras, Poethig (1987) named the first and outermost layer of cells of the shoot meristem the L1 cells, from which are derived all epidermal structures including stomata, trichomes and the wide variety of other surface appendages: L1 corresponds to the outer *tunica* of the earlier anatomists. Within the anatomist's tunica lies an immediately subepidermal cell layer, the L2, and further inner layers; those of L3 generate cells of outer and inner cortex and a centrally located lineage for the vascular system – together, the anatomist's original *corpus*. These lineages function in each part of the plant. Even at later stages of development, the internal L3 layer operates to determine floral meristem size and carpel number in the periclinal chimeras of tomato (Szymkowiak and Sussex, 1992).

Two examples here illustrate how chimeras have helped to elucidate the cross-talk between neighbour target cells of different lineages. From studies of pedicel abscission zones in periclinal chimeras of the mutant jointless (non-shedding) and normal wild-type (shedding) tomato, we know that it is the presence of wild-type cells in the outermost L3 layer of the pedicel cortex that determines whether or not cell separation across the zone can occur (Szymkowiak and Irish, 1999). Their presence in L1 or L2 positions alone cannot do so (see Chapter 6).

In the grapevine, Pinot Meunier (a dwarf-type phenotype) is a periclinal mutant of Pinot Noir. When the L1 and L2 layers of Pinot Meunier were separated and regenerated through somatic embryogenesis only those of L1 origin exhibited the dwarf growth of the Pinot Meunier (Boss and Thomas, 2002). Here we have an example of L1 determining the performance of the L2 population.

Evolution of cell types

The presence of cuticle, stomata or lignified tissue has been long accepted as an indicator that the possessor is a land plant. It is therefore of interest to speculate upon the potential for environmental monitoring, long-distance information exchange and molecular signalling that the evolution of an epidermis or a vascular system would have brought to the primitive land colonizer.

Some of the earliest indications of true land forms have been found in Silurian rocks. *Baragwanathia* exhibits a slender central strand of cells with annular thickenings in the walls that are attributable to tracheids, while *Zosterophyllum* shows, in addition to annular and spiral thickening, clear evidence of an epidermal cuticle. The presence of well-developed stomata are common to Middle Devonian fossils (*Rhynia Gwynne-Vaughani*) and phloem can be distinguished around the xylem elements of *Asteroxylon*. Subsequent development of flattened leaf-like structures in conjunction with increasing size and a more substantial transport-acting and mechanically supporting vascular system brought new challenges for an effective integration of plant parts and new requirements for coordinated signal-sensing between cells.

For differentiation to occur, a community of cells must somehow generate gradients between the different members of that community. We know that single cells, in particular zygotic cells, can be polarised by physical inputs (light in the case of a free-floating fertilized egg cell of *Fucus*), by the surrounding maternal tissue *in planta* or possibly even by gravity in the fertilized ovules of the higher plant. The physical environment of each cell and the subsequent organisation of cells into complex tissues was calculated by Crick (1970) to be a manifestation of the temporal maintenance of simple chemical gradients extending over as few as 50–100 cells – a number that approximates to a meristem in *Arabidopsis*. This distance of influence was then considered enough to send signals to cells as they escaped from a meristem and joined those of the enlarging tissue below.

A simple chemical diffusion gradient will convey signals in all directions to the surrounding cells, but for attaining the complex differentiation of the higher plant, a directionally controlled flow of information is critical for the maintenance of form and function. The possible evolution of one such directional flow has been studied in the sporophyte tissue of three orders of bryophytes: a hornwort (*Phaeoceros personii*), a thallose liverwort (*Pellia epiphylla*) and an erect moss (*Polytrichum ohioense*) (Poli et al., 2003). By following the auxin flow and its polarity in segments of tissue researchers found that hornworts showed a very low flux of applied IAA, with almost no difference in basipetal and acropetal movement and no inhibition by the auxin transport inhibitor NPA whereas the liverwort sporangial setae showed a higher rate of transport, still with no preference to a basipetal or acropetal directional movement, but with both directions subject to inhibition by IAA transport inhibitors. However, in the most advanced bryophyte tested, *Polytrichum*, auxin flux was high in the setae of young sporophytes with a basipetal polarity equal to that found in the coleoptiles of maize and was inhibited by NPA. This polarity was reduced in older sporophyte tissue by an increase in the extent of acropetal movement – an event similar to that found in the fully mature tissue of higher plants.

In each of these bryophyte sporophytic tissues, cell elongation was enhanced by the addition of auxin, and Poli et al. (2003) have proposed that these examples indicate that during the evolution of such primitive plant forms, there also arose from an initially simple or activated diffusion between cells, the polarity of a unidirectional auxin transport system. Based on the transport inhibitor data, this could indicate a mechanism comparable to the auxin efflux carriers encoded by PIN genes in *Arabidopsis* (Steinmann et al., 1999).

In an early study with gametophytes of the aquatic liverwort *Riella helicophylla*, Stange and Osborne (1988) had already established that growth of the cells was enhanced both by auxin and by ethylene with an additive 'super growth' similar to that exhibited by the Type 3 target cells of semi-aquatic higher plants. The polarity of auxin transport and certain auxin–ethylene interactions were therefore established as early events in the evolution of land plants.

The study of primitive plants provides fertile ground for exploring the origins of cell-to-cell interactions, and the present-day use of expression mutants represents a powerful genetic approach with which to elucidate the molecular mechanisms that control the pattern of plant cell differentiation. Only a selection of the many mutant studies described in the literature is discussed here for it is not our intention to provide a review of such work but rather to use examples to illustrate, in a molecular context, how the physical and chemical cross-talk between cell types can dictate overall plant performance. Inherent in the control of cell fates dictated by gene expression is the effectiveness of cell-to-cell signalling and the associated perception and response to these signals. Each cell, as it progresses through its journey of differentiation will act as a target cell for long-distance signals and/or positional information from and to its neighbours. It is to the development of the target cell concept with respect to differentiation and the changing nature of signal perception and response that this and the next chapters in this volume are addressed.

Meristems as stem cells

Meristems are sites of cells with unlimited cell division (Fahn, 1990). This anatomists' definition, though true in the broad sense, is not wholly precise as there are many exceptions to this general rule, not least of which is the death of root and shoot apices in the annual plant at the end of each season's growth. But as sites for generating a continuous flow of cells during the plant's lifetime, these two meristems are critical to the formation of every cell type the plant expresses. For the apical meristem the flow is always basipetal. Not so the root, where one side of the meristematic cell group is the source of all the cell types that make up the permanent body of the root, while those produced from the opposite side of the meristem form the highly specialized and short-lived cells that make up the root cap. Certain intervening cells between these two divergent groups of root meristematic cells are filled by a single centre of cells that has been called 'quiescent' (Feldman, 1976; Clowes, 1978) since although being part of a meristem they are essentially reservists to be called upon to divide only in response to conditions of stress. How far any of these quiescent cells is already programmed for its future response we do not know. Perhaps they are the few that do not have their target status already defined, though isolated quiescent centres of maize held in culture do retain the capacity to regenerate a whole new root rather than a different tissue (Feldman and Torrey, 1976). It has been calculated that in the primary root of maize for example that contains about 110,000 cells only some 600 are quiescent. In other species such as *Arabidopsis*, however, the centre can be as few as one or two cells.

Experiments of Racusen and Schiavione (1990) with developing somatic embryos of carrot at the torpedo stage revealed that root pole ends containing 10 percent of the total embryo can continue as roots, but they produce no shoots even though shoot pole ends at this stage have the flexibility to produce both. This indicates that cells at the root pole end are already dedicated to being root-type cells very early in the cell division programme of the carrot zygote, so root 'identity' had apparently already been acquired even though a specific target cell type had not yet been expressed.

This raises the question of how soon in biochemical terms, if not in anatomical visualization, a cell acquires specificity that denotes it as a target cell distinct from its neighbours and indicative of its future development. The numerous monoclonal antibodies that have been raised against arabinogalactan- and rhamnogalactan-linked pectin epitopes have facilitated the following of such changes, particularly in the composition of cell walls during differentiation and pattern formation (Knox, 1997). For the root, Smallwood et al. (1994) raised monoclonal antibodies to hydroxyproline-rich glycoproteins of the cell wall. Two in particular, JIM11 and JIM12, were studied in detail. JIM11 antibodies react with walls of the central root cap and the meristem and later at the cortex–stele boundary, progressing outwards to the body of the root cortex. Even later the JIM11 epitope is expressed in pericyclic cell files adjacent to the phloem and in the epidermis. In contrast, JIM12 is quite restricted. It is recognised in neither the root cap nor the meristem but reacts at the intercellular spaces formed at the junction of the oblique and radial walls in sectors of the pericycle opposite the xylem poles and later in the future metaxylem cells. This has many implications for the loci of new meristems in the siting and initiation of lateral roots. Further, the absence of JIM12 recognition from the walls of root cap cells and the absence of lignification may be significant as early diagnostic markers of a specialized gravity-sensing function to come and their differentiation as cap border cells with cell separation potential.

It is very evident that the pectin components of cell walls are rapid and sensitive reflections of changes in local signal inputs and changes in target status. For example, within 12 hours of exposure to low (3 percent) oxygen a banding pattern in cell walls of *Zea mays* roots was detected with JIM5 and was diagnostic of sites of future aerenchyma formation (Gunawardena et al., 2001b). The subculture of carrot cells at a low density and withdrawal of the synthetic auxin 2,4-D can result in an increase in expression of the JIM4-recognised epitope in most surface cells but most abundantly in those surface layers at the future shoot end of developing embryos. The transition to heart-shaped embryos occurs concurrently with the enhanced expression of the JIM4 epitope by groups of cells just below the developing cotyledons, at the junction of the future shoot and root. At this stage, the epitope is recognised on a single well-defined layer of cells at the embryo surface. On reaching the torpedo stage, expression of the epitope occurs also in two regions of the future stele and in cells of the cotyledonary provascular tissue. Using an anti-arabinogalactan protein (AGP) monoclonal antibody raised from a peribacteriod membrane of the pea, Pennell and Roberts (1990) followed the expression of the AGP epitope during the floral meristem development in pea plants. Whereas the antibody recognised all cells in vegetative meristems, in differentiating floral tissues there was no detectable presence of the epitope. In the stamen, for example, recognition disappeared from the four apical cell clusters that developed into the single pollen sacs containing a layer of tapetal cells and the sporogenous tissue, although the epidermal boundary of each pollen sac remained delineated by the antibody. That expression can be regained was shown when meiosis and the final haploid mitosis were completed in the male gametophyte; then the epitope was re-expressed at the plasma membrane of the pollen vegetative cell although it was not present in either the generative or the sperm cell.

The mutliplicity of AGPs that are formed by growing cells is seen from the panel of antibodies (JIM13, JIM14 and JIM15) that has been raised against two AGP fractions from embryogenic carrot cell media (Knox et al., 1991). JIM13 recognises an epitope in the plasma membrane of cells positioned at the root apex, and in particular those forming the epidermis and marking the region and axis of the future xylem. By contrast, JIM15 recognises all cells that are not recognised by JIM13. JIM14 recognises glycoproteins of low molecular weight in the cell wall which are also present in all cells in the root apex whereas JIM4 recognises L1 cells, but not L2 cells.

In a study of potato stolons and their development into structures of mature tubers, Bush et al. (2001) used an *in situ* immunological approach with antibodies raised to specific galactans (L5) or arabans (L6) to mark changes in pectic epitopes in the cell walls as tuberization proceeds (see Table 4.1) Clearly the dynamics of wall transformations and the multiplicity of their glycan components provides for an almost unlimited signal potential for target cell identification.

Flexibility and plant cell differentiation options

Meristem centres

The molecular identity of wall components laid down during the early development of a cell would appear therefore to be one of the factors that can dictate the flexibility of performance in later cell life. The origins of lateral roots, for example, have long been known to arise in the pericycle and opposite the xylem poles (Dubrovsky et al., 2000). In the tiny root of *Arabidopsis,* this is derived from an asymmetric division of two pericyclic founder cells in the same file. Continued asymmetric divisions of the daughter cells eventually provide a primordium that continues to cell cycle when division is suppressed in neighbour cells (Beeckman et al., 2001). The initiation of these lateral primordial centres is normally precisely spaced, but modifications (enhancement) of auxin availability can convert all xylem pole pericycle cells to primordial cells (Himanen et al., 2002). Auxin from the root apex affords a transcriptional control of genes in cells along the root axis. A down-regulation of the cyclin-dependent kinase (CDK) inhibitor of KRP2 is attributable to auxin; suppression of pericyclic cell division in isolated root segments can be relieved by the addition of auxin (Dubrovsky et al., 2000). Roots of a mutant that lack the auxin influx carrier and fail to accumulate IAA at the root apex also show reduced primordial meristem numbers. The presumptive auxin influx carrier AUX1 has been localised at the basal plasma membrane in cells of the root protophloem so it seems evident that an auxin-regulated cell-to-cell cross-talk is one of the earliest controls in lateral root meristem and root pattern formation. Further, the target cells for response are precisely positionally located with respect to their neighbours.

Whereas a molecular determinant in the cell wall may be the earliest marker for potential lateral root meristem initiation, flexibility at least to auxin control still exists for a limited time after which the pattern of pericyclic cell behaviour is fixed (Casimiro et al., 2003). These windows of opportunity may represent one of the most critically sensitive stages in the life history of any plant cell in the pathway of differentiation and development to its final target state.

Table 4.1. *Monoclonal antibodies raised against pectic glycoprotein epitopes and their preferential recognition in some specific target-cell types*

Monoclonal	Cells and epitopes	Reference
JIM4	Recognises arabinogalactans.	
	In primarily epidermis and pro-vascular tissue of shoot somatic embryos.	Stacey et al. (1990)
	Epidermal but not cortical cells of carrot root apical meristems in somatic embryos.	Knox et al. (1989)
JIM5	Recognises low esterified homogalacturonans.	
	In intercellular spaces of non-dividing cells in *Arabidopsis* root tips, absent in dividing cells.	Dolan et al. (1997)
	In style epidermal pectin determining lily pollen adhesion.	Mollet et al. (2000)
	At initiation of aerenchyma formation in maize roots.	Gunawardena et al. (2001b)
	Walls of carrot root intercellular spaces, not in epidermis or root cap.	Knox et al. (1990, 1991)
JIM7	Recognises more highly esterified pectin than JIM5.	
	In carrot root cortex and stele.	Dolan et al. (1997)
	Not in epidermis or root cap.	Knox et al. (1990, 1991)
	Binds to *Arabidopsis* seed mucilage and at root surface.	Willats et al. (2001a)
JIM11	In carrot seedling root meristem and cap.	Smallwood et al. (1994)
	Cortex/stele boundary, adjacent to phloem.	
	Intermediate filaments in carrot nuclear matrix.	Beven et al. (1991)
JIM12	In intercellular spaces, particularly pericycle opposite future metaxylem.	Smallwood et al. (1994)
	Not in root cap or meristem.	
JIM13, 14	Especially plasma membrane of root epidermis and protoxylem.	Knox et al. (1991)
JIM15	Recognises cells not recognised by JIM13.	Knox et al. (1991)
JIM19	Recognises glycoprotein antigens on epidermal guard cells and mesocotyl cells of *Pisum*.	Donovan et al. (1993)
JIM20	Recognises all JIM11 and 12 cells.	Smallwood et al. (1994)
PAM1	Recognises non-esterified galacturonic acid repeats, especially middle lamella of dividing cells.	Willats et al. (2001a)
LM5	Recognises $(1\rightarrow4)$-β-D-galactans.	
	In phloem sieve tubes and xylem secondary thickening in potato stolons, absent from mature tuber stele.	Bush and McCann (1999)
	In primary wall of tuber cortex, absent from middle lamella and cell corners.	
	Does not bind to plasmodesmatal regions.	
LM6	Recognises $(1\rightarrow5)$-α-L-arabans.	
	In potato tuber cortex, not at cell corners. Absent in loose suspension cultures of tobacco, so may be required for cell adhesion.	Bush et al. (2001)
LM7	Recognises partially methyl-esterified homogalacturonans.	Willats et al. (2001b)
	Distinguishes middle lamella from primary wall; binds to cell walls lining intercellular spaces in *Pisum* stems.	

The origin of founder cells for cambial meristems of the stem has also, in the past, been linked to the apical flow of auxin, and additions of auxin, together with cytokinin, are effective in modifying their formation *in vitro*. Experimental systems have certainly upheld this view and a control of cell cycle progression is implicated in *Arabidopsis*. At the molecular level a phospholipid kinase involved in the synthesis of phosphoinosotide signalling molecules is predominantly expressed in procambial cells of *Arabidopsis*. The reduction in polar auxin transport in the inflorescence stems of *ifl1* mutants leads to a block in vascular cambial activity at the basal parts of the stems and to a reduced expression of the auxin efflux carriers, PIN3 and PIN4 (Zhong and Ye, 2001). Further, a gene encoding a cytokinin receptor (WOL) is localised in procambial cells of roots and embryos and overexpression of another gene (ATHB) leads to an overproduction of vascular tissues from procambial founders (Ye, 2002). It is clear that a subtle hormonal control of procambial cell initiation and the subsequent cell division of the cambial cells exists, but how the products of these divisions are so precisely targeted along either phloem or xylem pathways of differentiation *in planta* remains to be discovered and is presently unknown in molecular terms. In contrast, conversion of cell cultures *in vitro* by the introduction of specific ratios of auxin and cytokinin is easily inducible. In *Zinnia* mesophyll cell lines, transdifferentiation of already expanded cells occurs directly to lignified xylem elements without intermediate cell divisions (Fukuda, 1997).

Options for cell enlargement

The progression from division to enlargement represents a major change in target status. Many years of research on cell elongation has produced good evidence for a number of recognisable target states with respect to the major plant hormones. Considering just auxin and ethylene, cortical cells of the pea stem, for example, enlarge in volume in response to the level of auxin perceived, but the lateral orientation of this enlargement is under regulation by ethylene; therefore the cell shape achieved reflects the combined influence of these two signals (Osborne, 1976; Figure 1.1). The growing shoot therefore appears shorter and fatter when the ethylene exposure of the cells is enhanced during the expansion period. This cell type has been designated Type 1 and is represented by those shoot cells that enlarge or elongate in response to auxin but not to ethylene (Osborne, 1979). Such cells form the majority in the population of any land plant.

Other enlarging cells respond in quite an opposite way to those of the pea shoot and can be found interspersed between the Type 1 cells that respond normally to auxin. Such cells are typical of abscission cells. These may be differentiated to their target state later than others in the shoot primordia, but they are usually fully functional as potential abscission target cells very early in the development of leaves, flowers, fruit and the organs that can shed. These cells are characterised by volume enlargement in response to a perceived ethylene signal but not to an auxin signal. In the abscission zone cells investigated to date, auxin actually represses the final stages of cell expansion just prior to abscission (for a more detailed discussion see Chapter 6). Such cells have been designated Type 2 target cells (Figure 1.1).

Another type of cell enlargement control was reported first by Ku et al. (1970) in a monocotyledon (rice) of the flooded paddy field. The coleoptile and leaf sheath cells were found to elongate in response to added ethylene but with no evidence of lateral swelling. It was already well established that mesocotyl cells, like those of their coleoptile neighbours, would enlarge and elongate in response to auxin (Imaseki et al., 1971), but the evidence that auxin-induced elongation could be enhanced if ethylene were present defined a target cell with quite distinct characteristics and one that differed fundamentally from Types 1 and 2. Such cells were classed as Type 3 (Osborne, 1977a, b). What subsequently became of great significance for ecological studies was the discovery that cells of many plants that colonised or were adapted to flooding or semi-aquatic conditions possessed cells in their stems or petioles that would elongate in response to either auxin or ethylene (Figure 1.1). This type of cell was present not only in monocotyledonous plants but also in dicotyledonous plants as well as in lower plants including ferns and liverworts (Osborne et al., 1996). Although all plants have cells that enlarge and elongate with auxin like those of Type 1, the Type 3 cells possess the additional feature of a much greater potential extension and enlargement when ethylene is also present. The differentiation of this type of target cell, the Type 3 cell with respect to ethylene and auxin, is central to the ability of submerged shoots of semi-aquatic plants to elongate to the water surface as ethylene accumulates within their tissues when underwater (for a more detailed discussion see Chapter 5).

In the integrated society of cells that forms a living organism, every cell is not required to perform the same function; the larger the organism, the greater the need for diversity and specialization and for specifically differentiated cells to act as signal perceptors and transducers on behalf of neighbour cells; but they also need the added capacity to convey information to long-distance communication networks.

Whereas groups of cells of similar target type are seen to perform in concert in a coordinated way (as in the many examples of the stem cortex), neighbour cells may also differentiate with dissimilar molecular markers that denote them as possessing quite distinct target states, which then designate alternative pathways of neighbour interaction and further growth.

The search for molecular markers

In evolutionary terms, the loss of adhesion between living cells, or the continued ability of neighbour cells to remain in cytosolic contact despite their differentiation to specific target states must have occurred early in the development of multicellular, multifunctional cell assemblies. It is not clear whether the liberation of spores from primitive land plants was the mechanical result of dehydration events or whether the cells were metabolically rejected as in the abscission processes of higher plants. Whereas root cap cells will lose contact with each other as a matter of course over a distance of only a few millimetres from the point of their formation, those of abscission cells remain attached to their neighbours for weeks or years until they receive the signal to separate. However, the fact that only the cells of the actual zone will separate indicates that they already have a target state that differs from their constantly self-adhering neighbours. Distinguishing this target state led researchers to an early quest for molecular markers that would designate when a difference had been established between a potentially separating abscission cell and its non-separating neighbours.

In the 1970s the first evidence for a specific abscission cell target state was demonstrated by Chee Hong Wong in an examination of the cell separation loci below the female flowers of the cucurbit, *Ecballium elaterium*. Unpollinated flowers abscind and premature abscission can be induced by ethylene but only at a precise stage of flower development. This occurs when groups of cells below the gynaecium have acquired a nuclear DNA content of 8C (Wong and Osborne, 1978). DNA replication in the absence of cell division is a common event in the cucurbits, and the 8C endoreduplicated condition was diagnostic of those particular cells that would separate from their neighbours and lead to normal shedding of the unpollinated flower bud or the inducible shedding by ethylene of either pollinated or unpollinated buds (Figure 4.1).

It was clear that the 8C content of nuclear DNA was not the marker of a potential separating cell in all tissues, for 8C nuclei were not found in the abscission zones of shedding mature *Ecballium* fruit, but it was the first molecular marker to identify the differential target state of adjacent cells that showed quite differing responses to ethylene.

That the abscission cell target state could be identified by the presence of particular polypeptides in cells that showed no nuclear DNA endoreduplication was subsequently clearly demonstrated for the leaves of both *Phaseolus vulgaris* and *Sambucus nigra* (McManus and Osborne, 1989, 1990a, b, 1991).

For this, proteins were isolated from abscission zone and non-zone adjacent tissues and fractionated by electrophoresis to seek peptide bands that were either enhanced or diminished between the two types of cells. Proteins from abscission zone cells that had separated following ethylene treatment were similarly fractionated.

Antibodies raised from polypeptides of interest were used for immuno-competition of the different protein extracts using immuno-affinity column chromatography. For *Sambucus nigra* an antibody raised against a 34 kDa polypeptide was shown to mark a protein present in abscission cells, both before and after abscission, that was absent from the cells of adjoining non-zone tissues.

This indicated a positional differentiation of ethylene-responsive target cells with an inherent potential for cell separation that could be recognised early in differentiation by zone-cell–specific antigenic determinants (McManus and Osborne, 1990a, b).

Similar experiments carried out with the leaf abscission zones of the bean (*Phaseolus vulgaris*) confirmed the preferential expression of a particular 68 kDa polypeptide marker in the target, ethylene-inducible, separating cells of the pulvinus-petiole zone, both before and after the separation event. Again, the non-separating cells of adjacent tissues did not register immuno-recognition of this protein component (McManus and Osborne, 1990a).

Ploidy level	Mean nuclear diameter	Response to C_2H_4

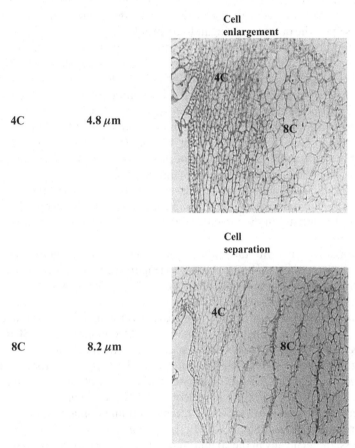

Figure 4.1. Endoreduplicated nuclear DNA levels mark cells responsive to ethylene in floral buds of *Ecballium elaterium*. Photomicrographs from Wong and Osborne (1978).

Further identification of the protein markers was sought in *Sambucus*. Monoclonal antibodies raised against the major peptide cross-reacted with a number of plant enzymes (peroxidase and α and β-glucosidase) and their associated oligosaccharides. The sequence Manα3 (Manα6) (Xyl β2) Man β4 Glc *N* Ac β4(Fuc α3) Glc *N* Ac was identified and the monoclonal to this N-linked oligosaccharide was used in further experiments to identify abscission, but not non-abscission cells *in situ* both before and after separation (McManus et al., 1988; McManus and Osborne, 1989).

Another interest to the target theme is the recognition of self and non-self as seen in pollen–stigma interactions and the signals generated between the two that determine whether or not a pollen tube will grow (Dixit and Nasrallah, 2001). This particular aspect of cell growth concerns the extension of pollen tubes *in vivo.* Most pollen can be induced to germinate on suitable media *in vitro,* but extension is highly restricted when a grain lands upon a stigma. The recognition of self, cell-to-cell compatibility and the subsequent permissive growth of the pollen tube between cells of the pistil towards the ovule is highly specific and depends upon the operation of ligand-receptor kinase interactions between these two different types of cells. Each is identifiable by target proteins. In *Brassica,* for example, the stigma expresses an *S*-locus receptor kinase (SRK) that binds a small pollen-expressed cysteine-rich protein, SCR. On the occurrence of this stigma–pollen (incompatible) recognition the papilli tissue of the stigma initiates a transduction cascade that blocks the hydration of self-pollen so that pollen tube germination is inhibited, the pollen dies and self-fertilization is prevented. In contrast, pollen-expressed receptor kinases (PRKs) permit tube growth in compatible associations in a number of plants so far studied (see Chapter 2 for details of the *S*-locus cysteine-rich proteins, SCRs).

The specificity of pollen- and stigma-expressed proteins and the specificity of their surface kinases are now seen as important intercellular recognition markers in the self-compatibility/incompatibility responses between pollen and stigma target cells.

In Brassica plants with the multi-allelic S-locus, a self-recognition process occurs as soon as pollen alights on the stigma (Figure 4.2). In the absence of compatible cell-to-cell recognition, the pollen tube fails to grow. Specifically, SCR/SP-11 peptides, of 74–83 amino-acid residues (47–60 after removal of the secretory N-terminal sequence), are secreted from the developing microspores in the tapetum to reside in the pollen coat exine layer. At pollination, the SCR peptide translocates into the cell walls of the stigma epidermal cells and there activates the stigma inductive processes. The putative SCR stigma receptor (marker) has been identified as the *S*-receptor kinase (SRK), a membrane spanning Ser/Thr protein kinase which, in common with other RLK protein in plants, has extracellular leucine-rich regions (LRRs), a transmembrane domain and a cytoplasmic Ser/Thr kinase domain (Stein et al., 1991; Goring and Rothstein, 1992). The SRK is expressed specifically in the stigma, and the occurrence of this protein is critical for a successful pollen–stigma interaction. It is now known that another associated protein, the *S*-locus glycoprotein (SLG) secreted by the stigma, can enhance the interaction. This glycoprotein is homologous to the SRK extracellular domain, but lacks the cytoplasmic kinase activity (Takasaki et al., 2000; Takayama et al., 2001).

In tomato, an interaction between the LAT 52 protein from the stigma and LePRK2 kinase located at the plasma membrane of the pollen is essential for successful extension growth of the pollen tube between the cell walls of the style (Figure 4.2). Here it is the LAT52 of the adjacent stigma cell that interacts with LePRK2 of the pollen, thereby controlling the growth of the pollen tube target cell (Johnson and Preuss, 2003). Transgenic tomato plants expressing anti-sense LAT52 not only prevent the proper growth of the pollen on and within the style, but also block pollen from these plants from proper germination *in vitro*. So, both stigma and pollen marker proteins are therefore evidence of, and a requirement for, a successful duet (Tang et al., 2002).

BRASSICA

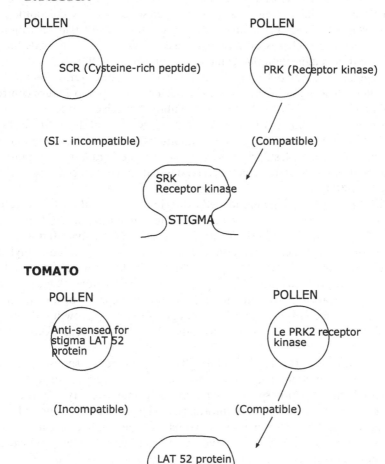

Figure 4.2. Receptor kinases and cysteine-rich peptide interactions can determine compatibility between stigma epidermal and pollen target cells. For experimental details for Brassica, see Brugiere et al. (2000); for tomato, see Johnson and Preuss (2003).

Cell performance and ageing in the target state

Many cells will retain their target status commitment for the whole of the cell lifetime of the plant. This is true of most cortical cells, but a cell undergoes constant dynamic change so the extent of both perception and response will not remain constant. Depending upon the performance of the rest of the plant and the signals the cell receives, a cell may up- or down-regulate the level of competence to perform as a specific target cell (see Chapter 7 for discussion of receptors). An example is the progressive reduction in the rate of elongation of a cortical cell as it reaches maturity despite the fact that auxin still elicits a response (e.g., proton extrusion). Fully extended segments of pea stems or maize coleoptiles, for example, will still release H^+ to an external solution through auxin-activation of membrane-located ATPases, even though the cell walls are no longer extendable by the extent of acidification that is induced (Venis et al., 1992).

The regulation of wall pH, either by addition of acids (short-term effects) or by a continued activation by auxin of the electrogenic proton pump, in turn controls a membrane potential across the plasma membrane which can exceed -200 mV (negative on the inside). So whether or not the target state of the ageing cell still responds to wall loosening under these conditions, the effects of auxin binding and auxin availability will still control events such as membrane-bound transport of proteins and attendant transduction cascades that lead to new gene expressions (Palmgren, 2001).

A direct link between auxin-induced extension growth and proton secretion is not, however, universal. In the semi-aquatic fern *Regnellidium diphyllum* additions of auxins do not enhance proton excretion and vanadate (an inhibitor of plasma membrane ATPases), although inhibiting both auxin and ethylene-induced growth in these cortical (Type 3) petiole cells, has only very marginal effects upon the net proton efflux. As this plant retains a permanently low apoplastic pH (5-6), it appears to by-pass the usual auxin-induced control of wall acidification via plasma membrane ATPases, although normal cell growth and proton release have been shown to operate for other Type 3 target cells (Ridge et al., 1998).

Most cells of the cortex progress through an enlargement response only once during their developmental ageing programme but other cells of cortical or epidermal origin may perform specific functions many times over. Type 2 abscission cells, for example, are receivers of information which is then acted upon by the cells themselves, either by remaining in contact or once only dissociating from their neighbours. Other cells, such as the gravity-sensing statocytes, perceive and convey gravitational cues both by long- and short-distance communication systems to other cells, retaining this privileged function of gravity sensor throughout the life of the plant (for further discussion see Chapter 6). Stomatal guard cells, too, maintain a green and functional condition even after the rest of the leaf has become yellow and senescent, and they open and close (expand and contract) many times during the life of the leaf.

Light intensity and quality of hormone signals (predominantly ABA) – sent either long distances from the roots or short distances from the adjoining mesophyll – regulate the specific turgor and volume changes of guard cells. Opening is determined by proton efflux driven by plasma membrane H^+-ATPases causing membrane hyperpolarisation with a balancing influx of K^+ and an enhancement of solute levels, mainly sucrose. Closing is essentially in response to water deficit and requires ion efflux from the vacuole compartment. As ABA levels rise in response to dehydration, cytosolic Ca^{2+} levels also rise. The Ca^{2+} elevations in turn inhibit both the plasma membrane proton pumps and the K^+ ion channel influx, while inducing a temporary anion efflux. The ABA-initiated rise in free Ca^{2+} followed by these rapid changes in membrane potential and the changed ion-channel directional flow leads to loss of guard cell turgor and stomatal closure (McAinsh et al., 1997).

The behaviour of guard cells with respect to expansion and contraction and their continued reversibility for cell size, indicates that the normal ageing changes to a permanent state of cell expansion do not apply to these specialized target cells. Furthermore, their extended survival and function, even when the surrounding neighbour cells of the leaf are senescent, indicate an absence of the usual programmes to cell death experienced by the rest of the mesophyll tissue (Schroeder et al., 2001).

Another example of developmental changes in the ageing state is seen in leaf senescence. Excised cherry leaves were found not to be maintained in a functional photosynthetic state by the addition of an auxin when tested in their expanding or fully expanded stages of growth presumably because auxin is not limiting during those stages, but they were at once functionally sustained, with senescence retarded, by auxin supplied later in the year while in their autumnal pre-shedding condition (Osborne and Hallaway, 1964). The gene expressions during cell separation in abscission zones of explant tissue are not necessarily the same as the gene expressions of much physiologically older abscission zones separating *in planta* (Del Campillo and Bennet, 1996). Particularly with regard to the terminal events of fruit ripening and abscission, the progress of the senescent changes that precede these on the plant are not duplicated in total when the organ or tissues are removed at earlier stages of development or are isolated from all but their near-neighbour short-distance signal inputs. The still unresolved and so-called tree factor which precludes the ripening of avocado on the tree (until very late and crop-wise unsaleable) is an example (Biale et al., 1954; Burg and Burg, 1962b): the harvested mature avocado removed from the tree-factor influence will certainly ripen, and ethylene speeds the process, but we do not know if the consortium of gene expressions that leads to this state are identical to those *in arbor*.

Senescence and cell death

Whatever may be a target cell progression through the life of a plant, it is evident that amongst the whole cell population, life spans are directly related to the target status and can vary greatly. Some of the earliest to programme to cell death are those destined to become routes for long-distance transport. The conduits of water and metabolites were the essential components of higher plant evolution and survival and are some of the earliest in the population to undergo a differentiation to death. Certain epidermal cells that develop target status to epidermal hairs or trichomes are likewise programmed to die early as part of their differentiation *in planta*. Not all epidermal cells, however, are of such attenuated life spans, for the epidermal LI lineage can produce some of the longest functional living cells in plants. Stomatal guard cells can survive green and fully photosynthetically active long after the rest of the leaf blade has become yellow, exhibiting all the symptoms of cellular disorganisation of a senescent and dying tissue.

In relation to the progress of senescence in individual plant parts, evidence of the signalling of reproduction to overall cell death in monocarpic species (in addition to experiments showing the retardation afforded by auxins, gibberellins and cytokinins with the acceleration of senescence by ethylene, abscisic acid and other senescence-inducing factors) have all provided an encyclopaedic background of important information gleaned over the past 30 years (Thimann, 1980; Noodén and Leopold, 1988). These now form the basis of the more recent genetic and molecular advances into how and why cells die. Again, we are still very ignorant of the molecular signals and immediate responses that initiate cell death programmes though we can recognise cellular markers that can identify a programme to death that is in operation. Also, we can follow in detail the sequence of cellular events along the pathway – for example, the expression of senescence-associated genes (SAGs).

The cues can either be wholly endogenous as in the case of procambial or root cap delineation or environmentally generated. Such cues are transduced into endogenous cues as in response to desiccation, changed light intensities or photoperiod. An example of a combined endogenous and environmental death-path signalling is seen in the nodes of the deepwater rice (*Oryza sativa*). Root primordia are initiated at each node but remain below the epidermis until such time as the node becomes submerged and subject to low oxygen stress. A precise localised death of a number of epidermal cells just above the primordial tip then follows which permits the emergence of the undamaged root through the gap formed in the epidermis. Though the initiation of the process is environmental, the endogenous signal is ethylene which increases under the lower oxygen tension of submergence. Nodal sections treated with ACC showed early signs of epidermal cell death within 4 hours which was inhibited by a concomitant addition of the ethylene receptor inhibitor, 2,5-norbornadiene (NBD) (Mergemann and Sauter, 2000).

Whichever the signal may be, cell death results from dysfunction and loss of integrity of chloroplast, mitochondrial and nuclear genomes. As a result of that, unwanted or ageing cells are removed by a signalled and regulated cascade of specific nucleases and proteases. Hydrolytic enzymes have long been acknowledged to alter during senescence by an up- or down-regulation of synthesis or through new gene expressions (Watanabe and Imaseki, 1982). The induction of specific serine and cysteine proteases and net protein loss have all been fully confirmed and indicate that plant cell senescence is not a random event (Thomas et al., 2003). Fragmentation of nuclear DNA of senescing leafy cotyledons of mustard (Osborne and Cheah, 1982) and during loss of viability of embryos of rye seed (Cheah and Osborne, 1978) has long been known, as shown by *in situ* labelling of end-groups of single-stranded breaks with 3'-OH-deoxynucleotidyltransferase and by electrophoretic fractionation of DNA of isolated nuclei. Not all cotyledon cells senesce at the same rate, those closest to the veins accumulate breaks more slowly indicating a positional control of senescence within the mesophyll and palisade. Environmental and internal insults to nuclear DNA are continuous throughout the life of cells, and DNA repair processes continually operate to restore the integrity of the genome. Failure to repair DNA damage and strand breaks leads to the progressive accumulation of unrepaired lesions which signal senescent decline. In the case of leaves, environmental signals are quite variable so organ life spans are equally variable. Where signals are essentially internal as in tracheid formation *in vivo*, the progress is highly orchestrated both in physiological and developmental time-frames. From very early plant physiological experiments it was evident that cell survival and the onset of senescence were under hormonal control. Chibnall's (1939) early observation of the extension of life of an excised bean leaf following rooting (now known to be attributable to the cytokinin flow from the root tips), and the substitution of roots by cytokinin in excised *Xanthium* leaves by Richmond and Lang (1957) are foundation experiments. Gan and Amasino (1995) placed the physiology on a molecular basis when they demonstrated that insertion of the isopentenyltransferase gene enhanced cytokinin production in transformed tobacco plants and delayed leaf senescence. While we remain ignorant of many (perhaps most) of the complex internal signalling mechanisms that direct temporal cell death programmes at the intimate levels of a few cells, the predictability with which they occur tells us that neighbour-to-neighbour cross-talk between different target individuals is a precisely organised exercise with an acutely sensitive perception and response.

It can be an instructive exercise to read the early papers by those who worked to understand the nature of cellular senescence in plants. In today's molecular scenarios of programmed cell death, the pinpointing of serine and cysteine as accelerators of senescence when supplied to excised oat leaves and the evidence for the synthesis of a protease with L-serine at its likely active centre (Martin and Thimann, 1972) were forecasts of the future. Thirty years later, Roberts et al. (2003) isolated and characterized a 59 kDa protease that is senescence-cell-specific from detached and yellowing wheat leaves. This protease possesses a 17-amino-acid sequence with a 65–75 percent identity to the highly conserved region of several plant subtilisin-like serine proteases. Previously to this, it is the cysteine proteases that had been directly linked to plant cell senescence (Granell et al., 1998). Both the cysteine proteases and this most recent senescence-specific serine protease of the cells of wheat leaves link to the cysteine proteases of the apoptotic caspase cascade of animal cells with their caspase-activated DNAse and the observed resultant loss of genomic integrity (Enari et al., 1998).

In developmental terms, the earliest examples of unwanted cell disposal are those that occur at embryogenesis when synergid and chalazal cells die and suspensor cells are subject to a precise programme of cell destruction. Such a programme has been followed in detail during the generation of somatic embryos in cultures of *Picea abies* following withdrawal of auxin and cytokinin and the addition of ABA (Smertenko et al., 2003) (Figure 4.3). Small cells with dense cytoplasm and mitotic activity develop adjacent to a group of vacuolating cells in which division is suppressed and the golgi activity increased. This polarised organisation marks the initiation of a programme to cell death in the vacuolated cells of the suspensor targets. The eventual nuclear DNA cleavage involves first the dismantling of the nuclear pore complex and the release of loops of 50 kb DNA

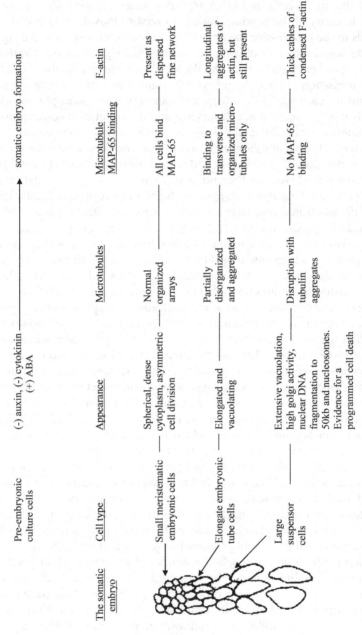

Figure 4.3. Cell and cytoskeletal markers denoting target states in developing somatic embryos from cell suspension cultures of *Picea abies*. For details of experimentation see Smertenko et al. (2003).

fragments followed by internucleosomal fragmentation. In the cytoplasm, there is a loss of microtubules, the formation of tubulin aggregates and the depolymerisation of actin fibres, which suggests that (as in animal cells) actin fibres function in the maintenance of the cytoplasmic matrix and hence cell survival. Further, a 65 kDa microtubule-associated protein (MAP-65) normally bound to the microtubules in functional dividing cells becomes disassociated from the depolymerising microtubules of the suspensor cells. A gradient of these responses has been recognised between the embryo centre and the adjoining lines of suspensor cells, in concert with the sequence of the suspensor cells towards death. This cell death sequence is evident both positionally and in temporal progress and provides an illustration of the differential target state that can exist even on adjacent cells destined to the same fate.

Many of the dead or dying and unwanted cells in animal tissues are engulfed and destroyed either by neighbouring cells or by phagocytic macrophages summoned by the liberation of signal molecules such as phosphatidyl serine from the disturbed inner plasma membrane bilayer of the dying cell (Savill et al., 2003). Macrophages carrying a phosphatidyl serine receptor (PSR) are attracted to the dying cell and there perform the operation of elimination. Failure to remove unwanted cells by PSR-deficient neo-natal mice was found to lead to serious abnormalities in development.

The complete ingestion of dying cells in this way is uncommon for plants, where instead, the internal dead cells become an integral part of the plant's anatomy. Typical examples are xylem cells and other lignified cells, those of the phloem and the many intricate but dead cells of the epidermis such as trichomes and hairs. The lysigenous formation of aerenchyma tissue may be an exception, but there too, much of the cell walls remain, even though the cytoplasmic contents are degraded. The large-scale elimination of senescing cells is achieved instead by the signal (ethylene) activation of specific groups of highly metabolically active abscission cells whose subsequent dissociation from one another leads to the shedding of all the dying tissues distal to them (fading flowers, senescing leaves, ripening fruit).

Is regeneration possible?

A number of tissues are known to 're-green' after yellowing, suggesting that certain aspects of senescence are reversible. Since plastids contain many copies of their circular genome which replicate within the plastid, it may well be that just one surviving DNA circle is sufficient to permit multicopy genome restoration and re-greening providing that the nuclear genome is also still functioning. The same argument may be true for the mitochondria. Not so for the nuclear genome, where unless the cell possesses endoreduplicated copies of the gene sequence the minimum number of copies they contain is either one (haploid cells) or two (diploid cells), and any long-sustained damage to these is likely to be lethal.

Repair processes of DNA are therefore critical for cell survival and the postponing of cell death. Much effort has been directed to the study of DNA processes in plants by either photolyase mechanisms for restoration of UV damage or for the excision repair events that are not light-requiring and monitor base-damage and DNA breaks. In the embryos of ageing seeds, DNA repair enzymes progressively lose function so that when water is again available for growth, germination and the onset of cell cycling are delayed and if the DNA damage is sufficient and the repair enzymes are already inactive, then the embryo dies (Osborne and Boubriak, 2002). This balance between genomic damage and repair is probably the most critical controller of plant cell death in any programme, but the exact sequence in which the events take place is dependent upon the target cell in which this imbalance progressively advances. There is no evidence to date, even in animal cells, that once DNA has fragmented to nucleosomes, there remains any chance for successful repair and continued survival.

Why life spans are so different amongst eukaryotic species remains an enigma. Even less understood are the survival times of individual tissues and cells within an organism. Whole plants of monocarpic species such as rye, wheat or tobacco have 1-year life cycles, but in normal circumstances the embryos of seeds of rye survive 3–4 years, those of wheat 30–40 years and tissue culture lines of tobacco (e.g., BY-2) may be immortal if no accident befalls them.

The origins of cell death in eukaryotes are commonly attributed to the acquisition of the multicellular state and the division of activity that the different cells then perform with the resulting diversification of development, function and life span. Recent investigations of certain unicells suggest that programmes to death are perhaps archetypal and that they have been inherited by metazoans and higher plants following viral incorporations of such genes into prokaryotic ancestors. When the unicellular photosynthetic alga *Dunaliella tertiolecta* was held in darkness, the disintegration of nuclear DNA to 50 kb then to 50 bp fragments occurred early in cell death, before degeneration of cytoplasmic organelles. These cells developed caspase-like protease activities homologous and immunologically similar to those of apoptotic animal cells that paralleled the process of algal cell death. Additions of caspase inhibitors prolonged cell life, and Segovia et al. (2003) suggest that these highly specific proteases were originally inherited from a common ancestor via ancient viral infections of bacteria, the bacteria then being engulfed by the originators of eukaryotic cells and thereafter maintained through evolution by all the different eukaryotic lineages. If this is so, then death is the inevitable outcome of the life form.

There remains a major question for which at present we have no satisfactory answer. How can some cells such as those of spores, windborne pollens, seeds and leaves of the so-called resurrection plants survive desiccation for days, months or decades, when loss of water is lethal to almost all others? Recently the growing root tips of certain germinating seeds were shown to possess a transient period of desiccation tolerance while still very small (Buitink et al., 2003). What exactly are the molecular strategies these cells have developed that retain the integrity of their genomes in conditions that lead to the complete disruption of cellular organisation and death in their neighbour eukaryotic tissues? The evolution of survival with limited water has been attributed to a number of possible mechanisms such as the organised accumulation of non-reducing sugars (Obendorf,

1997), specific protective proteins (dehydrins) (Close, 1996) or a variety of solutes that aid, on dehydration, the formation of a cytoplasmic 'glassy state' (Oliver et al., 1998) thereby limiting the mobility of water molecules within the cell. Perhaps the most intriguing and the most likely mechanism is the conversion of genomic DNA from the metabolically active B-form conformation to a metabolically inert A-form conformation by a highly specific genetically controlled sequence of new gene expressions evoked by the initial stages of cell dehydration. First shown for the developing and long-lived spores of *Bacillus subtilis* by Setlow (1992), lack of achieving this structural DNA transformation *in situ* may afford the critical mechanism that determines the onset of ageing and eventual cell death of every hydrated target cell. The ability to convert B-form DNA to A-form could there-fore be a marker for desiccation survival (for discussions of Setlow models, see Setlow, 1994).

5

Flexibility of Cell Types and the Target Cell Status

Every cell can be considered a target cell, with a status that is subject to change throughout its life until a state of terminal differentiation is reached. On this basis, every cell is slightly different from its neighbour with respect to position and signal response, so that at any one time each cell has a unique target status even though it is a member of an apparently uniform tissue. Although the number of signals that have been identified or described so far are limited, the number of responding target cell types in plants would appear to be unlimited.

The flexibility of an individual cell, or perhaps more correctly, the flexibility of a group of cells to give rise by repeated cell divisions to a whole new plant, is the basis of the concept that plant cells remain totipotent throughout their lives. Horticulturists have used this knowledge in vegetative reproduction following observations that many isolated plant parts will readily regenerate new individuals with all the anatomical and behavioural characters of the parent. Planting a cutting is one thing, where all the coordinating signals and target cells are, as it were, still in operational position. Propagation by pieces of tissue where lines of intertissue communication have been lost is quite another.

The question of how a community of cell types in a callus or suspension culture develops in an organised and temporal fashion into a meristem is essentially unresolved though certain clues give consistency to the concept that specific short-distance signals are operating between them. The early formation of vascular or stelar tissue close to the shoot and root meristem offers a potential source of positionally generated signal molecules. Here, we have considered the stele as a focal tissue for differentiation control, by the release of instructions to specific target cells adjoining the vascular centres.

In tissue culture, where cells are induced to multiply outside the organising society of the whole plant, lines of intercellular communication must be re-established and new target cell individuals must be generated in order to re-create a whole plant. Callus cultures or suspension cultures are not equivalent to the parenchyma cell types of the intact parent plant. Despite this, on suitable direction by auxin and cytokinin concentration, a suspension culture can be converted to over 60 percent lignified cells (*Zinnia*; Fukuda, 1994), a callus culture can be made to produce either a root or a shoot initial (tobacco; Skoog and Miller, 1957), or a group of cells can be induced to produce a somatic embryoid (carrot; Racusen and Schiavone, 1990). Once the root initial is present it has informational experience that determines it is a root, so neither excision nor manipulation abolishes this unless it is transferred back to tissue culture and the coordinated information state between the different cell types is again lost.

Coordination of a group of cells into a meristem leads to a curtailment of certain flexibilities. The question then arises as to which types of cells retain their flexibility in the intact plant and what are the degrees of flexibility that they represent. Clearly, cells such as xylem that differentiate to a terminal state of death within the community, or those that lose functional nuclei such as sieve tubes, must lose all options of further developmental change. A number of cell types, however, remain alive and retain their nuclei for long periods of time, but to our knowledge they are unavailable for further differentiation. Two examples of such living functional cells are the aleurone cells of graminaceous seeds and the abscission zone cells linked to organ shedding. Both of these have, until now, defied manipulative efforts to encourage them to express totipotency. They have failed to divide or to undergo any transdifferentiation to a new cell state. These examples are discussed further in Chapter 6.

In this chapter we examine examples of cells that *in planta* remain apparently unchanged for long periods but retain a flexible target status to respond to new signalling inputs and a capacity to perform new differentiation programmes. We have chosen cells that the anatomists classify as parenchyma – essentially epidermis, cortex and pith. However, although pith may be perceived as an undifferentiated parenchyma in a young plant it clearly loses this status and adopts a terminal death programme early in the shoot's life, as the dead pith of the anatomist's sectioning equipment has long portrayed. Why pith cells, which differentiate internally to the vascular tissue, should provide such excellent material for unlimited tissue culture when young, yet be programmed to cell death so early in life when *in planta* remains to be understood. Quite clearly its target status differs markedly from that of the long-lived cortex cells, which are differentiated externally to the vascular tract. Evidently, the differentiation of cortex external to the vascular tissue has significant positional target implications.

Whereas epidermis tissue initiates early in development a wide variety of positionally differentiated cells in the shoot (hairs, trichomes and stomatal pores) and in the root (the positioning of root hairs), only the cells of the cortical tissue retain the same flexible target state for the major part of the plant's life span. In other words, cells of the cortex retain a state of homeostasis for long periods, thus providing a 'cell bank' for perceiving and responding to new incoming signals with options for the greatest potential and flexibility for further differentiation.

Epidermis as a target state

As a target tissue, epidermis has the flexibility to produce probably the largest range of cell growth responses in the whole plant. Differentiated as the outermost layer of the meristem, epidermis performs the function of the plant's cellular overcoat monitoring change in the external environment and controlling shoot elongation growth, water loss and chemical and mechanical defence against pathogens. At such time in secondary thickening when an internal cambial layer is developed in the underlying cortex, the isolation, death and sloughing off of the cell layers external to this new cambium can lead to the permanent loss of epidermis in certain parts of the plant. The role of epidermis is therefore performed in its living state, unlike that of xylem and phloem, in which the major contribution to plant function occurs when the cells are dead.

Research carried out over the past decade has revealed epidermis as a highly complex tissue, under the control of many genes that provide distinctive properties and distinguish these cells from others internal to them. So far, it appears not to have been possible to culture epidermal cells as a specific cell type; furthermore, epidermal cells from upper and lower surfaces of dorsiventral organs are not necessarily the same. Evidence for intercellular informational networks from neighbour cells that can specify the differentiation of specific epidermal cell types has become a subject of intense interest to developmental biologists.

Flexibility in epidermal cells

Epidermis, as it emerges from the meristem as the tunicate L1 layer, very soon exhibits its remarkable flexibility, developing a distribution of still flexible, but already potentially designated, target cells open to further informational instructions from both the environment and neighbour cells. From the environment, we know that hypocotyls of *Arabidopsis* do not differentiate stomata unless they are exposed to light; but given exposure to light, the informational exchange between cells is open to the direction of a wide variety of signal molecules transmitted between them (Wei et al., 1994). Stomata, like trichomes, are considered in more detail in the terminal state of differentiation in the following chapter, but it is clear that they pass through a series of differentiation 'gates' before reaching their terminal differentiation condition.

Of interest in the present context is the intermediate stomatal target state that arises before guard cell precursors mature and become part of the functional stomatal complex. In tobacco leaf explants, cultured *in vitro*, oligogalacturonides have been shown to accelerate and synchronise the mitotic stage of the precursor guard cells and their conversion to the mature functional state. Neither final numbers nor distribution are altered but the rate at which the conversion takes place in the presence of oligomers of 9–18 sugar residues is more than doubled in the first 24 hours of treatment (Altamura et al., 1998). Oligogalacturonides of 1–8 were inactive and the response to the 9–18 polymers was repressed by the addition of auxin, indicating the dual regulatory role and control exerted on precursor stomatal development. What is the likely source of these oligogalacturonides *in planta*? The precursor guard cells upon which the fragments have their effects lie above the just-formed stomatal cavities. These cavities could result from a localised polygalacturonase activity, with the degradation of middle lamellae between the cavity cells being the likely source of the OGAs to which the stomatal precursor cells are then competent to respond.

In considering the multifunctional options open to the single sheet of cells that comprises the epidermis, it is not surprising that multigene controls have been identified that are specific to this target surface layer of the plant. Screening a cDNA library from epidermal peels of mature leaves of the succulent *Pachyphytum* (Clark et al., 1992) provided five cDNAs representing abundant mRNA transcripts exclusive to the epidermis of which three were then confirmed by *in situ* hybridisation to be confined to epidermal tissue. Two were located in all epidermal cells marking their common origin, but one was specifically enriched in the subsidiary cells of the stomatal complex.

The development of cell types with distinct differentiation pathways in cells that were initially part of the same sheet of tissue provides researchers with unlimited opportunities for investigation. From such studies important observations have stood the test of time. One of the earliest was the observation of Avers (1963) that the asymmetric division of an epidermal cell in the just fully expanded tissue behind the root tip in *Phleum* was the diagnostic feature which delineated that cell as a root hair initial. In stomata formation too, the first observable feature is an unequal cell division in which the smaller cell proceeds to the full stomatal complex (Kagan et al. 1992).

In *Arabidopsis*, epidermal cells destined to become root hairs are restricted to files of cells that adjoin two cortical cells beneath them, whereas the non-root hair files of cells are found in contact with only one adjoining cortical cell. Here, the difference in epidermal cell fate depends upon interactions with the number of cortical cells adjacent to them. Two genes, *GL2* and *TTG*, encode negative regulators of root hair formation so that mutations of these genes lead to roots with almost every epidermal cell developing as a root hair (Galway et al., 1994). This determination of a target state that is competent for a subsequent root hair development programme is clearly initiated by the epidermal cell positioning that occurs prior to final epidermal elongation. Another gene (*CPC*), which encodes for a small Myb-like DNA-binding protein, has now been shown to be additionally required for root hair cell formation and to compete with *GL2*, acting thereby as a regulator for promoting root hair formation. The balance between promoting and negating gene-directed programmes forces the response of the target epidermal cell. Of interest here is the evidence from Wada et al. (1997) that inhibition of the *CPC* gene and the expression of *TTG* and *GL2* result in inhibition of shoot epidermal trichomes as well as suppression of epidermal root hairs. Genes encoding transcription factors and elements of the ethylene signal cascade in *Arabidopsis* have indicated that ethylene (or ACC) may be a diffusible signal involved in the generation of these spatial patterns (Dolan, 1996).

Dorsiventrality in the epidermis

The upper and the lower epidermal layers of a leaf, although of common L1 origin, develop very differently and are visibly distinct in the mature leaf. Most obvious in land plants is the relevant paucity of stomata on an upper surface or their total absence on a lower surface, as in aquatics such as *Nympha* or in the New Zealand native tussock grass, *Festuca novae-zelandiae* (Abernethy et al., 1998). Another is the different extent of cuticle formation, but perhaps one that is less readily understood is the difference in the circadian rhythm of the petal epidermis response of *Kalanchoe*. Flowers open and close when the mesophyll cells between the two epidermal layers expand or shrink with a circadian interval of about 23 hours. The anthocyanin-containing upper epidermal cells are papillae-like and expand and shrink in concert with the mesophyll but those of the lower surface are thick-walled and bend only passively (Engelmann et al., 1997). The evidence for circadian turgor changes in one epidermis and not the other suggests either a targeting of functional oscillators to only one side of the petal or an inactivation of the oscillator by perhaps the mechanical suppression of its function. Whichever, the result is that only one epidermal side of the petal (the upper epidermis) exhibits the circadian response.

Big differences in epidermal cell sizes between developmentally upper and lower surfaces are noted on monocotyledon leaves. In barley, cells lying between the veins, for instance, the bulliform cells of the upper surface, reach approximately 200 μm while those of the lower surface can attain 2 mm, or more (Wenzel et al., 1997). To accommodate such differences, more cells lie over the veins than between the veins on the upper surface with the reverse on the lower surface. Such determinant patternings of epidermis indicate the range of potential sensing that the epidermal layer can achieve and the wide range of attendant target types that can be displayed.

The epidermis as a target tissue – Evidence from experiments in vitro

Mature shoot epidermis possesses a remarkable ability to respond to specific signals received from internal tissues and cells but has little capacity to respond alone directly to an added hormone. Chlyah (1974a), for example, found that single, excised epidermis layers of the rectangular stem segments of *Torenia fournieri* (a member of the Antirhinaceae) died within 48 hours and were incapable of differentiation unless placed back in contact with other cells or isolated with other subepidermal layers attached to them. In one respect, these epidermal cells resemble abscission cells and aleurone cells in that they resist culturing as an individual cell type, but they differ in the fact that they possess the ability to form any root or shoot cell type if provided with the appropriate contact signal from their neighbours. Thus *Torenia* epidermal cells can divide and develop to form buds or root cambia if sufficient subepidermal cells are isolated together with them and if appropriate levels of IAA or kinetin are included in the culture medium. Under these conditions, epidermis was found to be totipotent (Chlyah, 1974b). Because this epidermis can be removed from the stem tissue with relative ease, it is very suitable material for statistical studies. These studies include determining the number of cells in an epidermis that can divide, the controls exerted by position in relation to underlying tissues and very importantly in the target cell context, how far each epidermal cell can influence the performance of its neighbourhood epidermal community. In a study of bud formation in epidermal cells of stem segments of *Torenia*, the frequency of cell division centres was found to be non-random being lowest in cells overlying vascular tissue, and highest at the basal end (basipetally polar with respect to auxin transport). Using [^{3}H]-thymidine to label nuclei in S-phase, none of the 20 percent of non-stomatal epidermal cells labelled early after excision underwent division but, of later labelled cells, some were capable of forming centres for cell division indicating communication and an intercellular signal regulation in the determination of each epidermal cell fate (Chlyah, 1978).

Epidermal cross-talk

Epidermal cell communication and the concept of master cells directing the performance of neighbour cells derive from studies of epidermal stomata formation. Kagan and Sachs (1991), who called this 'epigenetic selection', investigated this in *Sansevaria* leaves, where half of the stomata initiated by near synchronous unequal divisions in files of epidermal cells failed to develop into mature stomata. Using computer-generated models, they deduced that formation and maturation of stomata does not depend upon the near-neighbour frequency of the initials alone, since all are initiated together in any area, but rather on another, as yet undetermined but epigenetic control that results in a final and non-random spacing. In other words, the mechanism of communication between the cells has a special significance in signal transfer.

Stomata are spaced so that at least one epidermal cell separates them. Following the first asymmetric division, the smaller epidermal cell divides symmetrically to produce the two guard cells, so each stomata is its own small clone. In *Arabidopsis*, neighbour cell feed-back loops have shown the expression of specific proteins as the stomatal clone develops. One protein, SDD1 (stomatal density and distribution 1), is primarily produced by stomatal meristemoids and guard mother cells. This subtilisin-like serine protease inhibits the expression of SDD1 in neighbour non-precursor cells. In the mutant, *sdd1* negatively regulates SDD1 production, so an abnormally high stomatal density results throughout the shoot. Overexpression of SDD1 from the CaMV35S promoter in a wild-type background reduces the number of stomata. Although SDD1 expression is not confined solely to epidermis and is present also in subepidermal tissues, it may well interact with other proteins such as the leucine-rich repeat receptor (TMM) that is confined only to the epidermis. Whatever interactions may eventually be revealed, it is clear that close interactivity between target cells is the basis of stomatal shoot patterning (von Groll et al., 2002).

Epidermis can therefore be seen as a target tissue in which signal responses may or may not require a neighbour cell association or cooperation with another cell type for expression; this can extend to communication between two genetically different cell types in the control of epidermal development. The formation of knots in maize leaves is such an example. When X-rays were used to create genetic sectors in the leaves of maize, a mutant knotted (*Kn-*) arose in which the leaf epidermis lacked the multiple cell divisions associated with the tissue below. The development of mesophyll knots was independent of whether the epidermis was of the *Kn* or non-*Kn* genotype but depended only on the presence of *Kn* in the mesophyll below (Hake and Freeling, 1986). However, epidermal cells clearly do have some autonomy *in planta* and are not always directed by their subcellular layers or influenced by their epidermal neighbours. In a graft-generated chimera of L1 of *Camellia sasangua* and the L2 and L3 of *C. japonica*, the epidermal cells of the petals are always those of the L1 *C. sasangua*, irrespective of which *Camellia* cells comprised the subepidermal layers (Stewart et al., 1972). In addition, epidermis may not exert a constraint upon the cell layers below. In leaves of the mutant cultivar of pea, *Pisum sativum* var. argenteum, in which the epidermis only loosely adheres to the adjacent mesophyll, removal of the epidermis by peeling at early stages of leaf expansion has no effect upon the subsequent growth of the cells beneath the peeled part. Mesophyll cell size and general leaf morphology remain similar to those of the unpeeled controls (Wilson and Bruck, 1999).

Epidermal outgrowths – Trichomes

Not only, it seems, are trichome cells specific in their origin and differentiation pattern within the epidermis, they can also express specific and independent biosynthetic pathways for secondary products. *L. pennellii*, for example, produces trichomes that form tri-acyl glucoses that are secreted from them to form a sticky impediment to aphid movement on the leaf surface. In graft chimeras of *L. pennellii* and *L. esculentum*, where epidermis arises solely from *L. pennellii*, these glucose esters are secreted irrespective of the cellular origin of the subepidermal tissues below (Goffreda et al., 1990).

More than twenty genes affecting trichome development have been identified in *Arabidopsis* where each trichome is the product of a single epidermal cell. A cessation of cell division but continuation of DNA synthesis to the 8C stage and abnormal (x2) cell enlargement causes the cell containing the endoreduplicated nuclear DNA to bulge outwards from the surface. This is the trichome stalk. The walls then thicken and the trichome is established. The *R* gene in maize (in common with the recessive *TTG* or *GLI* genes in *Arabidopsis*) leads to a reduction in the number of trichomes per leaf surface, but in maize the *R* gene also controls a root epidermal cell response, so there is also a reduction in the numbers of root hairs formed. *Arabidopsis* plants, homozygous for the recessive *ttg* or *gli*, are completely devoid of trichomes and, where chemical mutagenesis was used (ethyl methyl sulfonate, EMS) on plants heterozygous for *gli*, patches of trichome-free areas were formed in the leaves. Whether or not the glabrous patches were the result of the uncovering of the *gli* mutation, it is clear that the trichome-producing area cannot produce a substance that passes to and induces trichomes in the bare surface sites (Hülskamp et al., 1994). The suggestion is therefore that the primary function of *GLI* is at the level of the trichome-precursor epidermal cell and shows, if this is so, the likely temporal cell specific action for GLI. It indicates also that the trichome precursor cell is a highly specific target state in which the extent of nuclear DNA endoreduplication is a very early molecular marker.

The exact function of endoreduplication in determining a target state is still unclear. The presence of 8C cells at the base of the gynoecium in the female flower of *Ecballium elaterium* is a cortical nuclear DNA marker. In this instance it denotes a specific ethylene perception with ethylene-induced cell enlargement and separation that leads to premature ovary shedding at only these precise positions (Wong and Osborne, 1978; Figure 4.1 and Chapter 6). But here, too, the role for amplified DNA is unclear, as other abscission cells in this plant are not DNA endoreduplicated.

The experiments of Dan et al. (2003) with another cucurbit may offer a clue. They exposed cucumber hypocotyls to ethylene and noted that DNA synthesis and up to an eight-fold increase in DNA content per nucleus occurred in some 20 percent of epidermal cells, with no changes detectable in air controls. The endoreduplication of nuclear DNA content was not permanent, for on removal of ethylene rapid cell plate formation and cytokinesis restored DNA levels in the epidermal cells to the normal 2C level. This may be telling us that at least for the endoreduplicated state the condition of lagging plate formation has to precisely coincide in time with other signals presented to the target cell. The windows of opportunity for competence to perceive and respond may therefore be quite limited, as we already know from the photoperiodic and cell cycling regulation of flowering (see Bernier, 1988).

Epidermis and shoot elongation

Effects of hormones. It has long been known that the response of epidermis to hormonal additives can determine the growth elongation in segments of stem or coleoptile tissue. This was the basis of the famous 'split-pea' curvature test of Went by which a split stem would curve outwards on splitting through the greater turgor of the inner cells, but on placing in a solution of IAA, the ends would curve progressively inwards, the degree of curvature being dictated by the elongation induced in the epidermal cells (Van Overbeek and Went, 1937). Since then, experiments with peeled sections, epidermal strips and segments with inner cores removed have confirmed that auxin can induce neither elongation nor wall loosening in epidermis alone, and that the elongation event occurs in epidermis only when several subepidermal layers remain attached to it (Masuda and Yamamoto, 1972).

The concept that the less readily extendible epidermal layer acts as a constraint to the internal pressures exerted by turgor-induced forces of the cell layers below seems now to conform to the wealth of evidence provided by studies of elon-gating shoot and leaf tissues. A nice demonstration that this is applicable also to more isodiametric growth comes from a study of the enlargement of tomato fruit (Thompson et al., 1998). In those experiments, pericarp fruit slices taken from young and mature fruit, when put into solution, expanded outwards like the outward curving of the control pea stems in the Went pea test. In conjunction with this simple demonstration of the epidermal constraint they showed that a wall-loosening enzyme (xyloglucan-endotransglycoselase, XET) activity in the epidermis (but less so in the pericarp) was proportional to the expansion rate of the fruit until maturity, and only then was a putative wall-cross-linking enzyme (a peroxidase) expressed in the epidermis alone, indicating a final growth control by epidermis. This demonstrates again an independent target status of epidermis with respect to its pericarp neighbours.

In the isolated coleoptile segments of maize, auxin induces an increase in the plastic extensibility of the outer epidermal cell wall within 15 minutes. This wall loosening is associated with cytoplasmic changes that include deposition of osmio-philic granules at the wall's interface with the plasma membrane. Using probes for arabinogalactan proteins, Schopfer (1990) identified these granules by cytochem-ical light microscopy and linked their presence to the epidermal permissiveness to segment extension growth. Specific responses of epidermal cells are linked to other cytoplasmic characters, including microtubules. In most elongating epider-mal cells, microtubules are primarily located in the peripheral cytoplasm: their orientation transverse to cell elongation is associated with the similar direction of deposition of cellulose microfibrils whereas non-growing cells exhibit oblique or longitudinal arrays once they are mature and fully extended (Williamson, 1991).

Epidermis tubulin in coleoptiles of rye, however, seems to exhibit a specific location either within the nucleus or at the nuclear membrane, once the cells have become post-mitotic and started to elongate. Only then do the microtubules de-part to the usual peripheral cytoplasmic location, the earlier elongation events ap-pearing to take place before the microtubules migrate (Kutschera and Bett, 1998).

Whether this unusual difference in microtubule behaviour is linked to the particular control that the coleoptile epidermal cells exert over the internal tissue growth (and is therefore a marker of their target status) remains for more detailed exploration. Furthermore, the microtubule types may well be different in the different cell locations as has been demonstrated in animal cells (Byard and Lange, 1991).

Ethylene, which enhances the lateral expansion of shoot cells at the expense of elongation growth in Type 1 cells has tissue-linked effects on the epidermal cell layer; both peroxidase and extensin are specifically enhanced by ethylene in the extending regions of pea epicotyls (Cassab et al., 1988). It is clear that the different sides of an epidermal cell possess their own special organisations and structures. Not only is the orientation of the cell wall microfibrils differ-ent from that in neighbour subepidermal layers, but also the microtubules on the outer cell wall (of pea) epidermis have unique properties and fast rates of turnover (rhodamine-labelled porcine tubulin is incorporated into microtubule arrays within minutes) which may reflect a unique role for this face of the cell in environmental sensing (Yuan et al., 1992). In this respect, the outer tangential wall is also the pathway of traffic for fatty acids into surface wax and cuticular lipid, and an epidermal stearoyl-acyl carrier protein thioesterase is implicated in epidermal wax biosynthesis (Liu et al., 1995). In leaves of tobacco, other enzymes that may be related to defence functions are also specifically located in the epidermis. Basic isoforms of β-1,3-glucanases and chitinases are almost exclusive to epidermal cells in normal plants but, on treatment with ethylene, high levels (more than ten-fold) can be found also in the vacuolar compartments of all the leaf cells (Keefe et al., 1990).

These examples, as well as those related to touch and other signal inputs described in Chapter 3, suffice to outline the critical part played by epidermis as the coordinating interface between the plant and the ever-changing environment beyond.

The complexities of hormone and signal transduction that must be trafficked by the flexible epidermis of a young developing shoot or root and the protective functions that operate in the many inflexible epidermal cells types of the mature plant remain a challenge to current molecular understanding. The way that informational networks are generated by an epidermal cell and the outputs then selectively perceived by the target cells of cortical and inner tissues is presently a long way from elucidation.

Cortical parenchyma cells

Perhaps the cells that retain flexibility for the longest time in plants are the parenchyma-type cortical cells. But, make no mistake, not all cortical cells are the same in terms of their target cell status; this is displayed in the wide range of responses that they manifest to a multiplicity of signals.

One factor that limits interpretation of much of the work conducted with hormones is the difficulty in pinpointing the time at which cells become competent to respond to the signals that have been presented over extended periods of time. The individual roles of auxins and cytokinins in the regulation of cell division is an example. The original conversion of tobacco callus to the formation of buds or roots took many days (Skoog and Miller, 1957) so the exact timing of when either auxin- or cytokinin-induced competence of cells to respond to the hormone occurred is unknown. In an attempt to resolve this, Carle et al. (1998) examined the role of auxin and cytokinin in reactivating the cell cycle during the first 48 hours of tobacco mesophyll protoplast culture. Using hormonal delay and withdrawal studies they found that auxin (2,4-D or NAA) was required for the first 4 hours, with cytokinin (benzyladenine) not required until 10 to 12 hours, which is just 6 to 10 hours before S-phase. However, the accumulation of the cdc2 protein as a cell cycle marker for the onset of S-phase appears to be activated by auxin as well by cytokinin, and both are needed for full expression of *cdc2*. This indicates that each hormone probably controls a separate signal transduction pathway to the initiation of cell division with each pathway triggered at the same time.

A dual hormonal requirement of cortical cell growth is well exemplified in the ethylene-induced cell extension of stems and petioles occurring when certain semi-aquatic species become submerged below the water surface. The original discovery in *Callitriche* showed that the ethylene response is dependent upon the presence of gibberellin (Musgrave et al., 1972), though it was later shown that IAA was equally effective as the co-partner. In the semi-aquatic fern, *Regnellidium diphyllum*, cell elongation of intact rachi with leaflets attached was enhanced by either auxin or ethylene alone and growth could be more than additive when both were supplied together. Such cells were called Type 3 with respect to this unusual auxin/ethylene response. If the leaflets were removed, however, or segments of rachis used instead, the ability to respond to ethylene decreased with time from excision so that segments cut and kept for 24 hours in water no longer responded to ethylene alone. However, they could be caused to elongate again when an auxin was supplied and would grow even more when ethylene was also present. This has been called the supergrowth response (Ridge and Osborne, 1989). Adding an ethylene pulse first before the addition of auxin does not, however, cause more growth than that in auxin alone showing that both auxin and ethylene must function in concert in these cells though not through the same transduction pathway (Ridge et al., 1991). Although a model to account for this cooperative growth in these cells is described (Figure 7 in Ridge et al., 1998), we still have little understanding of how the wall-loosening dynamics of this auxin-plus-ethylene growth is achieved at the molecular level.

There are now numerous examples of the submergence-responsive Type 3 cells that elongate readily with ethylene. Some of these responses are remarkably fast (e.g., *Rumex palustris*) while others are relatively slow (e.g., *Ranunculus sceleratus*). The fast response of *Rumex* is known to be associated with an induction of new ETR receptor sites for ethylene on submergence which facilitates the rapid ethylene induced growth response (Vriezen et al. 1997). Furthermore, the rapid arrest of *Rumex* petiole elongation when the leaf again reaches the water surface has been correlated with the loss of trapped ethylene to the air and with the equally rapid (20 minute) suppression of further receptor induction when the internal concentration of ethylene to which the cells are exposed is reduced. Additionally, submergence induces the expression of the expansion gene controlling wall loosening in the flooding tolerant *R. palustris* (Vriezen et al., 2000) but not in flooding intolerant *R. acetosa*.

The behaviour of these cells is quite distinct from that of the pea stem, the maize coleoptile or indeed most higher plant parenchyma shoot cells, where auxin enhances rates of cell extension but ethylene does not. In fact, in pea stems ethylene alters the orientation growth from longitudinal extension to lateral expansion but with no detectable influence upon the final cell volume achieved (Osborne, 1976; see Figure 1.1). Elongation growth then appears arrested with a lateral swelling of the responsive region of the shoot. Such Type 1 parenchyma cells are the most common target cell type in plants.

Of course, not all shoot cells are highly responsive to the additions of auxin. The discovery of limited gibberellin biosynthesis in the dwarf mutants of maize and pea in which cell growth reaching normal extension could be achieved by the external addition of gibberellin tells us that the limiting signal for growth is not necessarily auxin. The complexities of wall loosening and turgor maintenance requires that many inputs are controlled and contribute to the final cell enlargement event. Nonetheless, either the addition of an auxin or the addition of a gibberellin to segments of tissue or intact plants can achieve the final release of growth restriction permitting the cell to take up more water and enlarge. It can reasonably be accepted that all the hormones are necessary players to some degree in the cell enlargement scene but only those that are at threshold or subthreshold levels can regulate the rate of growth (most usually auxin or gibberellin) or play a determining role.

The dual enlargement potential found in the cells of semi-aquatic plants in which auxin (*R. diphyllum*) or gibberellin (wild rice) or either auxin or gibberellin (*Callitriche*) are the co-signal partners with ethylene reinforces the concept that fine controls have developed within plants to match the environmental and genetic restrictions placed upon them. In all, however, the basic requirements for cell growth are the same: the need for a metabolic energy output to drive turgor forces, together with a means of cleaving those wall polymer bondings that maintain wall resistance. Both then permit the net movement of water into the cell and the enlargement of the vacuole.

A third type of flexibility in cortical parenchyma with respect to an auxin and ethylene response is represented by abscission cells (called Type 2). As far as we know, all abscission in dicotyledonous plants is set in train by ethylene produced in neighbour tissue and all the abscission zone cells studied to date show some degree of cell enlargement or increase in turgor during the progress of cell separation. Measurements made of the abscission zone cells in the bean show ethylene-enhanced cell growth and a suppression of that cell growth by auxin – i.e., the opposite of the hormonal auxin/ethylene interplay found in all other parts of the aerial shoot. This means that as a petiole or fruit stalk matures, the abscission zone cells already differentiated must keep pace in growth with those of the neighbouring Type 1 tissues, and presumably at this stage they respond to the same auxin and ethylene cell growth signals as their neighbours. Only when the Type 2 abscission cells perceive ethylene above a threshold level will they then start to enlarge further and initiate tissue and cell tensions with their adjacent non-zone neighbour cells (Wright and Osborne, 1974). The tissue tensions then set up between the two differently responding hormonal target cell types are critical to successful abscission. At this stage, water deprivation and turgor loss are major causes of a failure of plant parts to shed even when the process has been initiated. These very specifically located Type 2 abscission cells are few in number in any plant and are precisely positionally differentiated as a group or plate of cells between their normal Type 1 neighbours, most usually at the base of organs that will eventually fall from the parent plant.

It becomes suggestive to consider that Type 2 cells must undergo a target change from Type 1 at some stage during their maturation. Does this mean that ethylene perception is down-regulated once a zone cortical becomes a competent Type 2 cell? For most of their life span within the plant corpus, zone cells appear to be insensitive (unable to initiate cell separation) to the levels of ethylene normally encountered in the growing plant. However, once differentiated as competent Type 2 target cells, they can always be induced by *added* ethylene to start their cell separation programme; the increased ethylene formation by adjoining senescing leaves, fading flowers or ripening fruits is equally effective. Target change from Type 1 to Type 2 can take place very early in leaf development as experiments have shown. The tiny primary leaf of the *Phaseolus* bean that is already differentiated in the seed embryo cannot be induced to shed until several days after germination which appears to coincide with the first lateral root emergence (D.S. Thompson and D.J. Osborne, unpublished). The opening winter buds of *Sambucus nigra* or *Aesculus hippocastania* are also non-shedding until the leaf starts to expand (McManus, 1983; Osborne, 1989). The failure to respond to ethylene at the very early stage of leaf development performs a valuable function, for otherwise, the levels of ethylene produced at germination or bud break are sufficient to induce separation in competent Type 2 cells. The molecular and ultrastructural changes that convert Type 1 cells to Type 2 cells remain to be unravelled. During their early life, however, Type 2 cells appear to respond to many of the normal controls imposed upon Type 1 cells. They apparently offer no impedance to auxin transport as shown by classic auxin transport experiments in which freshly excised abscission-zone–containing petiole segments are compared with those of similar lengths of petiole only (Jacobs et al., 1966). Also, as the leaf petiole enlarges, the Type 2 cells keep pace with their Type 1 neighbours.

It is evident, however, that Type 1 cortical cells can retain a flexible status long after full expansion. In mature leaf petioles of *Phaseolus vulgaris*, excision of segments and suitable positional treatments with auxin or ethylene can cause localized groups of subepidermal cortical cells to transdifferentiate to Type 2 abscission cells, with the expression of the abscission-specific β-1,4-glucanhydrolase and ensuing cell separation (McManus et al., 1998; see Figure 5.1 and Chapter 6). Mature cortical cells of shoots (following decapitation; Webster and Leopold, 1972) and mesophyll cells of leaves (by fungal infection or wounding; Samuel, 1927) can also be induced to become Type 2. The ability of cells to retain long-lived flexibility and then to alter their target state for signal perception and response when fully mature, provides the plant with many performance options for survival.

In considering the age of cells and the potential for flexibility, we currently do not know if an apical meristem always consists of cells that are all Type 1. The actual time of conversion to a Type 3 cell, for example, is unknown, although the ethylene-responding cells of *Rumex, Nymphoides* or *Callitriche* are found both in very immature internodal and in petiolar tissues. From the many experiments carried out with *Nymphoides* (Ridge, 1992), it seems that the very youngest leaf petioles tested (1–2 cm) respond most readily as Type 3, while those that are further extended and more mature before submergence show a decreasing ability to exhibit the Type 3 target condition. This suggests that, in common with Type 1 to Type 2 conversions, an initial degree of cell expansion and/or cell division might well occur as a Type 1 response before the Type 3 target state is differentiated.

There is another aspect of the timing and change in flexible states as they impinge upon the course of plant development, namely, the *number* of cells that are caused to change their target status at any one time.

Zero days

2 days

4 days

6 days

8 days

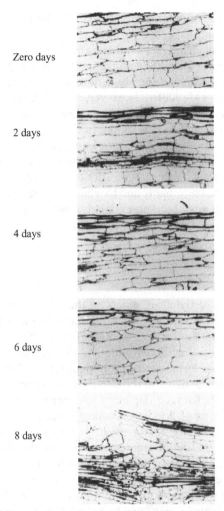

Figure 5.1. Photomicrograph depiction of the transdifferentiation of cortical cells of the petiole of *Phaseolus vulgaris* into functional abscission zone cells from day 0 to day 8 showing cell-to-cell separation at the newly formed zone. See McManus et al. (1998) for experimental details.

This is particularly marked in the conversion of Type 1 to Type 2 cells of cell separation zones. In the pulvinus-petiole junction of the bean leaf, a plate of one or two cells thick crosses from epidermis to epidermis – but it does not appear to include the epidermis. In *Sambucus*, this plate is composed of as many as 10 to 30 layers, and the number is not fixed. In *Ecballium* flower buds, pockets of Type 2 cells of different size and cell number develop, though no continuous plate is formed (refer to Figure 4.1). Given that each cell is an individual target cell and therefore is not identical to its neighbour (the stomatal guard cell pair may be an exception), it is understandable that the cell separation response to ethylene is also dependent upon the numbers of Type 1 to Type 2 cells that are converted and the positional conversion of each.

There is currently much interest in non-abscinding and delayed abscission mutants in plants. The non-abscinding lupin mutant, *Abs⁻*, is incapable of shedding its leaves and produces none of the abscission-specific cellulase associated with shedding (Henderson et al., 2001a). However, another lupin mutant, *delabs* (designated *Abs2*), shows delayed abscission together with low expression levels of the cellulase (Clements and Atkins, 2001). In *dab* (delayed abscission) mutants of flower shedding in *Arabidopsis*, we now know genes associated with this *dab* condition (Patterson and Bleecker, 2004). Although we might deduce that the mutation blocks the conversion of Type 1 to Type 2 in non-abscinding mutants, as yet we have little idea how many and to what extent the flexibility of Type 1 to Type 2 is controlled at any of the potential abscission zone sites of either *delabs* or *dab*.

Aerenchyma

The formation of intercellular air spaces in the cortical parenchyma of stems and roots is a common developmental programme in many plants, particularly those of aquatic or marshy habitats and is brought about by highly regulated species-specific patterns of cell separation and differential cell enlargements. A much faster event is the formation of prominent multicellular lacunae in otherwise conventional cortical tissue in roots and stems in response to stress, usually that caused by hypoxia (oxygen depletion to 3 percent) or attendant ethylene production in conditions of flooding. Cortical tissues retain their flexibility to achieve this response in any age of plant. In roots of maize seedlings, these lysigenous air spaces are initiated in certain cells of the mid-cortex within 12 hours of flooding and spread radially outwards to other cortical cells. This is followed by a progress to cell death and disintegration with the formation of aerenchyma pockets within 60 hours (Gunawardena et al., 2001a). The first observable changes are cytoplasmic, including plasma membrane invaginations and accumulation of vesicles between the plasma membrane and the cell wall, closely followed by chromatin condensation and fragmentation of nuclear DNA within 24 hours. In the final stages, the cells die, apparently apoptotically and the cell wall is degraded in clusters of cells together. The air space so formed leaves little evidence of the earlier presence of the once-living cortical cells because cell walls are wholly digested by a coterie of wall-modifying proteins amongst which cellulases, pectinases, expansins and xyloglucan-endotransglycosylase (XET) are all implicated (Jackson and Armstrong, 1999). Strands of living cells usually remain, however, connecting epidermis with the vascular supply, so it is clear that these particular cells are resistant to the death signals that kill their pre-selected neighbours. Both types of cells must therefore be present as different target cells, spatially distributed within the root cortex. Detecting the two types of cells before aerenchyma formation has proved difficult, but the putative aerenchyma target cells are known to expand more than those that are longer lived (Kawai et al., 1998). Studies using JIM5 and JIM7 antibodies to distinguish between low and more highly esterified pectins within the cell walls have failed to reveal pre-induction differences between the aerenchyma forming and non-forming cells, although wall de-esterification itself appears to be an early marker (within 12 hours) at the three-way corner junctions of cells that would become lysigenous (Gunawardena et al., 2001b).

Loss of flexibility with age

With respect to the capacity for meristems generated in the cortex to produce a range of cell types, we can speculate upon what appears as a loss of flexibility as the cortex ages. Cambial cells, for example, are a specific differentiation of the inner cortical parenchyma. These cells are not meristematic in the sense of apical meristematic cells, for they are committed to generating vascular tissues only, but in a highly polarised way. Although they are restricted in their products, these meristems persist for the life of the plant so although losing part of their flexibility *in planta*, they retain the capacity for indefinite cell division (see Fahn, 1990).

The continuous ring of the periderm that initiates bark formation and the limited pockets of cells that form lenticels are examples of centres of subepidermal cortical parenchyma that can be activated to meristematic activity, but have even less flexibility of product. In the numerous examples studied (see Fahn, 1990), the positional origins of lenticel meristematic centres in shoots are associated with stomata and stomatal signals, while in roots, the association is with lateral root positioning. The products of these late-formed shoot and root meristems are more limited than those of the earlier formed cambial cells, for they produce only parenchymatous cells, frequently suberized, but always of limited adhesion.

Where secondary abscission zones are induced across the cortical parenchyma of mature bean leaf petioles, the capacity for cell division appears entirely lost. But here, a flexibility to transdifferentiate to a Type 2 cell can still be expressed (McManus et al., 1998). It would appear that late-formed meristematic centres and mature cells do lose developmental options, while in lysigenous aerenchyma formation, the cells lose all options except an early programmed cell death (Gunawardena et al., 2001a).

This discussion of the flexibility of cortical cells has primarily centred upon their responses to auxin, ethylene, or stress, but it will be evident to the reader that other hormones and signals can equally play a part in the regulation of target cell change. As more information becomes available, the extent of target cell control in these cell-to-cell interactions should become clearer both at physiological and molecular levels.

Pith parenchyma

Each of the epidermal and cortical cell types so far described is a product of apical differentiation and external to the vascular system. However, many cells of parenchyma status remain within the confines of the pro-vascular ring and form a central core of pith tissue commencing at the apex of the shoot. Cell enlargement and cell divisions take place in pith cells to accommodate elongation of the epidermis and cortex. But, whereas the cells outside the vascular tissue remain living, though non-extending in the mature stem, those of the pith are relatively short-lived and, like the cells of the endosperm of cereal seeds, will undergo a process of water loss, ageing and cell death so that in the larger mature herbaceous plants (e.g., *Lupinus, Oenothera*), the majority of the diameter at the base of the stem is hollow resulting from the death of all the pith cells. The comparative short half-life of the pith compared with that of cortical cells external to the stele is a matter of some speculation. Both parenchyma cell types retain the flexibility to divide or to undergo transdifferentiation to other cell types in culture, but in pith this capability remains high only in young stems. The progression to cell death as the shoot elongates above them limits their potential as a source of totipotent cells for tissue culture and suppresses their ability to function as a source of new competent target cells.

Early pith cell death is not confined to stem tissue and occurs routinely in the petioles of many species. In the bean, for instance, the readiness of positionally directed cortical cells to convert to Type 2 secondary abscission zone cells along the length of an excised petiole segment depends in part upon the survival of the pith cells. The pith itself, however, appears not to have the competence to convert to Type 2 cells (McManus et al., 1998). Whether cortex conversion is determined by a signal emanating from the pith in its relation to the stele and cortex or whether the role of a living pith is to maintain that part of the petiole segment in a sufficient state of hydrated turgor for cortical transdifferentiation to occur is discussed in Chapter 6.

The formation of vascular tissue is closely linked to the development of leaf primordia and the delineation of the pith. The first signals to be perceived by the cells that constitute a provascular position are likely to be the auxin gradients generated between the meristem and the primordium. Genes that are concerned with the regulation of meristem development, the timing of leaf primordial initiation, vascular differentiation and the extent and temporal containment of the pith in relation to neighbour tissue could therefore play an intimate role not only in pattern formation but also in the subsequent target performance of all the parenchyma cells derived from that apex community (Steeves and Sussex, 1989).

The terminal ear 1 gene (*TE1*) of maize is one example (Veit et al., 1998). The mutant (*te1*) has compressed internodes between the leaves (this completely entraps the otherwise normal terminal male tassel). *te1* is thought to inhibit wild-type apices from acting as closely repeated organisers of leaf primordia; apices lacking *te1* suppression permit the development of repeated leaf organisation sites with the resulting compression of both cortex and pith.

This survey of the changing flexibility of parenchymatous cells as the plant develops has long been the domain of the anatomist, but with the wealth of genetic information at our disposal, we now have new opportunities to determine how specific suppression or activation of gene functions with age can operate to curtail or tailor new developmental expressions even in fully mature cells.

Lignified cells

Higher plant lignins have been studied in considerable detail (Lewis and Yamamoto, 1990; Fukuda, 1996) and exist within the wall structure as cross-linked polymers of p-hydroxyphenyl, guaiacyl and syringyl units in varying proportions. The general phenylpropanoid pathway, a shikemate pathway and certain tissue-specific lignin pathways have been identified (Lewis and Yamamoto, 1990) and much use is now being made of lignin mutants.

The importance of lignins and cell thickenings in the present context of target cells is the enhanced potential for cell-to-cell communication that they can provide. The development of an erect growing plant with a protective cuticle containing air passages directly to the external environment must have been a major factor in progressing toward the colonisation of the land. The ability to synthesise the wall-stiffening and wall-supportive lignins or lignin-like polymers goes hand-in-hand with this progression.

Traces of sucrose (at suboptimal levels for xylogenesis, 0.001 percent) stimulate xylogenesis in (lettuce) pith explants and also increase ethylene production, and it is interesting to speculate that the early transportation of inductive sucrose concentrations along elongated cell pathways in primitive plants could have been causal to the development of lignified elements appearing in the adjacent cells (Warren-Wilson et al., 1994).

In the lower plants studied so far, there could be a link between the unusual features observed in their alternative pathway for ethylene biosynthesis and the wall thickenings these plants possess. All produce ethylene, and one (the water fern *Regnellidum diphyllum*) that has been studied in greatest detail produces high levels of the gas which is involved in the regulation of cell elongation of submerged petioles (rachii). None produces ethylene from ACC and evidence to date indicates a lack of the higher plant pathway originating from methionine (Osborne et al., 1996).

Is there a connection between lignification or the chemical composition of wall thickening and ethylene production? Mosses and liverworts in general show no lignified tissues, although elongated cells adapted to transport activities are present in the central regions of thallose and foliose species. There is still controversy as to whether higher plant lignins, as a variable group of chemical constituents of secondary cell walls, are present at all in these lower orders. $^{13}C_{NMR}$ spectra have indicated an absence of p-hydroxyphenyl, coniferyl and sinapyl units and the presence of 3,5-hydroxy-polyphenols, which could be derived from an alternate branch of the phenylpropanol pathway (Wilson et al., 1989). Phenylpropanoids are potential precursors for ethylene and the switch between lower plant ethylene synthesis to that of the higher plant methionine-mediated pathway may be linked to a re-routing of the lignin-like pathway found in bryophytes and pteridophytes. The accepted regulatory pathways to secondary wall rigidification expressed in the tracheary elements of higher plants and in the transdifferentiation of parenchyma cells to tracheary cells in various cell cultures (e.g., *Zinnia*; Fukuda, 1994) would, to date, appear to differ from those of lower plants. The signalling potentials of these complex substances, specifically when the lignin-containing cells are part of a complex of differentiating or senescing cells, may eventually be seen to be of major importance.

The two examples we have to date (in bean and in *Kalanchoe*), in which abscission is blocked (even in the presence of ethylene) when vascular tissue is removed from excised abscission zone explants, and cell separation is re-initiated when the vascular tissue is replaced, are indicative of the signal function of the stelar tissue and perhaps also of its accompanying lignin polymers (Thompson and Osborne, 1994; Horton, R., pers. comm.). The implied feature is the signal that could be generated in stelar tissue during lignin degradation. In bean, we know that it takes about 24 hours for the abscission-inducing signal from the stelar tissue of a pulvinus to be generated and then passed to the abscission zone cells. There is no inducing stelar signal present in freshly cut non-senescing tissue. Always there is a lag period before the abscission-inducing signal is realized (Thompson and Osborne, 1994). The lag period could be the time required for lignin degradation to reach the threshold level of signal production to induce the abscission response (see Chapter 6).

The marking of a future cell type by the early anatomist, morphologist or embryologist was once a temporal or positional placement from the knowledge of what that cell would usually become in a developmental programme. Such was the case with xylem positional differentiation. New molecular analysis of gene expression and the identification of a cell that produces either the mRNA or protein product of that gene allows the target identification to be put back in developmental time to the earliest stage of target identification. One such example is the asymmetric first division of the zygote in *Arabidopsis* in which the larger cell differentiates to form the hypophysis and the suspensor that anchors the developing embryo to the ovule tissue and the smaller cell becomes the embryo with cotyledons, root and shoot meristems. Two genes *G564* and *C541* are first expressed in the suspensor cells but not the embryo cells, with this being evident as early as the four-cell stage of embryo development (Weterings et al., 2001). What is not clear, however, is the actual signalling mechanism involved that informs the zygote of these target cell differences. Whereas the free-floating fertilized *Fucus* egg differentiates into opposing poles upon stimulus from unilateral light (Jaffe, 1966) we must assume that the zygote orientation within the embryo sac can provide the differential signalling that determines which are the embryo cells and suspensor cells and thereby directing the biochemical pathways by which only suspensor cells are induced to produce the G564 and C541 products, these genes remaining suppressed in the cells that give rise to the embryo itself.

In meristems, differentiation of pro-stelar tissue in central cells just behind the apical tissue of root and shoot indicates how early both cell and tissue individuality and target status are introduced into the plant corpus and how soon the life spans of certain individual cells such as those of the vascular tissue are determined.

The much explored tissue cultures of *Zinnia* cells that readily convert by trans-differentiation into tracheary elements (Fukuda and Komamine, 1980) have been used to investigate in molecular terms how differentiation of vascular tissue may be directed (see Chapter 6).

6

Terminally Committed Cell Types and the Target Status

Cells that we see as permanently committed offer us the opportunity to follow their performance in both excised pieces of plant tissue as well as *in planta*. With a number of these it has been possible to establish with relative certainty the nature of their target status and the inputs of signals and signalling molecules that they can both perceive and respond to in predictable ways.

Also, it has been possible to follow associations with neighbour cells that influence the pathway to the committed cell state and to deduce certain of the cross-talk and physical communication that leads to a final differentiated condition. Two types of commitment have been considered. The first type is one in which the committed cells remain alive in the body of the plant and their function can therefore be called into operation by the perception of specific signals evoking a one time only response (as is the case with abscission or aleurone cells) or by the differentiation of a response mechanism that can be activated many times without loss of function (as in statocytes and stomata). The second terminally committed cell type to be considered is one that dies *in situ* amongst its living cell neighbours in the progress of the commitment, but then forms an essential component of the plant's structural architecture and overall function. The xylem cells of the vascular system are such a terminally committed example, playing an informational role in pattern formation while alive and operating as transport conduits and sources of lignin-derived or other signal molecules in death.

These selected terminally differentiated cells and their specific target status within the plant are by no means the limit of such types; pollen cells, collenchyma and epidermal root hairs are all candidates, but their progress to the committed state is less well understood, so they are not included in the present discussion.

Statocytes

With the emergence of land forms, a means for directional gravity sensing to maintain the erect plant habit became an essential component of the cell society. With sophisticated mechanisms for signal transduction to all plant parts it is evident that not all cells need to sense the gravitational field directly. Provided the signals from just a few gravity sensors can be conveyed to the whole community of cells, the vertical habit of root and shoot can be maintained. Furthermore, only the immature cells that still possess elongation potential are capable of the growth response that can bring back a deviant shoot or root apex to the vertical.

For the vegetative or flowering shoot of *Arabidopsis* it is clear that the single ring of cells, the endodermis, that surrounds the vascular cylinder is the site of gravity perception (Fukaki et al., 1998). These cells, called statocytes, develop enlarged starch grains or statoliths within the plastids (or chloroplasts) present in the cytoplasm. The mobility and density of these statoliths causes them to relocate always to the lowest part of the cell with respect to gravity, thus setting the signal for the directional elongation of the surrounding tissues. When placed horizontally, the cytoskeleton, plasma membrane or endoplasmic reticulum elements (possibly all three) within the sensor cell are agitated by the displacement of statoliths to the new lowest position. That side of the cell next to this lowest position emits signals that lead to enhanced rates of elongation growth of both the parenchyma and epidermal cells adjacent to that side. Statoliths falling in those endodermal cells whose statolith-receiving side abuts the vascular cylinder do not lead to enhanced elongation, hence by this anisotropic growth, the shoot is returned to the vertical; at the same time, the statoliths return to their normal position at the base of the endodermal cell. The *scr* and *shr* mutants of *Arabidopsis* that fail to generate a proper radial tissue pattern and lack a normal endodermis within vegetative or flowering stems demonstrate that the ability to properly recover to the vertical when displaced horizontally is dependent upon the development of functional statocytes (Fukaki et al., 1998).

It is still not entirely clear, however, how the signal transduction pathway operates. Do the statocytes constantly emit signals with respect to gravity? Do they emit only when their statoliths are perturbed? Do they emit only when their statoliths fall towards an outer cell wall facing towards the cortex and epidermis? Or do they not emit at all, functioning only as mechanical agitators of cytoplasmic actin fibres? It is reasonably assumed that the signal to differential cell growth in the two sides of the horizontal shoot is a difference in auxin (or another hormone) level generated by statolith movements. Experiments with the nodal regions of the flowering stalks of grasses, which can respond by bending within 30 minutes of a gravitational displacement, have shown that free auxin levels change in both upper and lower sides within that time, the upper side decreasing and the lower side increasing. These changes (both in auxin levels and bending response) are equally rapidly reversed again if the horizontal position of the nodal tissue is rotated by 180° with respect to gravity. There is no necessity for the upper and lower sides to communicate, for the changes in auxin levels and the growth responses occur equally if the node is intact or if it is slit longitudinally into separated upper and lower halves (Wright et al., 1978).

The differentiation of a shoot statocyte from its pro-statocyte parenchyma origins appears to be dictated by its position with respect to vascular initials. In the dicotyledon stem, this is immediately external to the protophloem, eventually providing a ring of endodermal gravity-sensing statocytes surrounding the vascular cylinder and central pith. In the monocotyledon, the gravity-sensing tissue of the flowering shoot is located in the hollow leaf sheath bases that surround the stem, with differentiation again localized to parenchyma cells adjacent to the phloem, but here it is the internal phloem. No ring of tissue is formed; instead groups of parenchyma cells undergo what appears to be a transdifferentiation to mature statocytes: the larger the vascular bundle, the greater are the number of neighbouring parenchyma cells that undergo the conversion. This suggests that each committed statocyte behaves independently and the summation of their signalling determines the extent and speed of the gravitational response. It is important, however, that although most shoot cortical cells have the capacity to accumulate some starch, the starch-containing chloroplasts do not develop into statoliths and do not attain the density to sediment. Only vascular bundle-associated cells become committed as statocytes with differentiated statoliths that are free to move in response to gravity. Whereas the normal photosynthetic parenchyma exports the daily fixed carbon as sucrose during darkness, at some stage of their development the chloroplasts of a potential statocyte lose this capacity and continue to accumulate starch, which even at tissue senescence is not hydrolysed.

In dormant potato tubers, the tiny bud meristems are rarely in a vertical position with respect to gravity. The formation of statocyte endodermal cells with enlarging starch granules occurs only after dormancy is broken and cell growth has started, commencing external to the protophloem of the scale leaf that is eighth in succession from the primordium. As the shoot grows, the endodermal ring is progressively closed and is complete at leaf primordium 9 (Macdonald, 1984).

Each sedimentable statolith in green plants originates from a cytoplasmic plastid – either an etioplast as in the potato tuber bud, or from a normal chloroplast in stems and in the leaf nodes of the grass stalk. These organelles possess their own genetic information; currently we do not know the signalling that initiates starch storage in the plastid and inhibits sucrose traffic back to the cytoplasm, but the result is a continually increasing size and density of the starch grains within an increasingly swollen plastid. Although endodermal starch grains are positionally recognisable in the pro-statocyte endodermal cells of potato tuber buds as early as leaf primordium 3 or 4, it is only when these plastids reach a critical density with respect to the viscosity of the cytoplasm that they move as a result of a change in the directional force of the gravitational field and hence function as gravity sensors.

The leaf bases of young flowering shoots of grasses also cannot respond by bending until their statoliths reach the size and density to move within the statocyte cytoplasm: only then the rate of the differential growth response is

Table 6.1. *Relationship between percent cross-sectional area differentiated as statocytes in the mature leaf sheath bases of the first nodes of the flowering stalks of different varieties of wheat and barley, and the angle of recovery to the vertical (θ) after 24 hours. Values are means ± standard errors, n = 9–11 (S. Dunford and D.J. Osborne, unpublished).*

	% of transverse area as statocytes	θ/24 hr in excised segments	θ/24 hr in excised shoots with ears attached
Barley (3 lines)	7.1 ± 1.2	44 ± 4	36 ± 9
	7.7 ± 0.5	46 ± 7	41 ± 8
	6.7 ± 0.9	32 ± 8	30 ± 3
Wheat (3 lines)	5.0 ± 0.8	27 ± 3	23 ± 7
	3.9 ± 0.7	25 ± 6	23 ± 5
	3.7 ± 0.7	24 ± 2	21 ± 4

determined both by the number of statocytes present and size of the statoliths (Wright, 1986). In several varieties of wheat and barley, studies have shown that a direct correlation exists between the rates of bending of horizontally placed flowering stalks back to the vertical position and the areas in cross-section of their fully differentiated statocytes per node. Varieties with a high proportion of statocyte area respond the quickest and so are the best plant lines to recover from lodging (Table 6.1; Figure 6.1).

In shoots, evidence presently suggests that each fully differentiated statocyte perceives gravity and transduces the perceived signal to a change in the level of a cell growth promoting hormone that is released by either plasma membrane or plasmodesmatal passage from the side of the cell upon which the statolith comes to rest. Once fully formed, the statocyte retains the terminal stage of differentiation and may be activated repeatedly by movement of the shoot from the vertical. Even at senescence and the death of surrounding cells, the starch-filled statoliths remain undiminished in size and starch content, though no longer functioning in an otherwise degrading cytoplasm.

Proximity to differentiating phloem is clearly one factor that determines the positional conversion of a shoot parenchyma cell to a statocyte both in the radial organisation of the dicotyledenous endodermal ring and in the monocotyledon discrete vascular bundle-associated cluster. What is not at all clear is the signalling that must take place from the phloem for a plastid to become a statolith.

Figure 6.1. a. Transverse image of gravity-responding nodal region of the vertical flowering stalk of *Avena fatua* showing stem, leaf sheath base and the sites of statocytes differentiated internally with respect to vascular bundles. Starch statoliths stained with iodine in potassium iodide. b. With nodes placed horizontally, statoliths sediment on the upper side to cytoplasm along the walls remote from the bundles (left) but sediment in close proximity to phloem cells on the lower side (right). c. Transmission electron micrograph of a statocyte displaying gravity-induced sedimentation of statoliths. a and b, bar represents 70 μm; c, bar represents 10 μm; g = gravity vector.

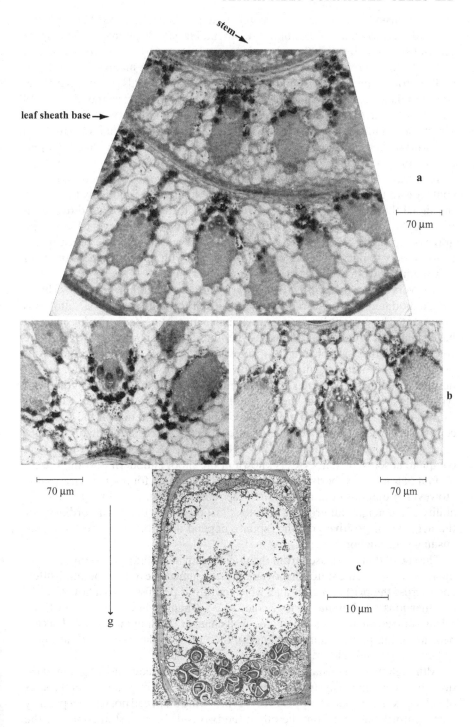

Gravisensing in roots is also a statocyte-dependent event, but here it is not the endodermis that differentiates sedimentable starch grains or develops a gravisensing role. Instead, these functions are deployed by the columella cells of the root cap. Therefore, when a root cap with its statocyte members is removed and the root is placed horizontally, the root no longer bends downwards to respond to the changed direction of the gravitational force and cannot do so until a new replacement root cap has been differentiated (Barlow, 1995). Further, the *scr* and *shr* mutants that disrupt both shoot and root endodermal formations in *Arabidopsis* and preclude graviresponses of the shoot do not preclude the perception of gravity and function of the root cap statocytes (Fukaki et al., 1998).

Root cap cells, originating as they do from their own meristem, do not have contact with the phloem, so a phloem-associated directive is unlikely to provide the signal that initiates a root statocyte. Equally distinctive in their differentiation are the root cap border cells that flank the columella statocytes. Unlike the gravity-sensing long-lived cells of an endodermis, all those of the root cap are progressively sloughed as independent single cells with a constant recruitment of new members from divisions of the root cap meristem.

It is very evident that effectiveness of a statocyte, whether of root or shoot, can be modified by the overall characteristics of the cell. Size of the statolith is important. The coleoptiles of the Amylomaize mutant of *Zea mays* for example have smaller statoliths than the wild-type; these sediment more slowly and the coleoptiles of these shoots are slow to respond to gravity. In the columella cells of the root caps of these same plants, however, the statoliths do not differ in size, density or sedimentation rates between mutant and wild-type although the wild-type roots are, like those of the coleoptile, less graviresponsive than wild-type (Moore, 1986). The reduced ability for particle sedimentation in the graviperceptive statocyte sensor is well documented for starch-deficient mutants of other plants (*Nicotiana*, Kiss and Sack, 1989; *Arabidopsis*, Kiss et al., 1997), and the complete absence of statocytes is known in the shoot endodermis of the Lazy-1 mutant of tomato (Roberts, 1984); this is one explanation for poor graviresponse. However, mutations in transmission of the sedimentation-evoked signal and the ability of the neighbour growing cells to adjust their growth rates accordingly for the negative or positive growth response depend upon the effectiveness of the ensuing transduction chain.

The signal that is released from the gravity-activated statocyte is still debated, but the pressure of the statolith movement is presumed to evoke a potential difference across the plasma membrane, which in turn is permissive to a flow of Ca^{2+} or other ions (Perbal and Driss-Ecole, 2003). Changes in auxin transport, the release of free auxin from conjugates, wall changes (both in proteins such as expansins and in polysaccharide linkages), then provide the basis for differential cell expansion and achievement of the graviresponse.

Although the role of statocytes in gravity perception seems no longer in question, these cells may play another role in the still debated mechanisms and routes of polar auxin transport. It appears that every living plant cell possesses a polarity that identifies one end from the other, the two ends being distinguished by the plasma-membrane-sited presence of specific influx and efflux carrier systems that control the preferential basipetal movement of the auxin molecule in conjunction with specific peptide transporters. The PIN gene family encodes a number of these proteins whose asymmetric localisation at the base of the cell is consistent with a directional movement of auxin in the cell from apex to base.

For a long time living cells of the vascular tissue have offered the attraction of a major route for auxin transport; particularly the endodermis by its close association with phloem. Several experimental facts are worth considering here that point to statocytes of both shoot and root as central target players in auxin movement. First, in the node of the flowering grass shoot, where the number of statocytes is high as a percentage of the total cell population (as much as 10 percent in TS section), normal basipetal transport of radio-labelled IAA occurs through excised young nodal segments (before statoliths have developed) whatever their orientation with respect to gravity. However, once sedimentable starch statoliths have formed, the polarity of auxin transport becomes flexible and is completely reversed if the nodal segment is inverted. The reversal of transport polarity can be detected within 20 minutes of inverting through 180 degrees; then by returning the segment back to the normal position with respect to gravity, normal basipetal polarity of auxin transport is restored (Wright, 1982, 1986).

As Wright (1986) showed, young rapidly extending leaf bases do not yet possess statocytes with sedimenting statoliths nor do they respond to gravity or produce enhanced levels of auxin on the lower side when placed horizontally. At some critical stage in the development of the flowering shoot, the statocyte progenitor target cells cease to export sugars and instead their chloroplasts accumulate starch. Once the chloroplasts will sediment in a 1-g gravitational field the node becomes not only graviperceptive and responsive, but the associated changes in free IAA formation and the reversibility of IAA transport polarity also take place.

The leaf sheath bases of flowering grass stalks may be particularly suitable for demonstrating flexibility in the polar direction of auxin transport only because of the very high percentage of statocytes that they contain. An acropetal auxin transport polarity was once reported for the flowering shoot of *Coleus* (Leopold and Guernsey, 1953). Although that of vegetative shoots was strictly basipetal, transport in the flowering shoot was similar in each direction. Unfortunately, there were no data for the statocyte populations of these tissues.

Interpreting certain auxin transport data for *Arabidopsis* roots suggests that statocytes might have a special role here also. The redistribution of auxin and the graviresponse is wholly dependent upon the root tip and the sedimentable statoliths of the columella cells. Inhibiting the transport of auxin that passes downwards from the shoot towards the root tip (i.e., acropetal movement with respect to the root tip) does not block a graviresponse, but blocking the basipetal auxin transport from the root tip backwards (i.e., from the statocyte-containing tip) does. Additionally, *Arabidopsis* with a genetic mutation (*eir1*) in graviresponse shows reduced basipetal transport (root tip towards shoot) but no change in acropetal auxin movement from shoot to root tip (Rashotte et al., 2000). Both results can be interpreted as indicating that the perturbated cytoskeleton of statocytes could be target sites for the redistribution of PIN efflux or permease influx proteins regulating directional auxin traffic at the plasma membrane (Friml and Palme, 2002). However, there is no evidence that statocytes are sites of auxin synthesis and the immunofluorescent localisation of the auxin PIN3 appears restricted to only a few of the root cap columella cells close to the meristem (Sievers et al., 2002).

The signals for differentiation of parenchyma to statocyte are clearly different in shoot and root, though the signals for plastid to statolith may be similar. As in the shoot, the number of statoliths per cell, the number of statocytes and the speed with which they sediment are determining factors for the sensitivity of the root geotropic response. Both in shoot and root, the response is by differential growth on the two sides of the elongating organ, engendered by a differential release of either elongation-enhancing hormones to the lower side of the shoot or, in the root, to an increase in the level of a cell-extending hormone on the upper side or an inhibitor on the lower side. The question of whether the same hormones are involved in both responses is still open to debate. It may well be that the increased level of a growth-inhibiting hormone such as abscisic acid in the lower side of the root effectively provides an additional means for determining the differential growth required for vertical positioning. But whichever way the growth response is achieved, it would seem that at least the mechanism for gravity perception through the interaction of a sedimenting particle with some part of the endomembrane system is the essential feature of the statocyte target cell in the higher plant.

The concept that gravity sensing depends upon an essentially similar basic mechanism in all living organisms has considerable appeal. Some organelle movement or particle sedimentation seems universal. The mechanism appears to have arisen early in evolution, and amyloplast sedimentation has been described for gametophytes and sporophytes in a number of bryophytes and in protonemal cultures of a moss (Walker and Sack, 1990). Calcium carbonate otoliths in the ears of fish, otoconia in mammals (Verpy et al., 1999), membrane-bound barium or strontium sulphate granules in the rhizoids of Characean algae (Sievers and Schmitz, 1982), and the movement of membrane-enclosed nuclei in their actin cradles in the stipes of fungal fruiting bodies (Moore et al., 1996) can all perform the function of internal cytoplasmic perturbation. Whether a pressure component, an electrical stimulus, the altering of calcium or other ion channels or modifications of plasmodesmatal opening is the subsequent event that leads to a regulation of anisotropic cell growth, they are all secondary to the positional factors that cause an initially parenchymatous cell to become a highly organised and terminally differentiated statocyte. (For a review see Perbal and Driss-Ecole, 2003.)

This chapter is devoted to cells with a terminal state of differentiation and which, to our knowledge, possess no further options for alternative developmental pathways. Statocytes are one example. It is now of special interest that although plant cells show clear polarity gradients, as in the controlling mechanism for active auxin transport, we now have evidence for terminal domain-defined target areas of specialised biochemical function within a single cell. Until recently, C_4 photosynthesis has been linked to the Kranz-type leaf anatomy, which consists of the two adjacent cell types for mesophyll and bundle sheath cells, each with their specialised organelle function that operates in concert in carbon assimilation. However, two members of the Chenopodiaceae with C_4/crassalacean metabolism have been shown to lack this cell-to-cell cooperation typical of the Kranz anatomy; instead, these two separate biochemical pathways operate within specified domains of a single cell (Voznesenskaya et al., 2003). *Borszczowia aralocaspica* and *Bienertia cycloptera* both have mature leaf cells with dimorphic chloroplasts, one specialized for CO_2 fixation in the C_4 cycle, the other specialised for donating the C_4 acids to the C_3 cycle. In *Borszczowia*, spatial segregation of the two chloroplast types is at opposite ends of elongate cells; in *Bienertia* the separate functions are partitioned between peripheral and central cytoplasmic compartments. These cells are not the result of a degeneration of the wall between two initially separate cells, and the chloroplasts originate from a common pool of these organelles. The signals that initiate the migration of chloroplasts with selective expression of organelle-encoded enzymes such as rubisco into one localised compartment of the cell remain for discovery. As indeed do the signals that convert some or all chloroplasts (or plastids) to statoliths in maturing statocytes.

Abscission cells

Unlike the situation of statocytes, where we know that the gravity target cell has been fully differentiated once the statoliths move and sediment in the cell, we are somewhat unsure exactly when the competent Type 2 abscission cell is differentiated. Under normal conditions of growth these positionally differentiated cells (as single or multiple cell layers below organs such as fruit, flowers and leaves that are eventually shed) do not immediately become sites of cell separation *in vivo*. Although competent cells can start to separate at once if they are exposed to an appropriate level of ethylene, this may not happen until long after they are differentiated and until the distal tissues adjacent to them produce sufficient ethylene from either wounding or senescence to reach the threshold level that will induce the separation response. This means that abscission cells may be present at the specific shedding sites for weeks, months or even years before their competence becomes translated to an operative event.

Morphological evidence suggests that the sites of "pre-abscission" cells are already distinguishable in the embryos of seeds of the bean (*Phaseolus vulgaris*). At the leaf pulvinus-petiole junction of the primary leaves, recognisable abscission sites are present in the mature seed, but in the early stages of germination no level of ethylene will induce cell separation or the induction of the essential zone-specific iso-form of β-1:4-endoglucanase required to achieve cell-from-cell loss of wall adhesion. This suggests that competence is not achieved until a response can be shown to applied ethylene. The level of ethylene to which competent cells respond by separating *in vivo* appears to be determined in part by the level of endogenous auxin, since auxin invariably represses cell separation in abscission. When increasing concentrations of auxin were applied to excised (competent to separate) abscission segments of *Sambucus*, it was shown that increasing concentrations of ethylene were required to induce cell separation (Osborne and Sargent, 1976). The balance between auxin repression and ethylene induction is precisely monitored by the abscission target cell once it has attained competence, but the temporal attainment of this competent state remains elusive and has still not been directly established in molecular terms for any abscission cell. Nonetheless, distinctive "markers" have been established that distinguish the ethylene-separable abscission cell from its non-separating neighbours, the first being the induction of cellulase shown by Horton (Horton and Osborne, 1967).

The first evidence for a marker for abscission competence was sought in the Cucurbitaceae, where endoreduplication of nuclear DNA is a normal acquisition of ageing. When the nuclei of discontinuous clusters of cells just below the bud of an immature female flower of *Ecballium elaterium* reached an 8C value, these buds could be induced to shed by ethylene. As long as the nuclei remained at 2C or 4C no cell separation could be induced (Wong and Osborne, 1978). However, because the separating cells of the mature fruit abscission zone were not 8C, the 8C condition in the young floral bud was considered only as a marker, but not the requirement, for those particular cells to reach competence to separate from their neighbours (see Chapter 4).

Differences in protein expression also indicate abscission cell competence by the presence of specific peptide determinants that are not detectable in neighbour non-abscission cells. These include a 34 kDa peptide present both before and after cell separation in the *Sambucus nigra* leaflet abscission zone (McManus and Osborne, 1990b) (see Chapter 4). Although it has been shown for other species that mature competent abscission cells express cell-specific protein markers, so far these methods have not been applied to the more difficult question of *when* these molecular determinants are first expressed during the differentiation of the competent abscission cell.

The induction of abscission-cell-specific wall degrading enzymes as part of the shedding process has received much attention in the past 30 years and many genes have been cloned. Both abscission-cell-specific cellulases and polygalacturonases are integral components of the new gene expressions needed in response to ethylene during the wall dissociation process, but the extent of each induction depends upon which abscission zone site is involved. In peach, for example, the induction of polygalacturonase activity in the fruit abscission zone is much greater than that of cellulase; the reverse is true for the leaf abscission zone (Bonghi et al., 1992). Furthermore, although these two cell-specific enzymes are certainly the major inductive requirements for all abscissions investigated in many species of plants, they clearly act as part of a cohort with other wall degrading glucanases that are either constitutive or up-regulated in concert. Henderson et al. (2001b) have suggested that the enzyme complex that operates at each particular abscission site is a reflection of the middle lamella or cell wall saccharide composition that must be hydrolysed or loosened to allow cell separation to take place.

But there is now another interesting aspect to the operation of a competent abscission target cell. If the vascular tissue is removed surgically from the distal and immediate abscission centre in an abscission explant of the bean, ethylene is no longer effective in inducing abscission nor is the abscission cellulase induced. Replacement of the vascular tissue restores the full abscission response (Thompson and Osborne, 1994), but if any phloroglucinol-positive-staining vascular tissue is retained in the explant for 24 hours, abscission is initiated and separation is normal. This indicates that by 24 hours after an explant is cut, some product of stelar degradation during ethylene-induced senescence of cells distal to zone cells is responsible for signalling an abscission sequence of events, and that in the absence of the stelar signal, ethylene alone is ineffective as the abscission inducer. The nature of this stelar product remains to be discovered, but it is, without doubt, one of the most important for activating a competent abscission target cell and for regulating plant shedding processes. The signal candidate could well be a product of lignin hydrolysis and it is of interest that a number of lignin products have, in the past, been shown to possess some abscission-accelerating activity. Coumarin and ferulic acid have both been reported, but with no substantive follow-up. If cell wall compositions are highly specific to certain cells (see Table 4.1) then it may be that acceleration due to any lignin breakdown products will also be highly specific to the particular abscission cell to which it signals. Good use could now be made of lignin biosynthesis mutants in future explorations of the nature of stelar degradation signals.

So far, it has not been possible to cultivate the highly vacuolated abscission cells *in vitro* once they have separated from each other. They are resistant to undergoing further cell division and although they can be kept alive for many days, they are not known to undergo further differentiation and so they eventually die (McManus and Osborne, unpublished). Whereas those abscission cells that are differentiated at the base of organs that eventually shed are positionally dictated early in morphogenesis, as pre-abscission cells, those that arise later in the plant's life, through cell damage or hormonal manipulation, can be the result of a transdifferentiation from pre-existing cells of mesophyll or cortex parenchyma; they do not require a preliminary division from a parent cell. Those of cortex parenchyma in leaf petioles and stems of the bean have been studied extensively and provide a model both for cell-to-cell interaction and cell-to-cell signalling.

In the young shoot, for example, decapitation leads to the senescence of the stump that remains, senescence progressing in the basipetal direction towards the nearest node. But before that node is reached, a visible junction develops between the yellowing tissue of the senescing cells and the still green tissue below. At this precise position in the stem, the cells of the cortex of the green (proximal) tissue undergo a morphological and physiological change to become competent abscission cells (Webster and Leopold, 1972) and produce the abscission-specific cellulase in response to the ethylene generated by the senescing tissue above them. Cell separation is followed by shedding of the senescent stump. Unlike the normally positioned abscission cells that may not be activated to separate until long after they acquire competence, those of secondary origin from fully mature cortical cells separate from each other immediately upon conversion as one continuous event.

Other secondary abscission sites in leaf mesophyll will also separate immediately upon formation. The "shot-hole" discs of leaf blade tissue of *Prunus amygdalus* infected with *Clasterosporium carpophilum* that loosen from the rest of the blade to leave a hole, occur at the junction of the senescent (infected) and green non-senescent (uninfected) cells (Samuel, 1927). Similarly, leaves of *Streptocarpus* that can be induced to shed the terminal parts of their strap-like leaves, particularly following kinetin treatments of the basal parts (Noel and Van Staden, 1975), also separate at the junction of distal senescing cells adjacent to green non-senescent cells below them. What was not established either in the bean shoot, in the "shot holes" of the leaf mesophyll or in the loss of terminal parts of *Streptocarpus* leaves was whether any cell division had occurred before the formation of cells that could separate.

Later experiments with petiole segments of the primary leaf of bean demonstrated conclusively that there is a direct conversion of the flexible cortical parenchyma cell to a terminally differentiated abscission target state (McManus et al., 1998). These secondary abscission zones can be positionally transdifferentiated in a segment in the presence of ethylene by the application of different concentrations of auxin to the distal end – the higher the concentration of auxin applied, the farther towards the base of the segment the new separation zone will form (see Chapter 5 and Figure 5.1). No cell division is involved, and no discrete zone is generated if ethylene is not added. Always the zone forms at a distinct green–yellow junction, the apical auxin-enriched part remaining green and adjoining a yellowing senescing tissue below. The polarity of this junction (apical green and basal yellow) is the reverse of the situation in the de-topped stump of the bean shoot reported by Webster and Leopold (1972), but distal senescence and basal non-senescence can be produced in the petiole segment by applying the auxin to the basal cut surface (instead of the apical end) as well as exposing the segments to ethylene. Then the apical part of the segment becomes senescent while the proximal tissue remains green and non-senescent. Again, it is precisely at the junction of these senescent and non-senescent tissues that cells are transdifferentiated to abscission cells in the green non-senescent cortical parenchyma.

It would appear, therefore, that cell transdifferentiation to the terminally differentiated abscission target state requires the presence of senescing cells immediately adjacent to non-senescent neighbours, irrespective of the physiological polarity of the tissue. Furthermore, this transdifferentiation occurs only in those cells such as leaf mesophyll or cortical parenchyma which have retained their flexible target state. This tells us that immediate cell-to-cell cross-talk is critical to local cell performance and can override other long-distance informational cues.

At first sight it may seem that the signals that positionally initiate the normal abscission cells are quite different from those that lead to their conversion by transdifferentiation from an already mature parenchymatous cell. Certainly there are no senescent/non-senescent junctions in the former. What there may be in common, however, are signals from the dying cells of the vascular tissue. Whether the differentiation and programmed cell death of protoxylem or xylem elements provides such a signal for the formation of primary abscission target cells at sites of vascular branching remains to be elucidated, but the involvement of senescence (including vascular tissue) is clearly relevant to the transdifferentiation of the secondary abscission cell.

Spontaneous mutants that lack the ability to shed have been found in many plant species and there is no single reason for the absence of abscission. Some mutants are insensitive to ethylene through sequence alterations in the amino-acid motif of the membrane-bound ethylene receptor. These *etr* mutants have been found in *Arabidopsis* and the never-ripe (*Nr*) tomato (van Doorn and Stead, 1997). By transforming wild-type tomato plants with a construct that contains an anti-sense partial copy of the *ETR1* gene, the blocking of abscission by failed ethylene perception has been confirmed (Whitelaw et al., 2002). However, transformation affects other phenotypic expressions in plant growth, including reduced auxin transport, which are not specific to the abscission target cell. Accumulations of auxin could also result in apparent loss of ethylene perception and failure to shed normally.

Auxin availability is not the only hormonal cause of modulated leaf shedding; in the reduced abscission mutant of birch, supplementing abscisic acid (ABA) is effective in restoring normal abscission indicating that competent abscission target cells are indeed differentiated but require ABA or ABA-induced ethylene for separation (Rinne et al., 1992). Altered expression of the expansin gene in cell wall proteins of *Arapidopsis* changes the separation competence of the flower pedicels. Unlike the abscission zones of the perianth, those of the pedicel are considered to be vestigial for they show no phenotypic differentiation. Nonetheless, expansin sequences (expansin 10) are normally maximally expressed in these locations and cell breakage by an applied force is reduced in anti-sense plants and enhanced in those with overexpression of sense transcripts (Cho and Cosgrove, 2000). This indicates that cell separation (or the cell enlargement that accompanies this) has an expansin-regulated component. However, we do not know what happens in the petal abscission zones of these transformants.

In the *Abs⁻* mutant of the lupin *Lupinus angustifolius* van Danja, loss of organ abscission, but with otherwise normal phenotype and senescence, is due to a lesion in a single recessive allele. Anatomically and ultrastructurally the abscission zone cells appear normal, but in response to ethylene, the mutant *Abs⁻* exhibits a much reduced induction of wall hydrolyzing enzymes with only limited wall-from-wall loosening and no shedding (Clements and Atkins, 2001). Other studies revealed that an abscission-cell-specific 50 kDa β-1:4-glucanhydrolase induced in wild-type is not synthesised by the *Abs⁻* mutant, even though other changes in wall hydrolysing enzymes in response to ethylene were normal (Henderson et al., 2001a). Together, the results with these abscission mutants suggest that absence of shedding can result from imperfect target cell differentiation that either does not permit ethylene perception or fails in achieving the proper transmission and response cascade to the perceived inductive signal.

The factors that regulate the onset of senescence in a plant cell have a multitude of effects not only within the cell itself but also upon the immediate neighbour cells. In the timing of abscission, the speed of senescence of distal tissue is a major determinant and is particularly evident in excised segments of tissues containing abscission zones such as those of the pulvinar-petiole zones (explants) of bean, where it is essentially the rate of senescence of the pulvinus that controls the timing of explant separation. Increasing amounts of auxin applied to the pulvinus decreases the rate of pulvinar senescence and delays the time to the initiation of abscission, while ethylene hastens both processes. Not surprisingly, therefore, genetic regulators that influence the senescence process are also effective in altering the timing of abscission. In *Arabidopsis*, the MADS domain protein AGL 15, plays a pivotal role in modulating plant ageing and development. In transgenic plants, in which constitutive expression of AGL 15 is enhanced, the perianth organs show retention of chlorophyll, delayed senescence and, in parallel, a delay of abscission for at least two weeks. No phenotypic or cellular differences were observed between wild-type and the transgenics and a possible interaction of AGL 15 with reduced ethylene perception was excluded by the dual cross of AGL 15 with a dominant *etr1* mutation. In these plants too, senescence was delayed and perianth longevity was retained in the presence of added ethylene (Fernandez et al., 2000).

For both the conversion of the parenchyma cell to a terminal differentiated abscission cell condition and the final expression of the terminal condition, the requirement of a neighbour cell senescence may be obligatory. In the differentiation of the primary abscission zone cell in the early development of the plant the hormonal inputs and gradients would be quite subtle from nearby senescing xylem or protoxylem cells; but for mature tissue, in which secondary zones develop in response to major neighbour tissue damage or overall tissue senescence the result is dramatic. Cortical cell transdifferentiation and the subsequent cell separation that then follows are closely linked events in time.

There is no doubt that ethylene has a major role in the overall abscission process, although the essential requirement for ethylene in the induction process of competent cells of *Arabidopsis* petals, sepals and anther filaments has been questioned by Patterson and Bleecker (2004). But ethylene is clearly not the only molecular player in determining where, how and when abscission will occur, for some product of senescing stelar tissue is also required. Additionally, the complete perception and signal transduction chain must be operative to permit the induction of the new gene expressions required to change cell wall characteristics from those of adhesive cells to those of cells that separate. We do not yet know in molecular terms the genetic changes that rule the conversion of the one target type (Type 1) to the other (Type 2).

Separation of the specific target cells of an abscission zone is clearly an event that can happen only once, but the question arises as to how far the process must proceed before it becomes irreversible. In the agricultural practice of loosening orange fruit from the tree by spraying with an ethylene-liberating or ethylene-inducing chemical to permit easier mechanical harvesting, it has long been known that fruit can be re-tightened to their pedicels if an auxin is supplied sufficiently early after the ethylene release (Cooper and Henry, 1971). Biochemical investigations with the petiolar explants of the bean have shown that the major inducible enzyme and associated ultrastructural events (dilation of the endomembrane system and discharge of vesicles to the plasma membrane from inflated golgi) that accompany the ethylene-induced cell separation can be reversed at an early stage by removal of the ethylene or by a treatment with IAA (Osborne et al., 1985). A 'turned-off' ethylene-induced programme can be 'turned on' again by the reintroduction of ethylene to the system. This tells us that the new gene cascade that is released by an appropriate level of ethylene can be repressed again when

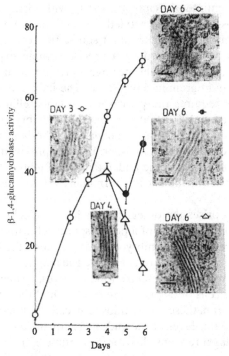

Figure 6.2. Reversibility of cellulase gene expression and dictyosome activation regulated in terminally differentiated abscission cells by auxin and ethylene signalling. See Osborne, 1990; ✧ continuous ethylene; ⬠ treatment with IAA from day 3; ● return to ethylene day 4.

the cells perceive and respond to an appropriate level of auxin, and can then be re-induced by the addition of more ethylene (see Figure 6.2). Similar events have been demonstrated for the abscission cells of *Sambucus nigra*, with immuno-confirmation by western analysis that the 1,4-glucanhydrolase that is switched on and off is indeed the abscission-cell-specific cellulase (McManus and Osborne, unpublished). This recognition of gene function reversibility *in vivo*, until a stage of no return is reached, indicates the sensitivity of response of a target cell to more than one signal input at any one time and the intimate regulatory cross-talk that must occur at the level of gene expression before the terminally differentiated cell is wholly compromised.

It is now evident that particular cell-to-cell associations, each with their own special target status, their particular hormonal cross-talk and their different rates of ageing must together determine the formation and final performance of the fully differentiated competent abscission target cell.

The aleurone

The developing embryo is the initial source of all target cells, so during the course of the formation of the new individual within the seed, many new identifiable cell types are differentiated in close proximity with highly specific responses to signal inputs. Suspensor cells, cotyledonary initials, root tips – all express protein molecular markers of their target states. In the mature seed, the earliest target cells to be recognised by a precise protein expression inducible in response to a defined hormonal message were the aleurone cells that surround the endosperm in graminaceous seeds. The induction of α-amylase synthesis in these cells requires the presence of a gibberellin signal. The initial demonstration by Paleg (1960) that removal of the embryo of the cereal grain of barley prevents both activation of amylase synthesis in the aleurone cells and the hydrolysis of adjoining endosperm starch provided the basis for the present spectacular advances in the understanding of gibberellin and ABA control of gene expression in the specific terminally differentiated target cells of aleurone.

The aleurone comprises the outer layer or layers of cells of the triploid endosperm formed from the fusion of a pollen sperm cell with two haploid nuclei of the maternal embryo sac. As embryogenesis proceeds, the inner triploid cells also develop particular positional target states. The majority become repositories for the starch and protein of the seed's food reserves, a small number develop functions as transfer cells or as protective cells around the embryo. During the stage of seed desiccation, these inner endosperm cells undergo premature death, unlike those of the outer layer of the endosperm or the embryo. Although this outer, aleurone cell layer remains alive for the lifetime of the seed, the cells never divide again. But they perceive and respond to gibberellin (the hormonal signal that Paleg showed emanated from the germinating embryo) by the synthesis and secretion of a number of gibberellin-inducible nucleases, proteases as well as the all-important starch-degrading α-amylase, without which the stored endosperm starch cannot be mobilized. ABA, synthesised during embryogenesis, represses the expression of the gibberellin-inducible enzyme proteins. The performance of an aleurone target cell with the on–off control of expression by gibberellin and ABA therefore resembles the dual control of the abscission target cell exerted by the interactive ethylene–auxin regulation of the specific inducible and essential cellulases and polygalacturonases.

The signals that direct the aleurone layer to become a tissue that is distinct from the rest of the endosperm, of which the cells are initially a part, are still unclear. All cells are at first triploid and of similar lineage and retain this or a further polyploid condition until death. Following fusion of the sperm and the two egg nuclei, nuclear division is not arrested but wall formation is suppressed. In this respect, the resulting endosperm coenocyte is therefore quite distinct from an organised meristem. At cell cycle arrest, the suppression of cell plate repression is lifted. Two MAP kinases, NTF6 and MMK3, are known to be associated with the instability of microtubules and failed phragmoplast formation and may operate in the termination of the coenocyte stage in the endosperm (Olsen, 2001). Within a few days, microtubular arrays are developed from the nuclear envelopes marking out individual nuclear domains. Cellularization and differentiation and the subsequent directional wall depositions are from thereon highly integrated events, both spatially and temporally.

The first factor to distinguish aleurone is the positional association of free nuclei with the embryo sac cell wall. This is followed by the enclosure of each nucleus within its own newly synthesised periclinal and anticlinal cell walls. Some earliest molecular markers for these aleurone cells are now known. The *Ltp2* transcript and a number of others are present soon after aleurone differentiation in barley is visible (Klemsdal et al., 1991) and the promoter of a barley aleurone-specific lipid protein transfer gene confers this aleurone-specific gene expression in transgenic rice (Kalla et al., 1994).

Wall enclosure of nuclei follows throughout the rest of the coenocytic endosperm to form a cohesive tissue whose function thereafter is dedicated to mobilization and to deposition of sugars and amino acids into specific starch and prolamin protein storage bodies. Two enzymes of the starch synthesis pathway, the ADP-glucose pyrophosphorylase of barley and the starch synthesis (SSI gene) of wheat are preferentially expressed in starchy endosperm and are the earliest molecular markers for starch accumulation. The aleurone outer cell layer also becomes a storage tissue, but into different bodies from the rest of the endosperm, containing many vacuolar proteins, lipids and polyphosphates of calcium and magnesium myoinositol (phytin). This storage distinction between aleurone and the adjacent cells of the rest of the endosperm is extremely well defined, metabolically and ultrastructurally (Lopes and Larkins, 1993) and is detectable from an early stage in embryogenesis. But we still do not know precisely how and why.

The initially uniform triploid endosperm is clearly subject to close-range positional signals from the maternal tissues that enclose the ovule. Of much interest is the nature of a positional molecular signal that specifically designates aleurone cells to the periphery of the endosperm. One suggestion has been the greater presence of a protein kinase-like receptor (Crinkly4) on the surface of the cells closest to the ovary wall, where a ligand-operated binding would activate the receptor and so provide for aleurone site specification (Olsen et al., 1998). Identification of the ligand and studies of aleurone-less or multilayer-aleurone mutants should eventually further elucidate the positional differentiation of aleurone. Another cell type of special importance to the accumulation of reserves and the eventual role of aleurone is the positional differentiation of endosperm cells with transfer properties close to the main vascular tissue that supplies the ovule. In barley, a transfer-cell–specific transcript *END1* identifies this location at the coenocyte stage of the endosperm and well before the development of an extensive endomembrane complex or wall outgrowths that characterise transfer cells (Doan et al., 1996). Other gene expressions (*BET1*) have been shown to mark these transfer cells in maize endosperm (Heuros et al., 1995). This positional signalling at such very early stages of seed development indicates the closeness of neighbour-to-neighbour cross-talk at the nuclear and cell levels and provides a further pointer to the directive role of vascular tissue in target cell differentiation.

Once they are differentiated as different target tissues, aleurone, transfer cells and endosperm have different fates. Whereas endosperm and transfer cells die before the seed is shed, those of the aleurone and the embryo develop mechanisms to stay alive despite desiccation to moisture contents of less than 5 percent and despite their quite distinct genomic DNA complement and their differences in ultrastructural identity. The factors that permit desiccation tolerance in the triploid aleurone and the diploid embryo are, at present, speculative. But whatever these factors are, they are not developed during differentiation of the body of the storage endosperm. However, both endosperm and aleurone possess a common feature: both are terminally differentiated and both undergo an organised death, although at quite different times. The starchy endosperm cells of maize, for example, like those of most graminaceous seeds, die toward the end of grain filling at cell desiccation (Young et al., 1997) while those of all aleurones degenerate while hydrated at germination and following the hormonal induction of α-amylase and associated hydrolases (Wang et al., 1996; Bethke et al., 1999). The initiation of programmes to cell death is now of major interest in studies of all living cells, but those programmes in plants, and particularly those of endospermic origin in graminaceous seeds, have already offered most rewarding material for such studies.

Endosperm death in both wheat and maize and in the early maize progenitor, teosinte, appears to be regulated by the levels of ethylene production and by ethylene perception. Following a peak in ethylene during early development, nuclear DNA starts to fragment to nucleosomal multimers associated with high endonuclease activity in mid and late stages of starch deposition (Young and Gallie, 1999, 2000). In the endosperm that is nearing seed-maturation, ABA delays the death programme by an ethylene–ABA orchestrated regulation. In the ABA-insensitive *vp1* and ABA-deficient *vp9*, mutants of maize, both show the expected progressive acceleration of cell death programmes. At germination, ABA acts to suppress the GA-induction of hydrolytic enzymes in aleurone, thereby essentially extending the aleurone life span. So ABA appears as the delayer of cell death in the endosperm cells as well as in the aleurone, though the inducer of cell death in the two types of target cells is different; for endosperm ethylene is the inducer, for aleurone it is GA.

Studies of mature isolated aleurone layers and single aleurone protoplasts indicate that each aleurone cell possesses its own identity, for the region closest to the embryo contains the highest proportion of cells responding to a gibberellin concentration equivalent to that from the embryo. Furthermore, not all the cells are induced to synthesise the new proteins in concert; a gradual recruitment of aleurone cells therefore takes place, with different lag periods before induction depending upon the different levels of gibberellin required for their activation (Ritchie et al., 1999).

The individuality of each aleurone target cell to the gibberellin inducer may not be attributable to a single cause. The availability of plasma membrane receptors for gibberellin is a first requirement for perception and transduction of gibberellin signals. High affinity GA-binding proteins have been isolated from aleurones and are present in other GA response plant tissues such as pea and *Arabidopsis*. GA deficiency mutants of *Arabidopsis* are many (Koornneef and van der Veen, 1980) but a GA response mutant, *gai*, that is not actually GA-deficient abolishes the GA induction of α-amylase when it (*gai*) is expressed in transgenic Bismata rice aleurone (Fu et al., 2001). G proteins, specific phosphatases and cyclic GMP are all now considered as important components of the gibberellin transduction pathway, and modifications in any of these can change the responsiveness of the cell (Lovegrove and Hooley, 2000). How does ABA function in the elegant balance between the two signals? In temporal terms, the presence of ABA prevents the expression of GA-inducible genes during embryogenesis, but at germination an ABA-induced protein kinase PKABA1 functions as a suppressor of α-amylase. Constitutive overexpression of the PKABA1 kinase very strongly inhibits a GA response, though overexpression in a null mutant of PKABA1 does not reduce or alter the GA induction of the α-Amy2 promoter or α-amylase formation (Gomez-Cadenas et al., 2001).

All controls are not immediately molecular for it has long been known that aleurone tissue of freshly harvested immature graminaceous seeds will not respond to gibberellin. However, when such seed (wheat) is dried or subjected to high temperature the ability of aleurone to then transcribe α-amylase mRNAs and synthesise amylase protein in response to added gibberellin can be unmasked (Evans et al., 1975). This physiological insensitivity appears separate from the genetic insensitivity of non-responding GA mutants (Cornford et al., 1986) but must eventually be attributable to changes induced in perception and transduction cascades involving both ABA and GA.

The acquisition of tolerance to desiccation at water deficit levels of -3 MPa is remarkably limited across living organisms, including plants. Although mechanisms to do so clearly evolved early in evolution, few whole plants, the 'resurrection plants' such as the grass *Eragrostis nindensis* (Gaff and Ellis, 1974), species of *Xerophyta* (Mundree et al., 2000) and *Craterostigma* (Bartels and Salamini, 2001) can do so. Of lower plants, many of the *Bryophyta* successfully dehydrate and survive. Certain limited parts of plants, however, are routinely dehydrated at maturity, particularly those of air-dispersed pollens and the embryos of seeds. In the gramineae, the attendant aleurone tissue is as desiccation tolerant as the embryo. The factors that provide for tolerance are still debated. One factor might be the accumulation of non-reducing sugars (or other solutes) that at water loss provide an immobile glassy state within the cytoplasm. Another is the production, during maturation, of one of the specific late embryogenesis abundant (LEA) proteins that are believed to function as protectants on binding to organellar membranes or by guarding against free-radical damage to nuclear DNA. It was once thought that such distinctions could be made between desiccation tolerant and intolerant cells or tissues but the exceptions are too many to fulfil any one rule. LEA proteins, for example, were once believed to be absent from the embryos of recalcitrant seeds that die if they are dehydrated. We now know that certain recalcitrant seeds do synthesise LEA proteins (Han et al., 1997). It is most likely that more than one mechanism and more than one protein or solute can offer cell stability to water loss and water re-entry. Where direct comparisons have been made within a single species between the tolerant embryo or aleurone and the desiccation intolerant tissues of the rest of the endosperm, there is good evidence to support a molecular control of cell survival. In barley, for example, one LEA group 3LEA (HVA1) protein is specifically expressed in embryo and aleurone (not endosperm) during the late stages of development, correlating with the acquisition of the seed's desiccation tolerance. Additions of ABA or imposed dehydration will induce HVA1 expression. Transgenic rice plants expressing this barley HVA1 gene are, overall, more tolerant to water deficits (Xu et al., 1996).

The differentiation patterns that take place during the development of endospermic-derived tissues provide an intimate example of positional information cross-talk leading to distinct gene expression in a limited tissue of initial coenocytic uniformity. There is no vascular differentiation within the endosperm so the only input to the tissue from this source must come from either maternal tissues that surround the embryo sac or the developing embryo. This may well operate with the positioning of the endospermic transfer cells that supply sucrose to the main body of the cereal endosperm during grain filling, for these cells arise not only close to the vascular tissue of the peduncle but also close to the embryo, even though the embryo at this early stage has not developed vascular elements. There is good evidence that hormonal signals operate on endospermic cells to achieve the spacial and specific responses that lead to their differences in performance. It is also clear that the cells acquire their designated roles very early in seed development. However, the precise mechanisms by which desiccation tolerance and gibberellin target status are generated in the aleurone but not in the other cells of similar origin are still a quest to pursue.

Following survival as desiccation tolerant cells, aleurone develops a second target state on rehydration. This second target state is fully terminal and depends upon the timing of the gibberellin-induced new gene expressions starting when the embryo germinates. The synthesis and secretion of gibberellin produced *de novo* by the embryo and a reduction in the ABA levels in the dead endosperm cells lead to a cascade synthesis of new enzymes (including α-amylase, proteases, lipases and nucleases) that on release then hydrolyse the starch, protein, lipid and cell wall materials that form the major endosperm storage reserves of the seed.

During this process, the aleurone cells expand and undergo considerable ultrastructural change; they do not divide but within days (depending upon the species) they are programmed to a disintegrative cell death as the growth of the young seedling advances.

Gibberellin perceived by a GA-receptor protein at the plasma membrane of the aleurone cell is transduced within a few hours to a derepression or (in the presence of ABA) a maintained repression of genes determining the new aleurone syntheses. In this respect these responses resemble those of abscission target cells during the turning on and off by the dual interactions of ethylene and auxin (see the previous section and Figure 6.2).

Much interest has centred on the involvement of G proteins in the GA transduction events. A heterotrimeric G-protein-receptor complex was proposed that induces the aleurone α-amylase mRNA transcription and translation (Jones et al., 1998b). A G-protein-agonist Mas7 enhances the α-amylase expression in isolated protoplasts and also the GA expression of an α-Amy 2/54:GUS construct. The presence of G-protein subunits in aleurone protoplasts was confirmed by PCR and northern analysis for a partial G_α subunit cDNA and two G_β cDNAs. In the *DWARF1* mutant of rice, which shows little growth response to GA and poor α-amylase induction in the aleurone, a sequence lesion was found in the G_α subunit

(GPA1) which again suggests that a functional G protein is required for successful GA signalling (Ashikari et al., 1999).

One of the first events to follow the addition of GA to aleurone is a fall in cytosolic pH and a rise in cytoplasmic levels of Ca^{2+}, which in wheat can be recognised within 2–5 minutes (Bush, 1996). These increases are prevented in the presence of ABA. In addition, calmodulin levels are enhanced by GA and reduced by ABA, suggesting further that the GA–ABA interaction could involve a Ca^{2+}/calmodulin control that regulates the expressions of the hydrolases inducible in aleurone cells.

A myb-type protein (GAmyb) is also newly transcribed in the presence of GA and may well be an essential component of successful signalling since in the absence of GA, transient expression of GAmyb will activate the transcription of an α-amylase promoter fused to the GUS reporter gene (Gubler et al., 1995). As with most other hormonally controlled effects in plants, a protein phosphorylation event may be required to elicit the GA-induced α-amylase production by the aleurone. If the phosphatase inhibitor okadaic acid is supplied to wheat aleurones, then α-amylase formation is blocked and the associated Ca^{2+} changes do not occur. Furthermore, the usual progress to an early aleurone cell death is delayed (Kuo et al., 1996).

Once gibberellin has been introduced to a receptive aleurone cell *in vivo*, the terminal events of the cell's remarkable life are set in train. It enlarges and vacuolates, the endomembrane system extends and dilates, secretory vacuoles from the golgi fuse with the plasma membrane, the Ca^{2+} and Mg^{+} in the phytin granules are released and the nuclear matrix becomes diffuse with the chromatin condensed. These are all ultrastructural signs of a senescing cell. There is neither cell division nor cell separation, each cell dies *in situ*, surrounding the already dead tissue of the then reserve-depleted endosperm. Additions of ABA to isolated aleurone cells or to their protoplasts will arrest these changes and extend the aleurone lifetime before death, but will not redirect the target state. If the embryo is removed from a just-imbibed seed, the aleurone in contact with the endosperm can stay alive for many weeks, in fact it behaves as the aleurone in an imbibed dormant seed behaves, for a dormant embryo does not release GA to activate the aleurone cells to new gene expressions. An imbibed but dormant graminaceous seed may retain a living embryo and aleurone for months, or possibly years, but once the process of germination has started, aleurone cell death is inevitable. The question that has caused considerable debate is whether or not the demise of the aleurone is programmed in a similar way to that of apoptosis in animal cells.

In barley aleurone layers and protoplasts, Wang et al. (1998) found the fully induced and secreting material contained nuclei with DNA that was fragmented to nucleosomal multimers, features of a typical apoptotic mammalian cell. The interpretation of these results has been questioned since one of the features of the GA-induced aleurone is the expression of new nuclease activities towards the end of α-amylase secretion, which could be sufficient to degrade the nuclear DNA during extraction procedures (Fath et al., 1999). TUNEL assays to detect free 3′-OH terminal ends to the DNA *in situ* confirmed that fragmentation occurs progressively *in vivo* and that the DNA of the nucleus becomes internally cleaved during the final few days of GA treatment. Although at least three new nuclease activities (33, 27 and 25 kDa) were detected in response to GA, after 2–3 days no evidence was found for the formation of nucleosome multimers although high molecular weight DNA was clearly degraded as seen in the TUNEL studies. What is very evident from these results is that loss of integrity of DNA with a rise in nuclease activity and a hastening of cell death occur when aleurone cells respond to GA. In contrast, the addition of ABA not only suppresses α-amylase formation, but also suppresses the rise in nucleases, arrests the endo-cleavage of nuclear DNA and extends the life span of the cells. Whether or not the endo-cleavage of DNA is random or to 180 bp nucleosomal multimers would seem to depend upon the experimental conditions *in vitro*, and may be even more dependent upon the conditions *in vivo* as has been demonstrated for DNA fragmentation patterns in ageing (dry) and accelerated aged embryos of rye (Boubriak et al., 2000). Whichever nucleases may operate, it seems clear that the terminally differentiated aleurone cell has no recourse to DNA repair or to continued survival.

Stomata and trichomes

Stomata

The behaviour of the two identical chloroplast-containing guard cells that comprise the stomata on the surface of higher plants are one of the most well-researched examples of terminally differentiated cells. Not only is their early delineation from epidermis into a special cell type positionally dictated so that they are never in neighbour-to-neighbour contact with another stomata, but also the signalling that leads to their expressed patterning lies under strict environmental and genetic controls (see Chapter 5). Once formed and functional, stomata are repeatedly operational throughout the life of the plant. Compared with other cells of the epidermal surface from which they are generated, they show a level of independence and selectivity in recognition of signal inputs to which other non-stomatal cells may or may not respond.

The physiological function of stomata is to regulate the size of the pore that forms when the adjoining cell walls of the two guard cells first separate; this provides the potential open conduit between the external environment and the intercellular spaces of the mesophyll or cortex below. In this respect, photosynthetic rates and transpirational water loss are determined by the turgor of the two guard cells at any one time and the size and shape of the pore that they then delimit. Pore control is not, however, performed by the guard cells alone. The non-stomatal epidermal cells that surround the guard cell pair are essential members of a complex hormonal and signalling group that facilitates the functional operation of adjusting the guard cells' size. Whether each of these stomatal groups themselves have independence *in vivo* is not clear, but it is now evident that small patches of a leaf surface, for example, can perform essentially synchronously while neighbouring stomatal groups fail to respond, indicating a localised collective communication and behaviour between them (for review, see Mott and Buckley, 2000).

There also appears to be a long-distance signal output from mature leaves to very young leaves that respond by regulating their stomatal density pattern. Reducing incident light by shading expanded leaves or increasing the CO_2 levels to which they are exposed both lead to fewer stomata forming in newly developing leaves, while lowering CO_2 exposure of mature leaves increases the stomatal density in the youngest immature leaves (Woodward and Kelly, 1995). Mutants of the *HIC* gene from *Arabidopsis* that fail to sense CO_2 levels produce large numbers of stomata. It appears that the HIC protein is guard-cell-specific and behaves as a 3-keto acyl coenzyme A synthase (KCS) so controlling wax synthesis. This KCS protein marker for guard cells can be detected as soon as the pair is formed. It was proposed by Holroyd et al. (2002) that alterations in epicuticular waxes could control the permeability of the guard cell extracellular matrix to diffusible molecules. In this way diffusible inhibitors of stomatal development could be responsible for determining both the patterning and density of stomata in epidermal layers, also the long chain fatty acids could act as transported lipid signals. Other mutants of wax synthesis genes, such as *CER1* that encodes a carbonylase producing odd-numbered carbon chain alkanes and *CER6* that encodes a fatty acid elongase for carbon chains longer than 28C, produce plants with phenotypically greatly increased stomatal indices (Holroyd et al., 2002).

These studies clearly indicate not only that terminally differentiated stomata possess specific epicuticular matrix markers in the form of specific wax compositions, but also that these compositions are themselves intimately linked to stomatal development. It has long been known that many environmental impacts such as light or CO_2 levels will feed back to different structural formations of epidermal cuticles and waxes (Martin and Juniper, 1970) so the terminally differentiated stomatal cells *in vivo* cannot be considered as a uniform target population.

Also, it is relevant that the many mutants of stomatal guard cell formation that have been studied include genes coding for kinases and proteases important in most signal transduction processes. The mutant *tmm* gene that disrupts correct spacing of guard cells from neighbour pairs is a probable membrane-located leucine-rich receptor-like protein with similarities to CLAVATA2 (Nadeau and Sack, 2002) and the *SDD1* mutant *sdd1*, which permits enhanced stomatal numbers above wild-type, encodes a modified subtilisin-like serine protease. The *SDD1* expression in wild-type stomatal precursor cells is thought to act in a feedback loop to ensure that the guard cell pairs do not develop next to each other (von Groll and Altmann, 2001).

Given the diverse multi-signalling that determines the distribution and positioning of stomata within the epidermal L1 meristem layer, it seems that the turgor responses and opening and closing mechanisms have remarkable commonalities. Once the terminally differentiated state is achieved, the most important function of all stomata is the sensing and response to changes in water potential.

Research into the control mechanisms that have led to the current understanding of how stomatal guard cells open and close many hundreds of times during the life of a leaf and yet survive green with functional chloroplasts long after associated cells in the rest of the leaf blade are yellow and senescent does not yet provide us with all the answers. Without doubt these cells are terminally differentiated since they neither divide further nor become transdifferentiated to another cell type. They retain flexible walls although those on the sides of the pore are usually thickened, and their communication pathways with other cells at the wall or at their limited plasmodesmatal connections are restricted since viral particles that readily invade neighbour cells fail to enter.

Whole plant or plant-part experiments established long ago that negative water potentials lead to stomatal closure, while blue light or CO_2 induce opening – responses that were measurable within minutes by gas-flow equipment through leaf tissue using a 'porometer' method, first employed by Darwin and Pertz (1911). More modern studies using patch clamping and fluorescence ratiophotometry have been conducted on epidermal peels from suitably amenable leaf surfaces or on isolated guard cell protoplasts. Two factors were shown to play a critical role in guard cell closure: an overall rise in the levels of ABA in the tissues when water potential fell (this preceded closure; Wright and Hiron, 1969) and an associated increase in the cytoplasmic levels of free Ca^{2+} within the guard cells (Schroeder and Hagiwara, 1990).

In order for a sensor cell to function as a fast and sensitive responder to an input cue that can then direct alterations in turgor, mechanisms for ion conductance control need to be open to rapid change. In *Arabidopsis*, the ABA-induced stomatal closing is seen as cytosolic. The Ca^{2+}-dependent and ABA-insensitive mutants *abi1-1* and *abi2-1* show limited ABA-Ca^{2+}cyt rise with limited pore closure (Allen et al., 1999). Rises in cytosolic Ca^{2+} can, however, be induced by other means including the addition of auxin (Cousson and Vavasseur, 1998). Then, the rise in Ca^{2+}cyt can cause and precede stomatal *opening*, so a Ca^{2+} regulation of turgor control of the guard cell appears to operate in opposing ways, suggesting that more than one Ca^{2+}-sensitive site exists. It is clear that stomatal closure requires an efflux of ions and loss of water from the guard cells and this efflux has features that are specific to this particular target cell and may be specific also to the different signals that the target cell perceives.

Much effort has led to the partial unravelling of which ionic fluxes operate during opening and closure and hence to potential turgor and pore control. It is abundantly clear that the plasma membranes of guard cells have certain unique selective properties. An ABA activation of K^+ channels and an intake of K^+ has been linked to stomatal opening, via a protein phosphorylation event and an H^+ extrusion driven by plasma membrane (H^+)-ATPases. The rise in cytosolic Ca^{2+} however can block the K^+ intake channels almost completely as in experiments with *Vicia faba* guard cells (Grabov and Blatt, 1999). The resulting long-term efflux of both anions and K^+ and plasma membrane depolarisation are contributors to loss of turgor and pore closure (Figure 6.3). It is still uncertain whether restoration of turgor and pore re-opening is the exact reverse of the closing controls. Alternative Ca^{2+} influx pathways may be present as stretch-activated channels and additional controls of K^+ efflux channels may be integrated in an oscillating fashion between the vacuole and the guard cell walls to produce the subtle continuous sensing that epitomizes the rapidity of response of guard cells to external and internal stimuli (MacRobbie, 2000).

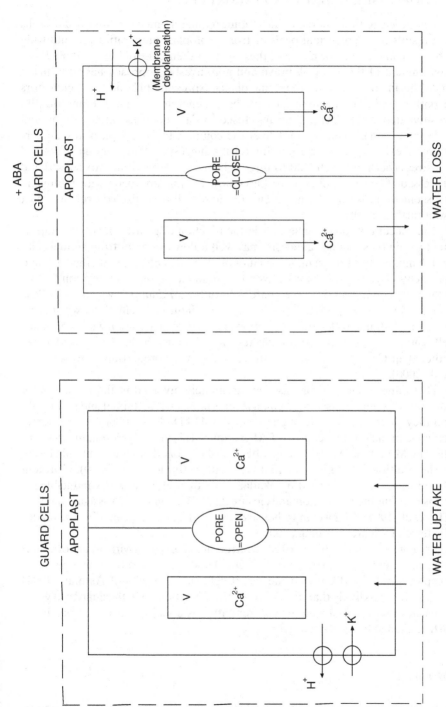

Figure 6.3. Diagrammatic representation of ionic fluxes in the control of stomatal pore opening and closure. V = vacuole

141

It is critical to the understanding of ionic regulatory mechanisms of guard cells that results from epidermal peels or from protoplast preparations should truly reflect the *in vivo* performance of these cells in their control of the intercellular environment of the plant. A brave and non-invasive approach has been made to do this in *Arabidopsis*. Transgenic plants expressing pH and Ca^{2+} indicators in both cytoplasm and apoplast have been generated. Using ratiometric pH-sensitive derivatives (At-pH luorins) fused to Ca^{2+} luminescent aequorins, and a chitinase signal sequence to deliver the complex to the apoplast, researchers obtained evidence from the ion flux results suggesting that osmotic stress and salt stress (both of which lead to turgor change) may be differently sensed and processed (Gao et al., 2004). To know how far this approach with multigene transgenics will reveal the intricacies of guard cell pore-size controls we must await further results.

The concept of ionic oscillations in the function of guard cell volume adjustment within these remarkable cells may well follow examples from animal cells. In T-lymphocytes, for example, calcium oscillations enhance the efficiency and specificity of gene expressions driven by proinflammatory transcription factors (Dolmetsch et al., 1998). In the guard cells of the *Arabidopsis* vacuolar ATPase mutant *det3*, the imposition of external Ca^{2+}-induced oscillations will rescue steady state stomatal closure, while imposing Ca^{2+} plateaus upon wild-type guard cells can prevent closure. This indicates that stimulus-specific Ca^{2+} oscillations, rather than Ca^{2+} plateau values, are critical to the pore closure events (Allen et al., 2000).

That kinases and phosphatases are intimately involved in the regulation of hormone or signal transduction chains seems universal for animal and plant cells, and they clearly play a role in guard cells. A 43 kDa MAP kinase is transiently activated in the positive control of ABA-induced closure of *Pisum sativum* stomata. The MAP kinase activity could be abolished, together with stomatal closing by the addition of the kinase inhibitor PD98059 (Burnett et al., 2000). If protein phosphatases were inhibited by okadaic acid, then in many ABA-inducible responses, including *Pisum* stomatal closure, the ABA-signalling was again blocked (Hey et al., 1997). (For a comprehensive review of guard cell signalling the reader is referred to Schroeder et al., 2001.)

The complexities of the guard cell controls indicate that as with other hormones (e.g., auxin and cell growth; Kim et al., 2000), there may be more than one signal receptor site for ABA within the target cell, as the results of Assman (1994) suggest. It seems likely that the highly intricate functioning of the guard cell when it reaches the terminal state of differentiation is a major factor that precludes further change to another target state.

Trichomes

In the choice of another highly distinct terminally differentiated cell with important surface functions in plant performance we refer to trichomes. Most epidermal outgrowths of shoot or root are short-lived and die soon after reaching the mature condition. In the shoot, their role becomes one of surface protector against hazards of environment, pest or pathogen.

Trichome initiation and trichome patterning are as complex as those for stomata. Many genes are now known that will determine cell fate, but whereas stomatal guard cells all adopt an essentially similar final morphology as a clonal pair amongst subsidiary supporting cells, trichomes exhibit a remarkable degree of flexibility in size, branching, final number of cells and degree of DNA endoreduplication (Szymanski and Marks, 1998). While still alive, different trichomes synthesise a wide range of cell-specific metabolic products. Salt-secreting trichomes as in *Atriplex* or *Avicennia* are initially rich in mitochondria, endomembranes and vesicles that actively secrete salt solution to the cell surface, while pores in the cuticle permit the release of aqueous saline droplets. Other trichomes secrete nectar, polysaccharide mucilages or proteases; each of these trichomes is structurally and metabolically a specific target cell type.

An interesting aspect of their differentiation control is that the same genes can regulate patterning for shoot trichomes and for root hairs but usually in opposing ways. In studies of *Arabidopsis* mutants, the MYB-class proteins WER and GLI are interchangeable, but whereas GLI is normally expressed only in shoot epidermis and is required for trichome formation, WER is normally expressed only in roots and hypocotyls and suppresses root hair cells and hypocotyl stomata.

Embryo polarity and the early self-identity of root and shoot poles is well established in zygotic and somatic embryogenesis so the interchangeability of these two genes could be seen as evidence of an evolutionary change in their regulatory function. As the distinction between root and shoot meristems became established in higher plants, more complex interactions with and between other genes are likely to have developed. The evolutionary history of WER and GLI may therefore provide footprints for the genetic designations of specific tissues and specific target cell traits. (For discussion, see Lee and Schiefelbein, 2001.)

Vascular tissue

Vascular tissue plays a central role in plant architecture. The development of wall stiffenings and lignin biosynthesis are as evolutionarily important to plants in their adoption of the erect habit as the bipedal transformations are to man.

The importance of vascular cells is far greater than that of mechanical support alone, affording, when dead, the conduits for all material (water and metabolites) transport from one part of the plant to another. They are also a central focus for positional information to each developing tissue with which they are in contact. Further, they can be considered as a signalling centre for all tissues in the immediate neighbourhood as well as the means for the conveyance, long distance, of hormones and signals to other parts of the plant.

The initiation of vascularisation *in vivo* takes place very early in cells just below the meristems, with a spacial organisation that is precise to each species. As living cells they clearly play a directive role in the polar transport of auxin. The concept that pro-vascular elements in some way direct a canalisation of auxin molecules along vascular pre-set pathways is attractive and supported by experiments that show that when these pathways are cut or interfered with, new, regenerated pathways are restored that reconnect the positions. Auxin treatments aid these reconnections indicating a positive feed-back control (Sachs, 1991, 2000). It is significant that vascular tissues occupy central positions in the developing plant so the vascular system can act as a positional reference point for surrounding cells, dispersing hormonal or other molecular signals by diffusion and by cell-to-cell trafficking independent of polar pathways of auxin.

Long ago, researchers demonstrated that cells of *Zinnia elegans* in tissue culture can be transdifferentiated into lignified xylem elements within a few hours by adding appropriate levels of a cytokinin and synthetic auxin to the medium (Fukuda and Komamine, 1980). The production of a 43 kDa nuclease (DNase/RNase) is induced in parallel with the cytoplasmic and nuclear autolysis of these cells (Thelen and Northcote, 1989). It is assumed that this glycoprotein nuclease first accumulates in the vacuole since breakdown of the tonoplast membrane precedes an immediate nuclear degradation associated with xylogenic differentiation. The accumulation of the mRNA for this cloned nuclease (ZEN1) precedes the nuclear breakdown by several hours (Aoyagi et al., 1998).

The cause of tonoplast membrane breakdown remains unknown in the trans-differentiating tissue cultures but the temporal sequence proposed by Obara et al. (2001) may well be similar to that of pro-vascular to vascular differentiation in meristems *in planta*. So far, *in planta*, our understanding of the signals that give rise to the orderly sequence of events that determine the species-specific pattern of root or stem vascular differentiation and the cell death that inevitably follows is still little understood.

In intact plants, vascular elements arise from procambial cells behind the meristems which first are distinguished by enlargement (both longitudinally and radially). Later, reorganisation of the cytoskeleton, golgi proliferation and deposition of the patterned lignified wall take place. Lignification, tonoplast disintegration, cell lysis and cell death are seen as coordinated events.

One key enzyme for lignification has now been established to operate both *in vivo* in *Zinnia* seedlings and *in vitro* in single cell cultures. A cell-wall–located basic peroxidase (pI 10.2) of 43 kDa and an N-terminal KVAVSPLS motif is expressed in both. This protein can therefore be considered as an early molecular marker for tracheary differentiation not only in mesophyll cells in culture, but also in the cambial-derived xylem vessels of the intact plant (López-Serrano et al., 2004).

The three cDNAs (TED2, TED3 and TED4) isolated by Demura and Fukuda (1994) that accumulate within 24 hours in a tracheary-induced *Zinnia* tissue culture have been shown to occur also in the vascular tissue of *Zinnia* seedlings with a distinct spatial and temporal expression. Cells with TED3 transcripts developed only into tracheary elements, but TED4 transcripts were found in vascular cells that became xylem parenchyma, and in fibres and tracheids. TED2 transcripts were throughout the xylem and at high levels in the immature primary phloem of roots and shoots (Demura and Fukuda, 1994).

Discovery of the dirigent glycoprotein that specifically utilizes E-coniferyl substrates to form lignans has provided new means to explore *in planta* early xylogenic targets in very young post-meristematic tissues of elongating stems of *Forsythia*. Metaxylem, for example, shows high recognition in *in situ* hybridisation with the dirigent protein anti-sera before any distinctive secondary wall layering can be observed, so providing another early marker of lignan initiation sites (Burlat et al., 2001).

What is not at all clear is the signalling that leads to differentiation of the pro-vascular initials from procambial cells immediately below the apical meristem. Mutants (*irx*) showing abnormal xylem phenotypes in *Arabidopsis* shoots have shown substantially lower levels and abnormal patterns of cellulose deposition in their walls. It has been concluded that the resulting irregular xylem formations with collapsing of the reinforced thickened cells results from a lack of the proper pattern of cellulose depositioning to direct the normal assembly of lignin and polysaccharides (Turner and Somerville, 1997). The actual amounts of deposition of lignin and non-cellulosic polysaccharides seem not to be altered, but the misalignment and misorientation of these components in the mutants may be responsible for the vascular abnormalities. The use of cellulose and polysaccharide synthesis inhibitors in *Zinnia* culture has provided evidence that alterations in the deposition of specific molecular components of hemicelluloses (xylans) precludes the normal self-perpetuating cascade of secondary wall patterning (Taylor et al., 1992) although inhibition of lignin biosynthesis does not necessarily inhibit the differentiation of tracheary elements in *Zinnia* cultured cells (Ingold et al., 1990). Together these results imply that *in planta* the differentiation of vascular tissue is determined by the very early differentiation of cells with specific polysaccharide compositions providing a vascular target status that then self-directs specific sites for wall and lignin assembly. They also suggest that very specific ratios in auxin and cytokinin could be generated in the apical meristem which then initiate xylogenesis by a similar molecular cross-talk that initiates induction of xylogenesis and subsequent death in isolated single cells *in vitro*.

The selection of terminally differentiated cells discussed in this chapter is by no means exhaustive and many others could be chosen. All the ones presented here will, however, give the reader an insight into the physiological and molecular specificities that contribute to a recognisable target cell condition.

7

The Mechanisms of Target Cell Perception and Response to Specific Signals

In previous chapters we outlined the target cell concept, identified the signals and hormones that a cell will encounter and discussed how types of cells can be identified as of particular target status whether during development or on reaching a terminal state of differentiation. Now, over the next chapters, we ask how cells actually recognise signals and question whether the target state dictates, or is dictated by, the mechanisms for signal recognition *in vivo*. The original description of a hormone, borrowed from the animal world, was a regulatory substance synthesised in one part of the organism and transported to another in which it is recognised and the effect of the hormone becomes manifest. Although the plant has sites of major synthesis of hormone signals and they are all known to be transported, all the evidence tells us that the majority of cells probably contain some level of each hormone and are constantly exposed to the hormones emanating from their neighbours. The plant, after all, is a coenocyte in which all living cells intercommunicate by plasmodesmata and by surface contacts at the cell wall. Of the many signals to which each cell is continuously exposed, why are certain of these perceived and responded to? Or, does a cell respond to all signals that are above a threshold level? If so, how is the threshold level determined and is it fixed or variable?

Animal physiologists deduced the existence of, sought and found receptor proteins on cell surfaces and within the nucleus. Hormone binding then directed the transcription of specific genes within the nucleus. In the example of prothoracic ecdysone in the insect epidermis, or in *Xenopus* metamorphosis or in other vertebrates, the subsequent coordination of response genes involves a series of secondary transcription hierarchies dictated, at least in part, by levels and changes in the levels of hormone titre in the circulating haemolymph. Is this how plants recognise hormonal signals? Can we distinguish how the different target cells might differ in their mechanisms of signal perception and response?

The emergence of the receptor concept in higher plants

That plant cells produce receptor proteins that recognise only specific hormones and signals was pursued in plants but with considerable difficulty. The first documented searches were for proteins that would bind auxin and auxin analogues; a number of proteins with high affinity for these molecules were identified, particularly in coleoptile tissues (Lembi et al., 1971; Hertel et al., 1972). As long ago as 1979, ideas that there were special binding sites for ethylene were pursued by Sisler (1979) and Jerie et al. (1979). Using ^{14}C-labelled ethylene, applied to various leaf tissues, and displacing bound labelled ethylene with unlabelled ethylene, Sisler (1979) concluded that the total number of binding sites for tobacco leaves approximated to 3.5 pmol per g of tissue. Further, the concentration of unlabelled ethylene required to then displace the bound labelled ethylene approached that required for a physiological response. Propylene, a physiological analogue of ethylene, would also displace the bound ethylene at physiologically active levels. Sisler (1980) partially purified an ethylene-binding component from the membrane pellet of mungbean hypocotyl tissue. Bean cotyledons were found to provide a better source of extractable ethylene binding sites than leaf tissue (Jerie et al., 1979; Bengochea et al., 1980), even though it is still not obvious what the role of ethylene might be in cotyledons. Cell-free systems were sought and, again using ^{14}C-labelled ethylene, a subcellular localisation of binding in cotyledon extracts of *Phaseolus vulgaris* was identified on both the ER and the membranes of protein bodies using both light microscope and electron microscope autoradiography (Evans et al., 1982).

It was not until the first report of *etr* mutants of *Arabidopsis* in which a reduced physiological response to ethylene was coupled with an 80 percent decrease in binding of ^{14}C-labelled ethylene when compared to wild-type plants (Bleecker et al., 1988), that significant progress with the characterisation of at least one class of ethylene receptor was made, and the stage was set for the genetic approach to receptor studies.

Previous chapters set out examples of neighbour-to-neighbour dictated interactions of different target cells and the evidence of their direction to different differentiation states by the hormone or signal levels that they encounter and contain. There are only two ways that these levels are initially controlled, either by inputs from outside the cell or by the balance of biosynthesis and degradation controls inherent within the cell.

It is tacitly assumed that all plant cells can synthesise the different hormones, unless the genetic constitution forbids a particular step in the biosynthesis pathway (as in many mutants). The extent of synthesis per cell is part of the target condition of that cell and this may not be a constant value, leading to great flexibility of internal signal and hormone levels. Large internal changes can occur when inputs are altered and this can happen on a massive scale mediated by long- and short-distance transport.

The next two chapters will not elaborate further on biosynthesis/degradation pathways, which are explored in Chapter 2, but rather will be concerned with the question, 'How do cells receive signals and hormones from outside the cell?'

They concentrate upon the role of receptors in the signal recognition process, and how the extent of receptor protein expression may be a critical control point for signal perception and response. Since cells control the level of receptors they produce as well as the levels transferred to the cell plasma membrane, the overall opportunities for control seem almost limitless.

Auxins and the receptor concept

We start with a discussion of how a receptor mechanism for target cell perception of input signals began with the search for proteins that would bind auxin molecules. The molecular mechanism of auxin action perhaps represents the most studied aspect of the molecular actions of plant hormones.

Over the past 30 years, there have been numerous reports of the identification of auxin-binding proteins in many plant tissues of many plant species. To navigate researchers to those proteins that may be genuine receptors amongst those proteins that through some aspect of structure can bind NAA or IAA simply by virtue of being low molecular weight acids, Venis and Napier (1995) proposed four criteria:

1. Binding should be reversible, of high affinity and of finite capacity.
2. The saturation range of binding should be consistent with the concentration range over which the physiological response saturates.
3. Binding specificity for different hormone analogues should be approximately in accordance with the relative biological activities of the compound.
4. Binding should lead to a hormone-specific biological response.

Since the days of Frits Went, the traditional biological materials for cell growth studies have been coleoptiles and the elongating submeristematic regions of shoots or roots, so it is to these tissues that we shall look first for an interpretation in terms of hormone receptors and the regulation of cell expansion. Physiological growth responses to auxins by different types of target cells are set out in Chapter 4 in the section entitled 'Options for cell enlargement' and in Chapter 5 in 'Cortical parenchyma cells'. In both Type 1 and Type 2 cells, auxin concentration in, or at, the cell surface plays a critical role in the response that results – a response possibly mediated by the level of an opposing ethylene production initiated by the concentration level of auxin present. For a Type 3 cell, found predominantly in plants of aquatic habitats, the positive growth enlargement initiated by both auxin and ethylene therefore becomes of special interest.

Each of these target cell types will be analysed with respect to a receptor-determined control of the hormone/signal repertoire. Some will seem topical, while for others the extent of current evidence is still fragmentary.

Identification and characterisation of auxin binding proteins

In many reviews, the characterisation of the auxin binding protein that is best seen as adhering to receptor criteria is the auxin binding protein 1, first identified in maize coleoptiles (Hertel et al., 1972). Many coleoptiles are highly responsive to auxins as Type 1 cells, and in these experiments, a synthetic auxin [^{14}C]1-naphthylene acetic acid (NAA) was shown to bind to membrane particles isolated from maize coleoptile tissue. For example, Ray et al. (1977) showed that NAA could bind to these membrane particles isolated from a 133,000 g membrane fraction, and subsequent Scatchard analysis revealed a binding constant (K_D) of 5 to 7 $\times$ 10^{-7}M NAA. Later, using ultracentrifugation within sucrose density gradients, Ray (1977) established that this high affinity site for NAA resided on the ER membrane (as determined by glucan synthetase II activity as an ER marker), and the binding protein was designated as the site I auxin binding protein (ABP). Many studies subsequently have reported on the isolation and characterisation of ABP1, originally in maize and then in other species, and for a more detailed account of the early characterisation of the protein, see Box 7.1.

Box 7.1 Characterisation of the ABP1 protein

Molecular characterisation of the site I ABP began with purification of a protein from maize coleoptile tissue using an affinity column comprising 2-OH-3,5-diiodobenzoic acid coupled to Sepharose (Lobler and Kambt, 1985). This 40 kDa protein separated as subunits of 20 kDa using SDS-PAGE, suggesting a dimer. The purified protein had a K_D of 5.7 $\times$ 10^{-8} M for NAA, which was similar to the site I binding sites identified in maize coleoptiles by Ray et al. (1977), and so was referred to subsequently as ABP1.

Three reports then followed, almost simultaneously, on the cloning of ABP1. Hesse et al. (1989) purified an auxin binding protein from maize coleoptiles using standard protein purification procedures and affinity column chromatography with NAA-Sepharose. Three isoforms of ABP were identified and the major protein was determined to be of 22 kDa with a binding constant of 2.4 $\times$ 10^{-7}M NAA, again corresponding to the site I auxin binding sites characterised by Ray et al. (1977). Using oligonucleotides to probe a cDNA library made to RNA isolated from coleoptile tissue of maize, Hesse et al. (1989) isolated an ABP cDNA designated axr[1]. In the second study, Tillmann et al. (1989), used purified anti-ABP antibodies to screen a λgt11 cDNA expression library made to maize coleoptile tissue and so identified and cloned the *ABP1* gene. Concomitantly, Inohara et al. (1989) also cloned a cDNA corresponding to ABP1 from maize coleoptile tissue. They designed oligonucleotide probes based on sequence information from ABP1 purified from whole maize shoot tissue including coleoptile, leaf rolls and mesocotyl (Shimomura et al., 1986).

Examination of the consensus sequence of these cDNAs revealed that the genes coded for a 38 amino-acid residue, N-terminal leader sequence for targeting to the ER, and a C-terminal KDEL sequence that implied retention in the lumen of the ER. A single N-linked glycosylation site was also identified, which was shown initially to be glycosylated with the high-mannose structure variants, Man$_9$GlcNac$_2$ (Hesse et al., 1989). This was confirmed by Henderson et al. (1997) to be the major structure, along with Man$_8$GlcNac$_2$ and Man$_7$GlcNac$_2$. The trimming of the Man$_9$GlcNac$_2$ structure, and examination of the proportions of $Man_8GlcNac_2$ and $Man_7GlcNac_2$ suggested to Henderson et al. (1997) that up to 15 percent of the ABP1 protein pool escapes the ER and proceeds to the golgi. Further, the occurrence of low amounts of complex glycans indicated that less than 2 percent of ABP1 is secreted.

Based on the structure of IAA, a putative auxin binding site was identified and, perhaps more significantly, Inohara et al. (1989) determined, using hydropathy analysis, that the sequence contained no hydrophobic domains of sufficient length to span a membrane.

It is now known that ABP1 is coded for by a small multigene family in maize and other higher plant species, although only one *ABP1* gene has been identified, using genomic Southern analysis, in *Arabidopsis* (Palme et al., 1992). Expression studies in maize have shown that genes encoding ABP1-like proteins are expressed in young leaves, coleoptiles and floral tissues but less so in older stems and leaves and in root tissues (summarised by Napier, 2001). Alignment of some of the ABP1 sequence with those from other plant species highlights the conserved nature of the protein (Napier, 2001). All sequences contain an N-terminal leader sequence to direct the nascent polypeptide to the ER, a C-terminal KDEL sequence that retains the mature protein of 163 amino acids in the ER, a conserved putative auxin binding site comprising $Thr_{54}-Phe_{65}$, and the conserved N-linked glycosylation site at Asn_{95}. In some dicotyledonous species, an additional conserved N-linked glycosylation site at Asn_{11} has been identified. Biochemical studies and deduction from the amino-acid sequence of maize reveals a dimer of 44 kDa comprising 22 kDa subunits, and while isoforms of the protein exist, the predominant protein is ABP1. Perhaps the most significant aspect of these primary structures is the absence of putatively hydrophobic membrane-spanning domains in any of the ABP1-like sequences identified thus far.

Genes from many species that encode proteins with identity to ABP1 from maize have now been isolated (Napier, 2001); but despite this progress with the characterisation of ABP1, the search for other auxin binding proteins continues. For example, more recently, Kim et al. (2000) characterised two 57–58 kDa isoforms of a soluble auxin receptor from root and shoot tissues of rice (later designated as ABP_{57}; Kim et al., 2001). The binding was specific to IAA, and the putative receptors could modulate the activity of a plasma-membrane-bound H^+-ATPase by direct interaction with IAA. This induced a conformational change in ABP_{57} (as determined by circular dichroism studies), for when IAA was bound to ABP_{57}, the affinity for the H^+-ATPase was enhanced (Kim et al., 2001). Interestingly, a second, lower affinity, auxin binding site may also be important in regulating ABP_{57} at higher IAA concentrations. When IAA binds to the second site, no conformational change in the protein is induced and the interaction between ABP_{57} and the H^+-ATPase is diminished. This implies that at low auxin concentrations, only the highest affinity primary sites are occupied, conformational changes are induced and putative signalling to the H^+-ATPase is initiated. As well, IAA is prevented from binding to the second site. As the concentration of IAA increases, however, more secondary sites become progressively filled, and the IAA effect at the primary site is correspondingly diminished. This is a possible mechanism with which to explain the wide dose-response curves to IAA in assays using plasma membrane vesicles from roots of rice, and the wide dose-response curve in experiments with excised tissue segments (see Chapter 5, Type 1 and Type 2 cells). Some experimental support for this role of plasma-membrane-bound ATPases in auxin-regulated growth has been provided by the studies of Rober-Kleber et al. (2003), although a direct interaction with the ABP protein was not investigated. In embryos of wheat, the occurrence of a plasma membrane H^+-ATPase has been shown to be regulated by auxin concentration, such that added auxin increases its accumulation (as determined by Western analysis). Further, the protein is distributed into the abaxial epidermis and tip cells of the scutellum, both of which are target tissues for auxin.

If ABP_{57} and other putative proteins are physiological regulators, including the plasma-membrane-bound ATPases, then we await, as we do for other binding proteins, further information as to their biological function in IAA-mediated plant growth and development. It now seems clear that there is more than one receptor site for auxin (plasma membrane and ER) and very probably more than one receptor protein, depending upon the target tissue. Following the isolation, characterisation and sequencing of ABP1, many studies now address the key question as to which of the cellular responses to auxin depend upon this putative receptor protein. The progress of these experiments from the earliest ones of Barbier-Brygoo and colleagues are outlined in Box 7.2, with evidence for the role of ABP1 in mediating cell expansion.

Box 7.2 ABP1 as a mediator of cell responses to auxin

The first evidence of a biological function of ABP1 arose from studies by Ephritikhine et al. (1987) who developed an assay for auxin action using tobacco mesophyll protoplasts. In this system, IAA-induced hyperpolarisation of the plasma membrane was measured using a micro-electrode impaled into the cell that recorded potential difference in response to ion channel activity as microvolt (Em) changes. They found that the extent of Em change in response to the same concentration of the synthetic auxin NAA for wild-type tobacco cells was greater than that from cells isolated from a NAA-insensitive mutant (originally described by Muller et al., 1985). Barbier-Brygoo et al. (1989) used the mesophyll protoplast assay to demonstrate that polyclonal antibodies raised to ABP1 protein purified by the authors inhibited NAA-induced membrane hyperpolarisation. In a later study, Barbier-Brygoo et al. (1991) showed that NAA-induced polarisation of the plasma membrane of tobacco mesophyll protoplasts from wild-type decreased by 50 percent when as little as 0.4 nM of anti-ABP IgG was added to the assay, but even less (0.04 nM) was needed to decrease the response by 50 percent in the NAA-insensitive mutant. However, the higher concentration of antibody (4 nM) was required to decrease the response by 50 percent in cells isolated from tobacco transformed with *Agrobacterium rhizogenes* (and so have an increased sensitivity to NAA). The authors also showed that adding ABP protein to the assay enhanced the sensitivity of the wild-type cells to NAA, suggesting that ABP can mediate the NAA effect *externally* to the cell.

The work of Venis et al. (1992) supported the concept that ABP is required for auxin-induced action at the cell surface. They made antibodies to a 14 amino-acid residue consensus sequence that comprises the IAA binding site (Arg_{53}–Thr_{66}; designated the D-16 antibodies). Again, using the tobacco mesophyll cell protoplasts, they showed that the antibodies could stimulate membrane hyperpolarisation without added auxin – that is, effectively simulating IAA activity. Using the D-16 antibodies from maize and antibodies made to the corresponding peptide from *Arabidopsis,* Steffens et al. (2001) showed that both antibodies could induce swelling of protoplasts isolated from maize coleoptiles and *Arabidopsis* hypocotyls without auxin treatment, so demonstrating interspecies conservation of the response. Taken together, these studies indicate that ABP1 mediates cell growth at the surface of the cell, since neither the ABP protein nor the D-16 antibodies can cross the plasma membrane.

Examination of the sequence data, however, suggested that ABP1 should be confined to the ER, and initial localisation studies supported this proposition. Using a specific monoclonal antibody, MAC 256, raised to purified ABP protein (Napier et al., 1988), and shown to identify the KDEL region of the protein (Napier and Venis, 1990), Napier and colleagues (1992) used immuno-gold labelling and located ABP1 to the ER. Further, the punctate nature of the staining suggested that the protein could be localised to subdomains of the ER. To support the immuno-localisation, Napier et al. (1992) also isolated ER membranes using sucrose density gradients and confirmed recognition by MAC 256. By expressing *ABP1* in a baculovirus vector and using monoclonal antibodies with immunofluorescence and confocal microscopy, Macdonald et al. (1994) showed that ABP1 was localised in the ER of the host insect cells; no ABP1, as far as could be determined, could be detected on the cell surface.

A cell-surface localisation was supported by Jones and Herman (1993) who used both cultured cells and maize seedlings to show that most of the ABP1 was indeed localised in the ER, but some could be detected at the plasma membrane and in the golgi vesicles. Staining in the golgi was confirmed using anti-KDEL antibodies, but in the cultured cells examined, some ABP1 was detected in the medium. Secretion of the ABP1 protein was interrupted, as predicted, using the golgi secretion pathway inhibitor, brefeldin A. The authors also noted that if the cultured cells were starved of 2,4-D, more ABP1 was secreted into the medium.

Further support for a plant membrane localisation of ABP1 came from Diekmann et al. (1995) who used polyclonal antibodies with silver-enhanced immuno-gold labelling viewed by epipolarisation microscopy to show that the protein is present on the surface of maize coleoptile protoplasts. Approximately 6,000 ABP1 proteins were present on each protoplast surface, and the addition of IAA caused the ABP proteins to cluster.

In the first use of molecular genetic approaches to support the assertion that ABP1 could mediate responses to auxin, Jones et al. (1998a) overexpressed the *ABP1* gene in tobacco and maize cell lines. In tobacco, the *ABP1* gene was transformed under the control of a tetracycline-inducible promoter into a genetic background expressing a tetracycline repressor on the assumption that constitutive overexpression of *ABP1* during the tissue culture procedures required to generate transformed plants might produce unusual phenotypes. Using auxin-dependent epinastic growth of leaves as an assay, they found that overexpression of *ABP1* conferred sensitivity to auxin in the normally insensitive mid-region of the leaf such that an increased angle of curvature, and an increased protoplast volume of cells in the mid-region of the leaf was observed. In further experiments, the *ABP1* gene, under the control of the CAMV 35S promoter was transformed into maize cells in culture (a cell type chosen because it does not contain any detectable levels of ABP1 or does not have a strict requirement for auxin to proliferate; Jones et al., 1998a). In these transformants, overexpression of ABP1 again conferred auxin responsiveness to the transformants as determined by increased cell volume.

The use of knock out mutants of *Arabidopsis* has revolutionised our understanding of the ethylene receptor (see later in this chapter), and so with only a single copy of *ABP1* in the genome of *Arabidopsis*, this approach seemed valuable for the determination of ABP1 function. The knock out mutants obtained using T-DNA insertional mutagenesis produced lethal homozygous lines (Chen et al., 2001b) with non-viable seeds and embryos arrested at the globular stage. However, these embryonic-lethal plants could be rescued by transformation with a functional copy of *ABP1*, indicating that ABP is an important requirement for normal development of the embryo.

Together, these studies provide quite compelling evidence for ABP1 located at the plasma membrane as a receptor protein in the transduction of many Type 1 responses to auxin, and with a 'storage' component in the ER. Whether ABP1 or another protein operates in Type 2 and Type 3 cells remains for future researchers to determine.

The effect of ABP1 in determining both auxin-induced expansion and division of tobacco leaf cells has been examined in detail by Chen et al. (2001a). They assessed the levels of ABP1, IAA and percentage of cells in G2 in both wild-type tobacco, and in leaves of plants transformed with the *ABP1* gene of *Arabidopsis*. They found that the level of ABP1 was highest at maximum cell expansion when auxin level was lowest, while at maximum cell division, the ABP1 level was low and auxin levels were high. In further experiments using BY-2 tobacco cells in culture, two dose-dependent responses to auxin were observed – at low auxin concentrations cells expanded, but at higher concentrations cells divided. Antisense suppression of *ABP1* expression in BY-2 cells reduced cell expansion but not cell division to the same degree. Chen et al. (2001a) concluded that the ABP1 receptor is responsive to low levels of auxin and mediates cell expansion, while cell division is regulated by higher auxin concentrations, possibly controlled through a different perception mechanism. It may be that the regulation of cell division in response to the higher concentration of auxin requires interactions with a heterotrimeric G protein (Kim et al., 2001), although the mechanism by which this might occur is yet to be determined.

An ABP1-dependent and an ABP1-independent pathway have also been suggested for the regulation of auxin-induced swelling by protoplasts of epidermal cells isolated from elongating internodes of pea (Yamagami et al., 2004). These authors concluded that the ABP1-independent signalling pathway is more responsive to lower concentrations of IAA and may be the pathway that operates in the control of growth in this tissue by constitutive concentrations of endogenous auxin. A role for G proteins is not excluded in this ABP1-independent pathway.

Auxin-mediated cell expansion and ion-channel controls. Because ABP does not appear to be a transmembrane protein (see Box 7.1), cell surface ABP must interact with (at least) one transmembrane protein. In stomatal guard cells of Vicia, K^+ channels are the targets of auxin action mediated via ABP1 (Thiel et al., 1993), and later work of Philippar et al. (1999) confirmed that expression of a gene coding for a K^+ channel in maize coleoptiles is regulated by auxin. Phillipar et al. (1999) cloned two genes coding for functional K^+ channels in maize, which they designated ZMK1 and ZMK2. During IAA-induced growth of the coleoptile, the expression of ZMK1 increased in good agreement with the kinetics of elongation. Further, in gravi-stimulated maize coleoptiles, the expression of ZMK1 increased and followed the gravi-induced auxin distribution. ZMK1 was expressed in the cortical cells of the coleoptile, but ZMK2 (which is not induced by auxin) was expressed in the vascular tissue.

Using patch-clamp techniques to examine further the relationship of ABP1 accumulation and K^+ channel current changes, Bauly et al. (2000) showed that overexpression of ABP1 paralleled an increased response of guard cells to auxin. They mutated the KDEL tag at the C-terminal of ABP and, using transgenic tobacco, followed the functional relationship between the pool of ER-localised ABP and the subset of protein that migrates to the cell surface. As predicted, overexpression of the ABP1 gene increased the responsiveness of the guard cells to auxin – that is, a lower concentration of IAA was required to stimulate K^+ channel activity when compared with guard cells from non-transformed plants. However, the localisation of the wild-type and mutated ABP proteins gave unexpected results. Mutation of the KDEL tag to HDEL did not interfere with the localisation of the ABP1 protein in the ER or the golgi stacks; neither did it alter auxin sensitivity of the guard cells, nor the occurrence of ABP1 on the surface of root cells as determined by immuno-gold electron microscopy. Mutation of the KDEL sequence to KEQL or to KDELGL, however, did cause the ABP1 protein to pass more readily across the golgi stacks; but again, no difference in the abundance of the ABP protein could be observed at the cell surface, nor was the sensitivity of the guard cells to auxin altered. The authors concluded that whereas overexpression of ABP1 alone is sufficient to increase sensitivity to auxin, this response is not regulated by the population of ABP1 receptors at the cell surface. It was possible, however, that small changes in the ABP population

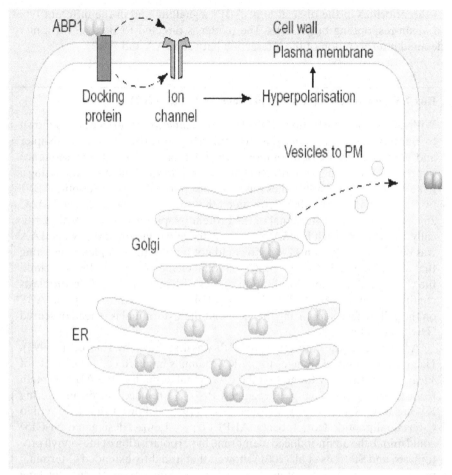

Figure 7.1. Proposed site of interaction of cell-surface ABP1 with a putative docking protein (modified from Timpte, 2001).

on the cell surface could exist between the different mutant forms; these could well be beyond the present limits of detection using immuno-gold. In these experiments of Bauly et al. (2000), auxin response was measured in guard cells, but the movement of the ABP proteins to the cell surface was measured in the root (due to technical difficulties when using shoot tissues). Nevertheless, overexpression of ABP1 does confer increased sensitivity to auxin and increased K^+ channel activity, in agreement with Philippar et al. (1999).

Present evidence seems convincing that the ABP1 protein does mediate the response to auxin in terms of regulating cell expansion. While such functional assays support an activity for ABP proteins on the external surface of the cell, the data of Bauly et al. (2000) call into question their overall significance. An intermediate 'docking' protein (see Timpte, 2001) has been suggested (see Figure 7.1), but the search for this potential downstream element remains as one of the priorities in the dissection of ABP1 signalling and in the different types of auxin-responding target cells. The reader is directed to Box 7.3 for a more detailed interpretation of the interaction between IAA and ABP1.

Box 7.3 Functional interaction between IAA and ABP1

With the first demonstrations of the biological function of surface ABPs shown by Barbier-Brygoo and colleagues (Barbier-Brygoo et al., 1989, 1991), Napier and Venis were aware of the importance of demonstrating some consequence to protein conformation or function in association with binding auxin. Using a panel of monoclonal antibodies raised against purified maize coleoptile ABP1 by Napier et al. (1988), they determined that two clones, MAC 256 and MAC 259, recognised the C-terminal of the protein (Napier and Venis, 1990). Critically, using a sandwich ELISA procedure, they demonstrated that when IAA was bound to ABP, then MAC 256 could not bind to the complex, suggesting that auxin binding induced a conformational change in protein that obscured the C-terminal. Immuno-localisation studies with maize coleoptile protoplasts confirmed that the MAC 256 monoclonal antibody failed to recognise ABP on the cell surface supporting the view that the C-terminal is indeed obscured (Diekmann et al., 1995).

A function for the C-terminal of ABP1 was first shown by Thiel et al. (1993). Using auxin-induced changes to the K^+ channels in guard cells of stomata of *Vicia* as an assay, they found that the C-terminal domain of the ABP1 protein from maize (Trp_{151}–Leu_{163}, designated $Pz_{151\text{-}163}$) replaced auxin in altering K^+ ion flow. Subsequently, Leblanc et al. (1999a) showed that a 15 amino C-terminal peptide from tobacco ABP1 (Trp_{153}–Leu_{167}, designated Nt-C15) could mimic the auxin-induced membrane hyperpolarisition of mesophyll protoplasts, and Steffens et al. (2001) showed that a slightly extended C-terminal peptide from the ABP1 protein of maize (Phe_{148}–Leu_{163}) could induce swelling in protoplasts isolated from maize coleoptiles.

The Leblanc et al. (1999a) experiments also showed that the entire ABP1 protein from tobacco, Nt-ERabp1, could induce hyperpolarisation of tobacco protoplasts without added auxin, an effect observed because the researchers used a homologous system (that is, tobacco cells treated with a recombinant tobacco ABP1 protein). They made a panel of monoclonal antibodies to recombinant Nt-abp1 (LeBlanc et al., 1999b), and showed that several recognised both active site (designated box A), and C-terminal regions of the protein suggesting that the native protein folded such that the auxin binding site came within a closer proximity to the C-terminal. While some monoclonal antibodies blocked auxin-induced membrane hyperpolarisation of the protoplasts, others, including clones that recognised both the C-terminal and auxin binding site *stimulated* hyperpolarisation without added auxin. By examining the kinetics of monoclonal binding to ABP1 using surface plasmon resonance, and relating this to biological activity, a conformational change in ABP1 was shown to be a requirement for the subsequent biological response.

The conformational structure of ABP1 protein has been modelled (Warwicker, 2001) and shown to be a β-barrel homodimer resembling the vicilin protein family. It is a metal ion-binding protein and may have oxalate oxidase activity. Using a tritiated photoactive auxin, azido-IAA to cross-link IAA to ABP1 by photolysis, Brown and Jones (1994) had earlier identified two residues in the C-terminal of the protein, Asp_{134} and Trp_{136}, that contribute to auxin binding. However, the later Warwicker model predicts that W_{44} participates in auxin binding, and that one of the C-terminal tryptophans may occupy the auxin binding 'pocket' in the absence of auxin – explaining perhaps how monoclonal antibodies can recognise epitopes in both the auxin binding pocket and the C-terminal domain. So, when auxin is bound to ABP1, a conformational change is assumed with the C-terminal then signalling the binding event. Whether this is directly to the membrane-localised ion channels or via a 'docking' protein is yet to be determined.

Aux/IAA proteins and auxin action at the gene level

While one thrust in understanding how auxin might signal a biological response focussed upon rapid ion channel changes and cell elongation via putative membrane-associated receptors, a second focussed on the significance of rapid auxin-induced gene transcription and how new auxin-induced gene expressions can occur within 5 to 30 minutes after auxin treatment. For example, on the addition of auxin to segments of pea or soybean hypocotyls in solution, an enhanced rate of growth, proton secretion and new mRNA synthesis are all detectable within 10 minutes (Walker and Key, 1982; Zurfluh and Guilfoyle, 1982; Theologis et al., 1985).

The first rapid auxin-response genes to be identified were in IAA-treated soybean hypocotyls (Ainley et al., 1988). Sequence comparisons have placed these genes in an *Aux/IAA* superfamily, and a large number of rapidly expressed auxin-induced genes have been identified in soybean, then in tobacco cell cultures and finally in *Arabidopsis*. These sequences can be grouped into gene families, including the *Aux/IAA* gene family, the *SAUR* gene family and the *GH3* gene family, but it is the *Aux/IAA* gene family that is of most relevance to our discussion here. The reader is referred to the review of Abel and Theologis (1996) for the early delineation of the other gene families.

A most significant observation concerning the rapid response genes to auxin is their expression in the presence of cycloheximide (in fact, cycloheximide induces their expression), indicating that new protein synthesis is not necessary initially for a response to auxin (Theologis et al., 1985). Genes expressed in the presence of cycloheximide and other protein synthesis inhibitors have, therefore, been referred to as 'primary response genes', while those that are suppressed by cycloheximide (and some members of the *Aux/IAA* genes are included here) are considered as 'the secondary response genes'. What is additionally intriguing about Aux/IAA proteins is the finding that they interact with certain members of a family of transcription factors, termed auxin response factors (ARFs) that reside on auxin-responsive promoter elements of the genome (AuxREs). The ARF proteins are activators of auxin-induced gene expression, but when Aux/IAA proteins become bound to the ARF proteins, then the ARF activation is repressed. A more detailed description of the Aux/IAA proteins and their interaction with the ARF protein family is given in Box 7.4.

Box 7.4 The function of Aux/IAA proteins in auxin target tissues

The first step in the elucidation of the role of Aux/IAA genes in plants was to examine their pattern of expression. Ballas et al. (1993) showed that the promoter region of at least one of the genes, *PS-IAA4/5*, has the signature auxin-responsive element, *Aux/RE*. Wong et al. (1996) created promoter:GUS fusions with two members of the Aux/IAA gene family, *PS-IAA4/5* and *PS-IAA6*, cloned originally from pea (Oeller et al., 1993). These constructs were transformed into tobacco and GUS staining was used to determine the expression of each gene. Both genes were shown to be expressed in the root meristem, at sites of lateral root initiation and in elongating hypocotyls – all target tissues for auxin. Further, *PS-IAA4/5* was expressed in root vascular tissue and in guard cells, while *PS-IAA6* was detected in the glandular trichomes and, significantly, in the elongating side of gravistimulated hypocotyls. Again, all these tissues are known to be auxin target tissues.

The function of the protein products of *PS-IAA4/5* and *PS-IAA6* has also been studied. Using [^{35}S]methionine pulse-chase and immune-precipitation experiments, Abel et al. (1994) determined that the half-life of the proteins was only 8 minutes (*PS-IAA4/5*) and 6 minutes (*PS-IAA6*), and GUS reporter gene: *PS-IAA4/5* and *PS-IAA6* translational fusions were localised to the nucleus. Comparisons of the deduced amino acid sequence of *PS-IAA4/5* and *PS-IAA6* with others in the *Aux/IAA* gene family identified four conserved boxes, termed domain I to domain IV, and preliminary modelling indicated that domain III at least should be involved in interactions with other proteins. Subsequently, Kim et al. (1997) used two hybrid analyses to demonstrate that the Aux/IAA protein PSIAA4 could form homodimers or heterodimers with other Aux/IAA protein family members tested, and that this binding occurred via domains III and IV. Of perhaps more significance is that the Aux/IAA proteins were shown to associate with IAA24, a protein with similarity to the auxin-response-factor (ARF) class of proteins. This association was confirmed by Ouellet et al. (2001) who worked with a series of semi-dominant *axr3-1* mutants (carrying lesions in the *IAA17* gene) previously characterised by Rouse et al. (1998). *IAA17/AXR3* is a member of the *Aux/IAA* gene family and these mutants contain lesions in domain II of the protein. Of the five intragenic revertant alleles described by Rouse et al. (1998), two contained second site lesions in domain III (designated *iaa17/axr3-1R1* and *iaa17/axr3-1R3*). Ouellet et al. (2001) showed that the IAA17/AXR3 proteins from these revertant mutants (that is – those with a lesion in domain III) could not heterodimerise with another Aux/IAA protein or, significantly, with two members of the ARF protein family, ARF1 or ARF5.

ARF1 was the first member of the *ARF* gene family to be characterised (Ulmasov et al., 1997a). In pioneering experiments, they established that four tandem repeats of the auxin response element (AuxRE) from the *GH3* promoter conferred auxin-responsiveness to carrot protoplasts using the GUS reporter assay. Then they showed, using gel shift assays and a GUS reporter system, that the ARF1 protein could bind to the AuxRE element in the *GH3* promoter, and that this binding conferred auxin-responsiveness. On sequencing the *ARF1* gene, conserved domains III and IV were found as identified on the Aux/IAA proteins. Further, studies with the *MONOPTERIS* gene from *Arabidopsis* showed that mutations in this gene interfered with the auxin-induced initiation of a body axis in the developing embryo, and subsequent sequencing showed it to be homologous to the *ARF1* gene. Hence, Hardtke and Berleth (1998) proposed that a function of the *ARF* genes *in vivo* was to regulate auxin responses.

This proposed relationship between ARF and Aux/IAA proteins and the subsequent auxin response had been examined earlier by Ulmasov et al. (1997b). Using a synthetic AuxRE element based on the *GH3* promoter, these workers found that this element could facilitate auxin-responsiveness in a transient GUS reporter assay in carrot cell protoplasts, or in stably transformed seedlings of *Arabidopsis*. The ARF1 transcription factor was shown to bind specifically to the AuxRE element and, in a yeast two hybrid assay, the ARF protein was shown to bind to a series of Aux/IAA proteins tested, including AUX22. Overexpression of AUX22 in the transient assay was sufficient to repress AuxRE driven GUS expression. These results led to the suggestion that the Aux/IAA proteins could *repress* expression of auxin-induced genes via complexes with the ARF proteins.

In a more detailed survey, Ulmasov et al. (1999a) examined nine *ARF* genes and determined, using a transient GUS reporter assay in carrot protoplasts, that at least one ARF protein repressed transcription in response to auxin, while four ARF proteins activated transcription in the presence of auxin. Importantly, these workers suggested that the activation or repression of transcription by ARFs is mediated by binding to Aux/IAA proteins via domains III and IV. This interaction was confirmed by Tiwari et al. (2001) who showed that the Aux/IAA proteins are active repressors of auxin-induced transcription by dimerising with the ARF proteins. For further details on the ARF protein family, the reader is referred to the review by Guilfoyle and Hagen (2001).

The validity of this interaction *in vivo* has been shown by examination of the *bodenlos* and *monopteris* mutants of *Arabidopsis* (Hamann et al., 2002). *Monopteris* mutants lack an activating ARF5, while *bodenlos* mutants contain a lesion in domain II of the *Aux/IAA* gene, *IAA12,* and both mutants fail to initiate the root meristem in the embryo – i.e., the target cells involved are auxin-insensitive. Hamann and colleagues showed that the BODENLOS and MONOPTERIS proteins can interact directly, and that both genes are co-expressed in early embryogenesis.

However, of relevance to the target cell concept is the question – if the binding of Aux/IAA proteins to ARF proteins becomes sufficient to repress a response to auxin, how are the levels of the Aux/IAA proteins controlled (and thus the auxin response in target tissues), particularly as the Aux/IAA proteins are known to be short-lived?

The first clue came from examination of the *axr3-1* series of mutants of *Arabidopsis thaliana* isolated by Rouse et al. (1998). Chromosome mapping studies determined that the *AXR3* gene was synonymous with the *IAA17* gene, a member of the *Aux/IAA* gene family. The *axr3-1* mutants contained lesions in domain II, one of the four conserved domains (designated I–IV) identified in Aux/IAA proteins (see Box 7.4). Later, Ouellet et al. (2001) determined that a mutation in domain II of the *IAA17/AX3* gene induced a significant (seven-fold) increase in the half-life of the mutant protein, when compared to wild-type. Thus when Aux/IAA proteins were stabilised so that they accumulated in the cell, the auxin-related phenotype in these cells become *less* responsive to the hormone. Using a different approach, Tiwari et al. (2001) showed that mutations in domain II of the Aux/IAA proteins that they tested resulted in an increased repression of GUS activity from a AuxRE:GUS construct when the construct was co-transfected with a mutant Aux/IAA protein into carrot cells that were treated with auxin.

These observations by Ouellet et al. (2001) and Tiwari et al. (2001) provided support for earlier observations by Worley et al. (2000) on the rate of degradation of Aux/IAA proteins. Worley et al. (2000) had shown that an N-terminal region spanning domain II of an Aux/IAA protein of *Pisum sativum*, PSIAA6, directed a low protein accumulation when expressed as a translational fusion with a luciferase reporter gene. However, single amino-acid substitutions in domain II of this region (equivalent to those found in two alleles of the *axr3* mutation), caused a 50-fold increase in protein accumulation. From these results, Worley et al. (2000) concluded that mutations in the *Aux/IAA* genes causing mutant phenotypes can result from changes in the extent of Aux/IAA accumulation, with rapid turnover of the protein being necessary for a normal auxin response.

In a similar study, Ramos et al. (2001) showed that mutations in domain II increased the half-life of PSIAA6, and that the addition of commercially available specific peptide-based proteasome inhibitors (designated MG115 and MG132) increased the accumulation of the wild-type Aux/IAA proteins. Such results suggest that a proteasome activity may regulate the turnover of Aux/IAA proteins, via interactions with domain II. A proteasome involvement has now been confirmed by several studies and this evidence is summarised in Box 7.5.

Box 7.5 Regulation of Aux/IAA protein degradation and the auxin response

An early clue to the mechanism of Aux/IAA protein degradation arose from the characterisation of the *tir1* mutants of *Arabidopsis* (Ruegger et al., 1998). These mutants lack a variety of auxin-regulated growth processes including hypocotyl elongation and lateral root formation. The TIR protein contained a motif that has identity to F boxes, and since other F-box proteins were implicated in ubiquitin-mediated degradation processes, the authors speculated that the auxin response might also depend upon the modification of a key regulatory protein or regulatory proteins by a ubiquitin-associated pathway. Reugger et al. (1998) examined the relationship of *TIR1* with another auxin mutant, *axr1*, and showed that the *tir1* mutants acted synergistically with the *axr1* mutant so implicating a ubiquitin-mediated degradation pathway in the auxin response. The *AXR1* gene was shown originally to code for a protein with identity to the E1 class of ubiquitin ligases (Leyser et al., 1993). Subsequently, AXR1 has been shown to form a heterodimer with the ECR1 protein, and together they activate the ubiquitin-related protein RUB1 (Del Pozo et al., 1998). Later, Gray et al. (1999) identified components of the SCF-ubiquitin ligase complex in *Arabidopsis* consisting of yeast Skp1p and Cdc53p homologues, designated ASK and AtCUL1 respectively (thus SCF comprises Skp1p, Cdc53p and the F box). They showed further, using the two-hybrid assay, that TIR interacts with ASK and AtCUL1 to form the SCF^{TIR1} complex. Mutations in the ASK or TIR1 proteins resulted in a decrease in auxin response (e.g., auxin inhibition of root elongation was reduced), while overexpression of *TIR1* inhibited primary root elongation and promoted lateral root initiation – that is, it promoted the response to auxin. Del Pozo et al. (2002) have subsequently demonstrated that the AtCUL1 protein is indeed modified by the AXR1-ECR1-activated RUB1 protein.

The key connection between ubiquitin-mediated proteolysis and the degradation of Aux/IAA proteins was shown by Gray et al. (2001). A functioning SCF^{TIR1} complex is required and these workers showed that SCF^{TIR1} physically interacted with two members of the Aux/IAA family, AXR2/IAA7 and AXR3/IAA17, via domain II on the Aux/IAA proteins. Perhaps most importantly, they showed that auxin stimulated the binding of the SCF^{TIR1} complex to the Aux/IAA proteins – that is, promoting the degradation of these repressors of auxin-regulated genes. The observation that auxin increases the rate of degradation of Aux/IAA proteins was confirmed by Zenser et al. (2001). Using a 13-amino-acid consensus sequence from domain II of the Aux/IAA protein PSIAA6 fused to LUC, workers created an auxin responsive reporter gene fusion and transformed it into *Arabidopsis*. Zenser et al. (2001) showed that the level of the reporter gene fusion protein was reduced within 2 minutes after application of auxin, and that increasing the level of auxin increased the proteolytic rate. The reader is referred to the review of Leyser (2002) for a more in-depth dissection of auxin signalling and the role of the ubiquitin-mediated degradation pathway.

How does auxin exert its cellular effects on target tissues –
A working model

Currently, many researchers have proposed working models as to how auxin modulates the transcription of the auxin-regulated genes. A consensus view appears to suggest that the ARF proteins permanently occupy their target DNA

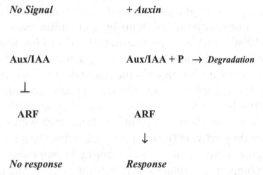

Figure 7.2. Conceptual overview of auxin signalling in plants. P = phosphorylation, ⊥ = repression of signalling

binding sequences, designated the auxin-regulatory elements (AuxREs) regardless of the cellular auxin concentration. At low auxin concentrations, Aux/IAA proteins are stable, but short-lived, and they dimerise with the ARF proteins and block transcription of the auxin-responsive genes. At higher auxin concentrations, the Aux/IAA proteins are themselves destabilised, potentially via auxin-induced kinase activity and subsequent proteasome-mediated degradation. The ARF proteins then dimerise and bind to the AuxREs which then permits transcription of the auxin responsive genes (see Figure 7.2 for a conceptual overview).

Significantly, as one class of these auxin-responsive genes are the *Aux/IAA* genes, a mechanism for a feed-back repression and modulation of the effect of auxin presents itself. Any model is complicated by the fact that at least 25 members of the *Aux/IAA* gene family, and 23 members of the *ARF* gene family are known in *Arabidopsis* alone. These proteins form specific combinations, and given that the expression of members of each gene family is also developmentally regulated (for example, Wong et al., 1996; Ulmasov et al., 1999b), the multitude of possibilities for this mechanism regulating auxin responses in different target tissues at different developmental states is vast.

It should also be noted that while most *Aux/IAA* genes are induced by auxin, there are notable exceptions. Rogg et al. (2001) isolated a mutant of *Arabidopsis*, *iaa28-1*, that displayed reduced apical dominance and lateral root initiation – i.e., an apparent auxin-reduced response phenotype. The *IAA28* has been isolated, sequenced, shown to be a member of the *Aux/IAA* gene family and found to be expressed preferentially in roots and in the stem tissue of inflorescences. Further mutant analysis has shown that, in common with other members of the *Aux/IAA* gene family, the IAA28 protein in wild-type plants of *Arabidopsis* will repress genes that cause auxin-induced lateral root initiation. However, unlike other members of the *Aux/IAA* gene family, the transcription of *IAA28* is not regulated by auxin. This shows that expression of the Aux/IAA repressors of auxin-induced genes may or may not be induced by auxin, again, increasing the multiplicity of signal combinations by which repressors can be regulated by changing levels of hormones in a target tissue.

For the final part of the mechanism by which auxin operates it is necessary to examine how the hormone regulates the proteolytic degradation of the repressor Aux/IAA proteins. The most promising approach appears to be that the Aux/IAA proteins become phosphorylated and therefore become competent to form a complex with SCF^{TIR1}, a component of ubiquitin-mediated proteolysis (Kepenski and Leyser, 2002). Phosphorylation of target proteins prior to interaction with the SCF complex occurs in mammalian cells, thus encouraging the search for suitable auxin-stimulated kinases in plants. Examples of such candidates include an auxin-stimulated MAP kinase, identified in roots of *Arabidopsis* (Mockaitis and Howell, 2000). Another is the PINOID protein with identity to serine-threonine protein kinases. Mutations in this protein negatively regulate auxin signalling (Christensen et al., 2000). It is intriguing to speculate, therefore, that such kinase-dependent activation may provide the link between perception of the hormone by binding proteins, and the activity of downstream receptor kinases. In this regard, it may be significant that ABP$_{57}$ soluble receptors identified by Kim et al. (2001) in rice will activate heteromeric G proteins.

For the second part of this chapter we will examine the mechanism of ethylene perception in its target tissues.

Ethylene perception

That plants can perceive and respond to an external source of ethylene has been recognised since the turn of the last century (Neljubov, 1901) but the mechanisms by which target cells perceive and distinguish ethylene between the many signal inputs that reach them from adjacent tissue is now becoming more tractable in molecular terms. It is accepted widely that the ETR-like family of proteins represent one class of ethylene receptors. These proteins may not be the sole class of ethylene receptor nor the sole mechanism by which the hormone is perceived by its target tissues. Nevertheless, a description of the ETR-like protein family illustrates a widely emerging theme in plant hormone receptor biology: namely, that the action of the hormone is to relieve a normal state of repression – a theme already developed in the previous discussion on auxin perception.

As three particular examples of ethylene target cells, we choose abscission cells (Type 2) and the induction therein of specific enzymes, the cell growth control that operates in the directional orientation of auxin-induced expansion in Type 1 cells, and the cooperative auxin-plus-ethylene–enhanced elongations of Type 3 cells.

Evidence that receptor proteins perceive ethylene

The first member, *ETR1*, of a gene family that was seen to function as a true receptor for ethylene was identified in mutant screens of *Arabidopsis* by Bleecker et al. (1988). Using suppression of the elongation component of the ethylene triple response to select plants that still elongated in the presence of ethylene, researchers identified phenotypes in a screen of 75,000 seedlings derived from the M_2 generation of EMS mutagenized seeds of the Columbia wild-type (Haughn and Somerville, 1986). These mutants displayed continued hypocotyl elongation in the dark in the presence of 5 ppm ethylene. Three lines were identified, and one of these, designated *etr* (for ethylene triple response), was characterised further. The *etr* mutant (later designated *etr1-1*, see Chang et al., 1993) was shown to lack other normal responses to ethylene; it exhibited no inhibition of root and hypocotyl elongation, no decrease in leaf chlorophyll content, no induction of guaiacol oxidase activity in leaf and stem tissue, and no stimulation of seed germination nor ethylene-induced reduction in ethylene biosynthesis (as determined by ethylene evolution from excised leaf tissues). Genetic analysis revealed that this lack of ethylene response was due to a dominant mutation, and the authors speculated at that time that the dominant lack of response to ethylene could mean that the *ETR* gene functions in the absence of ethylene to suppress the expression of ethylene responses.

Also using the triple response screen, Guzman and Ecker (1990) identified another ethylene insensitive mutant, *ein1*, that was shown to be allelic with *etr1* (and later designated *etr1-3*, see Chang et al., 1993). This screen also identified an *ein2* mutant, the gene product of which occupies a strategic position in the ethylene transduction pathway (see later discussion). Guzman and Ecker (1990) used the technique of double mutant crosses to determine the epistatic relationship of the mutant alleles. In this first analysis, *ein2* was shown to be downstream of *ein1/etr1*.

In all, 4 allelic *etr* mutants were identified and designated as *etr1-1, 1-2, 1-3* and *1-4* (see Chang et al., 1993), and all displayed a genetically dominant lack of response to ethylene. Chang et al. (1993) were the first to sequence the *ETR1* gene and to determine that the lesion in each of the four *etr1* alleles occurs in the reading frame of a gene coding for a protein with similarity to histidine kinase two-component signalling proteins in prokaryotes. Briefly, the typical two-component system comprises (i) a membrane localised histidine kinase with an extracellular domain to perceive the ligand input, and (ii) an independent response regulator that contains a receiver domain containing a critical aspartate residue and a signal output that activates transcription (see later in this chapter for a more detailed description of these ubiquitous signalling systems). Each *etr* mutant characterised was shown to be a miss-sense mutation that altered only one amino acid in one of three hydrophobic (membrane-spanning) domains (designated I to III) in the N-terminal region of the protein. By transforming wild-type *Arabidopsis* with *etr1-1*, Chang and colleagues produced a dominant ethylene-insensitive phenotype akin to the *etr1-1* ethylene-insensitive mutant described by Bleecker et al. (1988).

Using a modified mutant screen of *Arabidopsis* in conjunction with double-mutant analysis, Kieber et al. (1993) identified another downstream element of the *ETR* gene family, designated the constitutive triple response (CTR) protein. To achieve this, they mutagenised seed lots by treatment with EMS, diepoxybutane or X-rays and screened for plants that displayed the triple response in the absence of applied ethylene (that is, the opposite of the screen used by Bleecker et al., 1988). From more than 10^7 seedlings, 18 fertile lines were obtained and these were grouped into two classes: class 1 were those in which the mutant phenotype could be rescued with inhibitors of the key enzymes in the ethylene biosynthetic pathway, either aminoethoxyvinylglycine (AVG), an inhibitor of ACC synthase, or α-aminoisobutyric acid (AIB), an inhibitor of ACC oxidase; class 2 were those mutants in which the phenotype could not be rescued by either of the ethylene biosynthesis inhibitors.

Measurements of ethylene production of class 1 and 2 mutants showed that two of the class 1 mutants overproduced ethylene, but they were not allelic to the first ethylene-overproducing (*eto*) mutant identified, *eto1* of Guzman and Ecker (1990). Thus the Kieber mutants were designated as *eto2* and *eto 3*, where *eto2* produced twenty-fold more ethylene than wild-type, and *eto3* produced a hundred-fold more. Four class 2 mutants were also identified, and all four fell into a single complementation group designated *ctr1*. The *ctr1-1* mutant was diepoxybutane-generated, *ctr1-2* resulted from X-ray mutagenesis, while *ctr1-3* and *ctr1-4* mutants were generated by EMS treatment. Morphological analysis confirmed that all these *ctr1* mutants, both as seedlings and adult plants, displayed a phenotype identical to wild-type plants treated with 1 ppm ethylene.

With the identification of the *ctr* mutants, two ethylene-induced genes (*EI305* and chitinase) were shown to be expressed constitutively in the *ctr* mutant background, but in wild-type plants they were expressed only after treatment with 100 ppm ethylene. Kieber et al. (1993) then isolated the *CTR1* gene by T-DNA tagging, and sequencing revealed that the reading frame would code for a protein of 90 kDa molecular mass with no obvious membrane-spanning regions. The protein had highest identity to a serine/threonine protein kinase most closely related to the Raf protein family from eukaryotes (41 percent identity in the kinase domain). Finally, by using double mutant analysis, Kieber et al. (1993) showed that the *CTR* gene was downstream of the *ETR* gene but upstream of the *EIN2* gene characterised by Guzman and Ecker (1990).

Since these early studies by Bleecker et al. (1988), Guzman and Ecker (1990), Kieber et al. (1993), and Chang et al. (1993), the identification and characterisation of the nature of ethylene perception in plants has been greatly extended. There are now five members of the ETR-like proteins that have been characterised in *Arabidopsis;* ETR1 and ERS1 (the subfamily I receptors) and ETR2, ERS2 and EIN4 (the subfamily II receptors). The basic features of each ETR-like protein is shown in Figure 7.3, and their identification and characterisation as members of the ETR-like family is described in Box 7.6.

Characterisation of ETR function

The initial studies on ETR1 included assessment of its function as an ethylene binding protein. When *ETR1* was transformed into yeast, it was found to localise to the plasma membrane (Schaller and Bleecker, 1995), and ethylene binding (using $^{14}C_2H_4$) was shown to occur in these membrane fractions (for details of ethylene binding to ETR1, see Box 7.6).

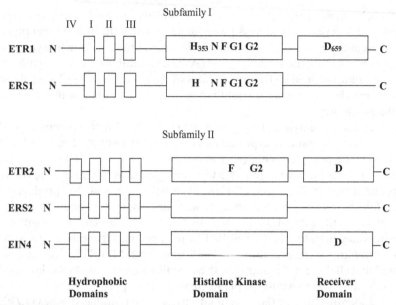

Subfamily I

Subfamily II

Hydrophobic Domains   **Histidine Kinase Domain**   **Receiver Domain**

Figure 7.3. Diagrammatic representation of the key features of the ETR-like receptor family of *Arabidopsis thaliana*. Using ETR1 as the reference sequence, the potential site of autophosphorylation in the histidine kinase domain, H_{353}, is indicated, as well as four other consensus motifs: N, F, and the G1 and G2 boxes that are found in bacterial histidine protein kinases. The critical asparate residue, D_{659}, on the receiver domain is shown, with the subscript numbers denoting amino-acid residues in ETR1. The presence of these critical residues on other protein members of the family is indicated. The reader is referred to Hua et al. (1998) for a detailed comparison of the protein sequences.

Box 7.6 Identification and isolation of ETR-like proteins

The use of EMS mutagenesis was employed successfully to isolate ETR1 (Bleecker et al., 1988; Guzman and Ecker, 1990), and further analysis of EMS mutants of *Arabidopsis* yielded two more homologues of *ETR*. Roman et al. (1995) characterised the *ein4* mutant and through detailed genetic analysis determined that *ein4* and *etr1* were upstream of *ctr1*, although the *EIN4* gene was not sequenced as part of this study. Sakai et al. (1998) isolated a line, designated *etr2-1*, that showed elongated hypocotyls and roots in the presence of 5 ppm ethylene and displayed a dominant phenotype for ethylene insensitivity in terms of etiolated seedling elongation, leaf expansion, and leaf senescence. Double mutant analysis placed the gene upstream of *CTR*. The *ETR2* gene was cloned and sequenced and shown to have 65 percent similarity with *ETR1*, with the lesion in the *etr2-1* mutation ($Pro_{66} \rightarrow$ Leu) occurring in the membrane-spanning hydrophobic domain I of the N-terminal region of the protein (Figure 7.3). Further, Sakai et al. (1998) transformed wild-type *Arabidopsis* with a genomic sequence incorporating the *etr2-1* mutant gene and showed that these plants had an identical ethylene-insensitive phenotype when compared with the EMS-generated *etr2-1* mutants.

Studies using more direct molecular approaches identified two further members of the *ETR*-like gene family. With *ETR1* as a probe, Hua et al. (1995) isolated the *ERS1* gene from a genomic library of *Arabidopsis*, then introduced a miss-sense mutation into domain I of the N-terminal of the gene, $Phe_{62} \rightarrow$ Ile (identical to the miss-sense mutation in *etr1-4*). The mutated gene was transformed into wild-type *Arabidopsis* and conferred dominant insensitivity to the transformants. Double mutant analysis also showed that the *ERS1* gene was upstream to the *CTR* gene. In further work, Hua et al. (1998) used *ETR2* as a probe of genomic DNA of *Arabidopsis* and isolated *ERS2* and *EIN4*. Sequence analysis of both *EIN4* and *ERS2* revealed that both genes had higher similarity to *ETR2* than with *ERS1* and *ETR1*. The *ein4* mutation had been identified previously by Roman et al. (1995) and shown to confer a lack of response to ethylene, and to operate upstream of *CTR1*. Three allelic mutants were sequenced, *ein4-1*, *ein 4-2* and *ein4-3* and again miss-sense mutations were discovered to confer single amino-acid changes in the hydrophobic transmembrane-spanning domains in the N-terminal domain ($Ile_{84} \rightarrow$ Phe in domain II of *ein4-1* and *ein4-2*; $Thr_{117} \rightarrow$ Met in domain III of *ein4-3*). The $Ile_{84} \rightarrow$ Phe mutation in *ein 4-1* and *ein4-2* is similar to that introduced by Hua et al. (1995) into *ERS1* and that which occurs in *etr1–4*. A mutation was also introduced into the *ERS2* gene: $P_{67} \rightarrow$ L – a mutation similar to that which occurs in *etr2-1* (Sakai et al., 1998). With the mutated gene transformed into wild-type *Arabidopsis*, analysis of the transgenic plants produced revealed that dominant ethylene insensitivity was conferred to the transformants. Again, double mutant analysis showed that *ERS2* is upstream to *CTR1*. In total, five genes encoding the ETR-like ethylene receptor have now been characterised in *Arabidopsis*: *ETR1*, *ETR2*, *ERS1*, *ERS2* and *EIN4* (features of each protein are set out in Figure 7.3).

Mutational screens identified *ETR1*, *ETR2*, and *EIN4* as genes that mediate the response of *Arabidopsis* to ethylene. For members of the gene family that were identified by molecular screening (*ERS1*, *ERS2*), their ability to confer ethylene insensitivity was confirmed by overexpression in the wild-type background. However, parallel to these approaches, direct biochemical means have been used to characterise the relationship between ethylene and the ETR proteins.

Using yeast transformed with the *ETR1* gene, Schaller and Bleecker (1995) showed that the ability to bind ethylene could be conferred to the yeast transformants; the binding could be inhibited by the competitive ethylene binding inhibitors, 2,4-norbornadiene and *trans*-cyclooctene. In contrast, yeast transformed with a mutant form of *ETR1*, *etr1-1* ($Cys_{65} \rightarrow$ Tyr; a substitution in the hydrophobic membrane-spanning domain II) did not show ethylene binding, nor did another lesion in domain II ($Cys_{65} \rightarrow$ Ser) or in domain III ($Cys_{99} \rightarrow$ Ser). Finally, truncation of the ETR1 protein revealed that ethylene binding was limited to the N-terminal hydrophobic-domain–containing region of the protein (residues 1–165). Subsequently, Hall et al. (2000), also using the transformed yeast assay, showed that the ERS1 protein can bind ethylene. In common with ETR1 (also tested by Hall et al., 2000), binding was inhibited by 1-MCP.* Together, these studies demonstrate that both ETR1 and ERS1 proteins can bind ethylene, and a mutation in these protein sequences leads to a dominant ethylene-insensitive phenotype.

The fact that yeast could be transformed with the *ETR1* gene and then shown to bind ethylene is critical to the ethylene receptor concept, as it establishes this protein as an ethylene binding protein. Additionally, homology to the histidine kinase two-component signalling molecules in prokaryotes also presents the possibility of the ETR protein operating as a receptor via downstream kinase-mediated signalling (Chang and Meyerowitz, 1995). Typically, the two component signal complex comprises two elements – a sensory histidine kinase and a response regulator. The sensory histidine kinase typically contains an N-terminal input domain and a C-terminal kinase domain with an invariant histidine residue. The response regulator possesses a receiver domain with an invariant aspartate residue and a C-terminal output domain. From studies with bacteria, we know that the sensor domain binds the ligand and thus induces autophosphorylation of the invariant histidine mediated by the histidine kinase domain. The histidine kinase then phosphorylates the invariant aspartate residue of the response regulator, an event that is then transmitted to the output domain. In some two-component systems, including those identified in *Arabidopsis*, a phospho-relay intermediate is involved in the transmission of the phosphorylation signal from the sensor domain to a remote response regulator (Urao et al., 2000).

Genetically, the *etr* mutants were all dominant mutations and so it was not readily apparent as to whether the loss of ethylene perception represented gain-of-function or loss-of-function in terms of receptor signalling. Then, in an elegant series of experiments, Hua and Meyerowitz (1998) produced a series of recessive loss-of-function mutations corresponding to *ETR1, ETR2, ERS2* and *EIN4*. For *ETR1*, four were generated (designated *etr1-5, etr1-6, etr1-7* and *etr1-8*) with a G → A transition at Trp_{563} (which resulted in a stop codon in *etr1-5* and *etr1-8*) and at Trp_{74} (which resulted in a stop codon in *etr1-7*), and in *etr1-6*, a single base pair transition in the intron putatively interrupting splicing. Therefore, each mutation gave rise to a truncated protein that, as determined by Western analysis, did not accumulate in the mutant plants.

Hua and Meyerowitz (1998) then made a series of double (*etr1-6,etr2-3; etr1-7,etr2-3; etr2-3,ein4-4; etr1-6,ein4-4; etr1-7,ein4-4*), triplicate (*etr1-6,etr2,ein4-4; etr1-7,etr2-3,ein4-4*) and quadruple (*etr1-6,etr2-3,ein4-4,ers2-3*) crosses with these loss-of-function mutants of *ETR1, ETR2, ERS2* and *EIN4*. The expectation was that successive losses of the competence to respond to the hormone would ensue (as more members of the gene family were 'knocked out'). However, the phenotypes were not as expected. Successive crosses that knocked out further members of the gene family produced phenotypes that were reminiscent of the ethylene response. In fact, the quadruple mutant plants displayed severe hypocotyl shortening, and failed to flower – a phenotype that occurs in plants treated with high concentrations of ethylene. The authors concluded that the role of the receptor

* 1-MCP = 1-methylcyclopropene

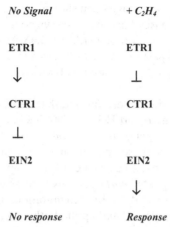

Figure 7.4. Conceptual representation of ethylene signalling mediated via ETR1 and CTR1 in *Arabidopsis thaliana*. ⊥ repression of the signalling function.

must be to act as a negative regulator of ethylene action such that if the receptor was non-functional, then a phenotype displaying an ethylene response resulted. Since the phenotypes of the loss-of-function mutants are opposite to the phenotypes of their corresponding dominant alleles, these dominant mutants (e.g., *etr1-1*) are gain-of-function mutations and the receptor protein must be held in the signalling active state – that is, repressing the ethylene response. Thus a non-functional receptor in terms of ethylene binding confers the observed ethylene-insensitive phenotype (e.g., *etr1-1, 1-2, 1-3* and *etr1-4*) because the hormone cannot stop the receptor (ETR) signalling to the CTR repressor. Correspondingly, in the loss-of-function mutants (e.g., *etr1-5, 1-6, 1-7*), the loss-of-function receptor cannot signal (to CTR) to repress the ethylene response (irrespective of ethylene binding) because the loss of function receptor is truncated and apparently does not accumulate. The phenotypes of these plants therefore exhibit a constitutive ethylene response.

It is now widely accepted that the negative regulator is the CTR protein, since mutations in this protein produce a constitutive ethylene response phenotype (see Figure 7.4 for a conceptual view of this signalling). A more detailed description of the cellular events succeeding the ETR-CTR interaction is given in Box 7.7.

Box 7.7 Downstream signalling from the ETR-CTR interaction

In the original screen of Kieber et al. (1993), four mutant alleles were identified. Two of these (*ctr1-1* and *ctr1-4*) had single amino-acid miss-sense mutations in the kinase domain of the protein and two coded for proteins with no kinase domain. All four mutants displayed the same phenotype and so it is possible to speculate that abolition of the kinase moiety from the protein abolished the negative regulation by CTR. It should be noted that abolition of kinase activity had not at this stage been shown for *ctr1-1* and *ctr1-4*. In other organisms, these Raf kinases have been shown to act as mitogen-activated protein (MAP) kinase kinase kinase (MAPKKK) with downstream MAPKK and MAPK elements as part of a signalling cascade to EIN2, and recent evidence suggests that such MAP kinases are also involved in ethylene signalling (Ouaked et al., 2003).

The earliest genetic evidence established that EIN2 operated downstream of ETR and CTR (Guzman and Ecker, 1990; Kieber et al., 1993). Guzman and Ecker isolated 25 independent lines from a population of EMS mutants of *Arabidopsis* and six of these showed at least a three-fold difference in hypocotyl length in response to ethylene when compared to wild-type. One dominant allele was designated *ein1-1*, and five recessive mutations were designated *ein2-1, 2-2, 2-3, 2-4*, and *2-5*. In each mutation, the plants display a lack of response to both endogenous and applied ethylene. Using *ein2-1*, Alonso et al. (1999) were the first to clone and sequence the *EIN2* gene. They determined that the protein consisted of 1,294 amino acids which gave rise to a 141-kDa protein with a dimorphic structure comprising a 461-amino-acid residue N-terminal with 12 (hydrophobic) transmembrane domains, and a 833 amino acid predominantly hydrophilic domain. Alonso and colleagues confirmed the N-terminal localisation to membranes by translation of *EIN2 in vitro* using a canine pancreatic cell-free system. The N-terminal sequence has homology with the Nramp family in proteins that are divalent cation transporters in many organisms from bacteria to humans. In critical expression experiments with transformants that displayed the ethylene-insensitive phenotypes of the *ein2* mutants, Alonso et al. (1999) determined that expression of the C-terminal region of EIN2 in the *ein2-5* genetic background (designated *ein2-5*:CEND1 constructs) did produce transformants with a constitutive ethylene response phenotype, akin to the *ctr1-1* phenotype. Further, the examination of expression of a number of ethylene-induced genes tested (*At-GST2, basic-chitinase; At-EBP*) showed that each was up-regulated in the *ein2-5*:CEND transformants without ethylene. Finally, the expression of the C-terminal portion of EIN2, CEND in an *ein3-1* mutant background is sufficient to activate downstream ethylene responses in an EIN3-dependent manner. It is now known that EIN2 signals to downstream transcriptional activators including EIN3 (Alonso et al., 1999).

The significance of EIN3 proteins in the ethylene response has been shown in *Arabidopsis* by Guo and Ecker (2003), who found that EIN3 protein levels increase in response to ethylene; an induction that was dependent on the presence of the ethylene receptors (ETR1 and EIN4), CTR1, EIN2, EIN5 and EIN6. However, in the absence of ethylene, EIN3 is degraded via a unbiquitin/proteasome pathway with two F-box proteins, EBF1 and EBF2 identified. Overexpression of either F-box protein causes EIN3 degradation and the ensuing insensitivity to ethylene demonstrates that the ubiquitin/proteasome pathway must act to negatively regulate plant responses to ethylene.

In terms of a biochemical dissection of this system, the initial cellular localisation studies performed in yeast confirmed the targeting function of the N-terminal hydrophobic membrane-spanning domains (Schaller and Bleecker, 1995). However, Chen et al. (2002) determined the subcellular localisation *in planta*. Using a combination of sucrose density-gradient fractionation and immuno-gold electron microscopy, they showed that in *Arabidopsis*, ETR1 associates with the ER membrane and not the plasma membrane. Using deletion analysis, Chen et al. (2002) also showed that the N-terminal membrane-spanning domains are sufficient for this targeting.

In studies on the nature of signalling from the ETR1 protein, Gamble et al. (1998) were the first to demonstrate that the histidine kinase domain (residues 164–609 of the ETR protein; see Figure 7.3) when expressed as a GST-ETR1 protein fusion could undergo autophosphorylation *in vitro*. No phosphorylation of the fusion protein was observed when the invariant histidine (the autophosphorylation target residue) was mutated to Gln ($His_{353} \rightarrow$ Gln). However, Gamble et al. (2002) also introduced mutations into the etr1–1 gain-of-function mutant protein to abolish kinase activity. To do this, substitutions were introduced into the G2 box of the histidine kinase ($Gly_{545} \rightarrow$ Ala; $Gly_{547} \rightarrow$ Ala), a region that is critical for kinase activity (see Figure 7.3). Nevertheless, when the (double) mutant was expressed in a wild-type background or the *etr1-7* background (with no functional ETR protein), perhaps unexpectedly, an ethylene-insensitive phenotype was still obtained. The authors had reasoned that the etr1-1 mutant receptor was 'locked' into signalling to CTR, repressing the ethylene response because it could not bind ethylene. However, it also showed that if the signalling capacity is abolished (in the double mutant), the repression of the ethylene response still continued.

This apparent non-interruption of etr1-1 function in ethylene repression with the abolition of the kinase activity highlights a potentially curious observation that emerged once the sequences of all five members of the ETR-like gene family were compared (Hua et al., 1998). That is, members of subfamily I (ETR1, ERS1) contain all the critical residues in the histidine kinase domain for function, whereas the subfamily II receptors (ETR2, ERS2, EIN4) do not (see Figure 7.3). Indeed, Chang and Meyerowitz (1995) showed that if the His_{353} in the histidine kinase domain or the Asp_{659} of the receiver domain were mutated, the etr1-1 protein could still confer a lack of response to ethylene.

To examine the role of the histidine kinase in ethylene signalling in more detail, Wang et al. (2003) transformed a loss-of-function double mutant, *ers1-2,etr1-7* (that displayed a severe constitutive ethylene phenotype) with each receptor driven by the *ETR1* promoter. As expected, accumulation of ETR1 and ERS1 restored normal growth to the transformants, but ETR2, ERS2 and EIN4 did not. However, transformation of *ers1-2,etr1-7* (or *ers1-2,etr1-6*) with a mutant of ETR1 ($G_{515} \rightarrow$ A, $G_{517} \rightarrow$ A, two mutations in the G1 box), so inactivating the histidine kinase, also restored normal growth (Gamble et al., 1998). Responsiveness to ethylene was also restored as determined by inhibition of root growth and hypocotyl extension. The interpretation of these findings is that the histidine kinase signalling from ETR1 may not be required for ethylene receptor signalling, but that this function may be carried out by a novel, as yet uncharacterised intermediate response regulator or by CTR directly.

Of significance perhaps to these findings is the observation that CTR has been shown to physically interact with the ETR proteins (Clark et al., 1998; Cancel and Larsen, 2002). Using a yeast two-hybrid assay, the N-terminal of the CTR protein was found to form a physical association with the His domain of the subfamily I receptors, ETR1 and ERS1 (Clark et al., 1998) but only a weak association to the subfamily II receptor, ETR2 (Cancel and Larsen, 2002). Further deletion analysis showed that the catalytic binding domain of CTR (which, in raf-like kinases from other organisms, has been shown to interact with GTP binding proteins) is not involved in the ETR–CTR association, suggesting that the GTP binding region is left free to signal in the CTR protein (Clark et al., 1998). Significantly, CTR1 has also been shown, using sucrose-density-gradient fractionation, to be localised to the ER membrane (in common with ETR1), but since CTR1 has no apparent membrane-spanning domains, it is likely that its proposed physical association with ETR1 may explain the common subcellular location (Gao et al., 2003).

In other downstream signalling experiments, Moshkov et al. (2003) demonstrated an ethylene up-regulation of the activity in several monomeric GTP-binding proteins. The activation is ethylene-dependent as determined by receptor-inhibitor experiments with 1-MCP (which inhibits ethylene perception), and in *etr1-1* mutants an even more attenuated activation was observed. A number of the ethylene-up-regulated monomeric G proteins were, as expected, also up-regulated in the *ctr1-1* genetic background, although not a complete match with the ethylene up-regulated profile. Nevertheless, the evidence looks promising that G protein activation is an integral part of ethylene signalling, although its relationship with the ETR-CTR-MAPKK cascade is yet to be determined.

Evidence that the expression of receptors is developmentally regulated in plant tissues

From the genetic evidence from *Arabidopsis*, it seems clear that the ETR-like family of proteins are *negative* regulators of the ethylene response. This has been tested directly in tomato, in which six homologues of the *Arabidopsis* gene family have been identified and designated *LE-ETR1* to *LE-ETR6* (reviewed in Klee and Tieman, 2002). In common with *Arabidopsis*, at least one member, *LE-ETR3* (originally the *NR* [*never-ripe*] mutant; Wilkinson et al., 1995), lacks the response regulator, and at least three (*LE-ETR4* to *LE-ETR6*) lack the complete set of conserved amino-acid residues in the histidine kinase domain. Ciardi et al. (2000) overexpressed the *NR* gene in a wild-type tomato background and determined that the transformants displayed a reduced response to ethylene, when compared with wild-type. The *NR*-overexpressing plants did not display the same degree of inhibition of hypocotyl elongation in seedlings or stem elongation at 9 weeks old, as the wild-type plants.

Still using tomato, Hackett et al. (2000) transformed *Nr* mutants with an anti-sense *NR* gene and demonstrated that fruit developed normally in the transformants and were not delayed in ripening, as in the non-transformed ethylene-insensitive *Nr* mutant. Such observations support the view that the function of the receptor is to repress the ethylene response, and so the higher the accumulation of the receptor in the target tissue, the more the response to the hormone becomes attenuated.

In evaluating the relationship between levels of ethylene receptor and target status, we can now review studies that have monitored the expression of homologues of the *ETR* gene family during fruit ripening in different plant species, as well as in other developmental processes.

Fruits as ethylene target tissues. In the earliest experiments, a homologue of *ETR1* that was expressed constitutively during fruit ripening was isolated from tomato by Zhou et al. (1996) and designated originally as *eTAE1*. However, a second homologue, *tETR,* that showed higher identity to the *NR* gene when compared with *ETR* was developmentally linked and was expressed highly in fruit tissues at the breaker stage, but its expression decreased as ripening progressed (Payton et al., 1996).

Lashbrook et al. (1998) followed three members of the multigene family of tomato (*LE-ETR1, LE-ETR2* and *LE-ETR3* [*NR*]) during ripening. Of these, *NR* was expressed primarily in fruit tissues but its expression displayed a discrete developmental programme, being first at a high level in the ovary, then at a decreasing level post-anthesis before increasing again at the ripening stage. Interestingly, *NR* expression followed the pattern of ethylene evolution from ripening fruit. An induction of *NR* expression by ethylene (which could be blocked by the ethylene perception inhibitor, 1-MCP) was shown first by Wilkinson et al. (1995) and confirmed by Nakatsuka et al. (1998). In a later study, Tieman and Klee (1999) determined that *LE-ETR4* was, in fact, the highest expressed receptor in ripening fruit, although, and unlike *NR,* it is constitutively expressed.

Such differential expression of *ETR* homologues has been examined in other fruits. In musk melon, *Cucumis melo,* the *ERS1* homologue, *Cm-ERS1,* increased in the pericarp during fruit enlargement, then decreased as the fruit matured before displaying a slight increase in expression at ripening (Sato-Nara et al., 1999). In contrast, expression of the *ETR1* homologue, *Cm-ETR1,* paralleled the climacteric ethylene production of fruit ripening (although the role of ethylene was not examined further using specific inhibitors). The accumulation of the Cm-ERS1 protein during early fruit development was confirmed using antibodies (Takahashi et al., 2002).

In peach, a homologue of *ETR1* designated *Pp-ETR1* was constitutively expressed during fruit ripening, while an *ERS1* homologue, *Pp-ERS1,* was up-regulated in parallel with ethylene production – an up-regulation that was inhibited by 1-MCP (Rasori et al., 2002). This differential expression in which the *ETR1* homologue is constitutively expressed and the *ERS*-type receptor is up-regulated by ethylene has been observed also in passion fruit (Mita et al., 2002). Here, expression of *Pp-ERS2* is enhanced; an induction that is inhibited by the ethylene action inhibitor, 2,5-norbornadience (NBD). Therefore, in the fruit species examined, expression of the *ETR* gene family is not constant but at least one member is expressed constitutively, while at least one other is under regulation by ethylene. Further, as Bassett et al. (2002) observed, the *ETR1* transcripts are differentially processed and display differential expression in peach fruit thus adding a further degree of complexity to the regulation of the *ETR* gene family. The question then becomes whether such differential receptor expressions are common to other ethylene-regulated developmental processes.

Sex determination and receptor abundance. The expression of three ethylene receptor homologues, *CS-ETR1*, *CS-ETR2* and *CS-ERS*, has been examined in cucumber plants (Yamasaki et al., 2000) where exposure to ethylene has long been known to promote the formation of female flowers (Takahashi et al., 1983). Although the presence of *CS-ETR2* and *CS-ERS* was highest in gynoecious plants when compared with monoecious plants, and the expression of both these genes was increased in response to ethylene treatment, expression of *CS-ETR1* remained essentially constitutive during the development of female flowers showing that expression of this gene can be tissue or target selective.

Organ abscission and the expression of ethylene receptor genes. The expression of *eTAE1* has been compared during leaf and floral abscission of tomato (Zhou et al., 1996). In leaf abscission zones, expression decreased after 48 hours of ethylene treatment, but in zones of the flower, where expression of *eTAE1* was highest prior to ethylene treatment, no change occurred. Payton et al. (1996) determined that expression of the *tETR* (*NR*) gene increased specifically in the abscission zone of the flower 8 days after emasculation. No expression of *tETR* was detected in tissues on either side of the zone either at 0 days or 8 days. In further experiments on tomato floral abscission, expression of *LE-ETR1*, *LE-ETR2* and *NR* were shown not to change in relative intensity during floral abscission (Lashbrook et al., 1998). Such experiments serve to illustrate the specificity and abundance of receptor gene expressions to particular target cell types.

Using anti-sense technology to down-regulate the expression of *LE-ETR1* in tomato, Whitelaw et al. (2002) examined the phenotypes of the transformants. The most common was delayed abscission and a slightly reduced plant size. As may be predicted, fruit ripening was not affected nor was the expression of the *NR* gene, which is induced during normal fruit ripening (Wilkinson et al., 1995; Lashbrook et al., 1998; Nakatsuka et al., 1998; Hackett et al., 2000). Interestingly, seedlings did show a normal triple response, in common with the observation of Tieman et al. (2000) who characterised a severe ethylene response phenotype in anti-sensed *LE-ETR4* plants, indicating that *LE-ETR4* may play the dominant role in the gene family (see later discussion). Nevertheless, the delay of abscission by the down-regulation of a single member of the *ETR* family as observed by Whitelaw et al. (2002) is significant. The authors express the view that this may be the opposite of the result expected as one hypothesis is that a *decrease* in receptor abundance should make the tissue *more* sensitive to the hormone.

But as Lashbrook et al. (1998) have shown, there is more than one receptor expressed in the tomato flower abscission zone, suggesting that the sensitivity of the tissue could be determined by the composite expressions of more than one receptor, particularly as Whitelaw et al. (2002) observed that the expression of *LE-ETR2* and *LE-ETR3* (*NR*) was not affected by the down regulation of *LE-ETR1*.

A multi-receptor control may be common to abscission. In passion fruit, expression of the *ERS1* homologue *Pe-ERS2* increased during the formation of the separation layer, while expression of the *ETR* homologue remained constant (Mita et al., 2002). Finally, in peach, expression of the *ETR1* homologue *Pp-ETR1* does not change in the fruitlet or leaf abscission zones in response to ethylene treatment. However, the expression of a *ERS1* homologue, *Pp-ERS1*, is up-regulated by propylene treatment (an ethylene substitute) (Rasori et al., 2002) indicating again the possibility of a corporate control by multi-receptors in particular target cells.

Receptor expression during growth and senescence. Payton et al. (1996) examined expression of *tETR* during petal senescence in tomato and determined that the levels were high in the early senescent stages but were later down-regulated. Lashbrook et al. (1998) followed the expression of *LE-ETR1, LE-ETR2* and *NR* during senescence in tomato leaves and found that the *NR* gene displayed the highest expression of the three, but the levels did not change as senescence proceeded.

The abundance of ethylene receptors characterises elongation in the Type 3 cells of *Rumex palustris*. This semi-aquatic species shows a pronounced response to flooding by increased ethylene production, which, in association with gibberellic acid, causes petioles to elongate (Voesenek et al., 1993; see earlier discussion of target status in Chapter 5). To examine changes in the levels of ethylene receptors during the growth response, an *ERS* homologue was cloned, designated *RP-ERS1*, and its expression determined during flooding-induced growth (Vriezen et al., 1997). Northern analysis revealed a significant increase in flooded plants after exposure to elevated levels of ethylene (although ethylene inhibitors were not used to validate this induction). Expression of *RP-ERS1* was shown to increase also in response to elevated carbon dioxide levels and to low concentrations of oxygen – that is, all conditions found as a consequence of submergence.

Does receptor abundance confer a target status to the cell?

The appreciation that binding of ethylene to its receptor may function to relieve a repression of ethylene responses requires a reversal in our thinking regarding the significance of (at least) receptor abundance. If the genetic evidence proposed by Hua and Meyerowitz (1998) and others is supported by subsequent biochemical analysis, then a tissue expressing a high number of receptors would be expected to be less sensitive to a given concentration of ethylene – that is, a higher concentration of the hormone would be necessary to evoke the response. Conversely, a tissue that is very responsive to the hormone should express few receptors.

However, an examination of expression studies to date has not borne out this apparently simple relationship, although there are many possible reasons for this.

1. The sole reliance on the abundance of receptors in a particular tissue as determined by expression of the *ETR*-like gene family may not represent the true pool of functional receptors. Similar studies using antibodies to the receptor proteins may add more certainty to such correlations, but they will still not tell us if the presence of a receptor protein is a direct measure of receptor function.
2. We consider it to be highly significant that the notion of receptor compensation has arisen from transgenic plant studies. In tomato, the down-regulation of the *NR* (*LE-ETR3*) gene using anti-sense approaches does not produce plants with a significantly altered phenotype, but the up-regulation of *LE-ETR4* presumably restores the appropriate degree of tissue sensitivity (Tieman et al., 2000). For this to occur, the degree of tissue perception at the level of the receptor must be monitored in any target cell by an as yet unknown mechanism. The identification and dissection of such a mechanism will, in turn, be crucial to our understanding of the target cell concept in plants.
3. Associated with the concept of functional compensation outlined in (2) is the emerging view that there is a hierarchy in the *ETR* receptor family in (at least) tomato and *Arabidopsis*. When the expression of *LE-ETR4* was reduced in tomato using anti-sense approaches, the transformants exhibited an extreme constitutive ethylene phenotype that included epinasty of leaves and stems, enhanced floral senescence, and accelerated fruit ripening (Tieman et al., 2000). This severe phenotype is not observed when expression of the *NR* gene is similarly down-regulated or if any one of the five members of the *ETR* family of *Arabidopsis* is down-regulated. It is, however, reminiscent of the triple or quadruple crosses of the loss-of-function mutants of *Arabidopsis* (Hua and Meyerowitz, 1998), except that the responses in tomato could be reversed by the application of silver thiosulphate or 1-MCP (either of which will block the ethylene response). This supports the general view that these changes are due to the transformants being extremely ethylene insensitive, which, in turn, suggests that *LE-ETR4* is the potent negative repressor of the ethylene response in tomato, and that changes in the expression of this gene lead to the major changes in ethylene sensitivity.

A similar hierarchy may also operate in *Arabidopsis*. Using the loss-of-function mutants of *ETR1, etr1-7*, Cancel and Larsen (2002) showed that these plants displayed an extreme sensitivity to ethylene (and propylene). The responses could be inhibited by pretreatment with the ethylene action inhibitor, silver nitrate, and were not caused by an overproduction of the hormone. The *etr1* single loss-of-function mutant (*etr1-7*) was the only mutant that was shown to be highly responsive to ethylene. It was not observed in the other single loss-of-function mutants tested (*ein4-4* and *etr2-3*), suggesting a dominant role for *ETR1*. Unlike the observations with *LE-ETR4,* lack of the highly sensitive phenotype in the other loss-of-function mutants tested was not through a compensatory increase in *ETR1* expression, since other loss-of-function mutants did not show this. Rather, the results indicate that *ETR1* may be unique in being a very potent regulator of CTR activity. In support of this, two hybrid studies have shown previously that the ETR1-CTR association is much stronger than the ETR2-CTR or ERS1-CTR associations (Clark et al., 1998; Cancel and Larson, 2002). Further, a triple loss-of-function mutant, *etr2-3;ein4-4;ers2-3*, exhibited a triple growth response that could be wholly abolished by treatment with silver nitrate, and since this mutant is not an ethylene overproducer, Cancel and Larson (2002) conclude that the class II ethylene receptors (ETR2, ERS2, EIN4) may not be significant regulators of CTR – rather the prime regulator is ETR1.

It is clear from other (non-receptor) studies that responsiveness to ethylene in a target tissue is a difficult concept to define biochemically. For example, the Type 2 leaf abscission zone cells of the bean, *Phaseolus vulgaris,* are widely regarded to be sensitive to ethylene such that when the hormone is perceived, all the cells in close proximity to the zone respond by the production of specific wall hydrolases (as well as other enzymes) but the specific cell-to-cell separation event takes place only in Type 2 cells. Actually, the reverse may be the case, with the Type 2 cells being much *less* sensitive than their immediate neighbours (see Chapters 5 and 6). Using induction of guaiacol oxidase activity as a marker for the different cell types, McManus (1994) showed that both the zone and the surrounding pulvinus and petiole tissue all respond to ethylene by an increase in total enzyme activity, but it was a different spectrum of peroxidase isoforms that was induced in each tissue. The stage of development is also critical. Immature fruit, for example, are not responsive to ethylene in terms of being induced to ripen, nor can the hormone induce expression of the *NR* gene in immature fruit of tomato (Wilkinson et al., 1995). However, immature fruit do show ethylene responses in terms of a sequence of new gene expressions (Lincoln et al., 1987).

Studies in which the expression and abundance of receptors are examined during different physiological events do underline the fact that the ethylene receptor gene family is under tight developmental control and it may be that a simple relationship between tissue sensitivity and the level of expression does exist. But other trends have also emerged. During fruit ripening, many species exhibit an up-regulation of at least one member of the gene family concomitant with a rise in ethylene production and, in many species, this increase in expression has been shown to be itself ethylene-dependent. During ripening of tomato, peach, passion fruit and muskmelon, *ERS*-like gene expression increases in parallel with an increase in ethylene production (Lashbrook et al., 1998; Hackett et al., 2000; Rasori et al., 2002; Mita et al., 2002; Sato-Nara et al., 1999). Such increases are also observed in *Rumex* in response to flooding (Vriezen et al., 1997). In contrast. during tomato, peach and passion fruit ripening, an *ETR1*-like homologue appears to be ethylene-independent such that the overall expression remains constitutive (Zhou et al., 1996; Rasori et al., 2002; Mita et al., 2002), although it should be noted that one *ETR1*-like homologue did increase in ripening muskmelon (*CM-ETR1;* Sato-Nara et al., 1999) and in mango fruit (*METR1;* Martinez et al., 2001).

Therefore, in addition to the developmental control of the expression of all genes in the *ETR* family, the endogenous formation of ethylene itself can also influence the expression of a subset of its own receptors. So far this has always been seen as an up-regulation of the receptor and (in theory) a down-regulation of the sensitivity of the target tissue to the hormone. But when seeking to understand the significance of receptor abundance and target cell sensitivity, we need to consider this aspect of ethylene responsiveness further before the role of receptors can be fully ascertained. What must be remembered is the close interaction that exists between the opposing growth responses to auxin and to ethylene of almost all plants and the control that each hormone exerts upon the biosynthesis of the other and also upon the rates of turnover and degradation of the other. These are interactions that must, therefore, accommodate the performances of all the auxin and ethylene receptors.

Hormone Action and the Relief of Repression

In the previous chapter, we considered the perception of auxin and ethylene in different target tissues. While these hormones were treated separately, the essential mode of action of ethylene and auxin is to relieve a pre-existing repression of response (see Figures 7.2 and 7.4). It is now clear that this mode of action of hormonal signals is widespread amongst plants. In this chapter we look at three further examples, the cytokinins, gibberellins and brassinosteroids, and again examine the evidence for perception of these signals via binding proteins or receptors in different target cells. We additionally examine the evidence that these developmental cues operate through the relief of pre-existing repressions of molecular responses in each cell type examined.

Cytokinin perception in the context of receptors and target cells

Two spectacular events in plant development are attributable to the action of cytokinins. The first, is the conversion of the cells of a callus culture into the organisational complexities of a shoot meristem. Cytokinins act not alone, but in concert with auxin, the ratio of one to the other being critical for optimal organ development. The second is the maintenance of the non-senescent state in specific tissues. Here, cytokinins act as repressors of cell death programmes in many target cell types, most notably those of the leafy tissues of herbaceous plants. Seemingly, there is a requirement for cytokinins primarily synthesised in the root meristems for the retention of metabolic function in the green shoot. Whenever the levels of cytokinins fall, either through excision of the leaf from the root source, or by competition from the more demanding and fast-growing younger leaves above, retention of photosynthetic capacity and metabolic function can be preserved (in most, but not all, circumstances) by the addition of one of the naturally occurring or synthetic cytokinins (see Chapter 4).

As the search for the auxin and for ethylene binding proteins intensified, and isolation of genetic probes for their receptors progressed, those working with cytokinins also sought evidence for similar mechanisms to explain the signal perception and response pathways for molecules that were more structurally akin to those of genomic and transcriptionally dependent DNA and RNA. In fact, because of their substituted purine structures, the natural cytokinins were once thought, wrongly as it turned out, to be breakdown products of nucleic acids, particularly tRNAs (see Chapter 2).

The search for the pathway of cytokinin perception has, initially, followed well-recognised biochemical approaches to identify cytokinin binding proteins. One of the best characterised of these was identified first by Brinegar and Fox (1985) in the embryos of wheat. The protein, designated CBP-1, was shown to be a homotrimeric protein consisting of three identical subunits of 54 kDa. The characteristics of this protein closely followed those of seed storage proteins of wheat in that it accumulated rapidly during grain filling, was localised in tissues surrounding the embryonic axis and had structural similarities with storage proteins of the vicilin type (Brinegar et al., 1985). CBP-1 had a low binding affinity for the isoprenoid cytokinins and some specificity for cytokinins bearing an N^6 aromatic side chain. Although there is good evidence that CBP-1 is a binding protein, more recently it has been proposed to act as a sequestering protein regulating the availability of free cytokinins to target tissues in the developing and subsequently germinating embryo. Kaminek et al. (2000) identified a naturally occuring, highly active cytokinin with an aromatic N^6-(3-hydroxybenzyl)adenosine side chain in wheat grains which may be one in which the concentration is regulated by binding to CBP-1. The identification of a similar CBP in oat grains with a higher affinity for N^6-benzyladenine (BA) when compared with zeatin indicates further that the sequestration of cytokinin in regulating embryo development and germination may be a widespread phenomenon (Kaminek et al., 2003).

The CBP-1 is a soluble protein of the cytoplasm, but it is evident that cytokinins can induce physiological changes when applied externally to plant cells. In mosses, for example, cytokinins play an important role in inducing bud formation (Saunders and Hepler, 1983) and 6-benzylaminopurine will stimulate Ca^{2+} influx with a K_m of 1 nM when added to protoplasts of moss protonema cells (Schumaker and Gizinski, 1993).

To determine if the cytokinin-binding proteins were membrane-localised, Brault et al. (1999) used an affinity probe comprising [9R] zeatin riboside conjugated to IgG from goat, and then sought any protein that complexed with anti-[9R] monoclonal antibodies. Binding proteins were detected in the membrane fraction of cultured cells of *Arabidopsis* and proteolysis confirmed that the binding component was proteinaceous. A range of biologically active cytokinins (zeatin, isopentenyladenine and isopentenyladenosine) competed for binding with the zeatin riboside, as did the anti-[9R]Z monoclonal antibody. Most significantly, Brault et al. (1999) showed that binding of the conjugate was correlated with the exponential growth phase and maximal cell cycling rate of cultured cells of *Arabidopsis* (a known cytokinin target state), with maximal binding at the mid-point of the exponential growth phase (at 3 days) and decreased binding again as the cell number reached a plateau.

Although cytokinin binding proteins have been identified, a recognisable mechanism for the transduction of the binding event through to intra- or intercellular signalling is yet to be fully identified. Since characterisation of the ethylene receptor owes much to the use of mutants of *Arabidopsis*, particularly the early triple response mutants that led to the discovery of the *ETR* gene family, early attempts to identify cytokinin receptors also utilised mutants of *Arabidopsis*. Indeed, some were found with altered responses to cytokinin. For example, it has been long known that low concentrations of cytokinins will induce the triple response in seedlings of *Arabidopsis* through cytokinin-induced ethylene production. Thus by screening for mutants that did not display the triple response in response to cytokinin but did in response to ethylene, researchers identified the *cin* series of mutants (Vogel et al., 1998). The *cin5* mutant was shown to have a lesion in the ACC synthase gene, *AT-ACS5*, thus disrupting ethylene production and induction by cytokinin (Vogel et al., 1998). Other *cin* mutants have been characterised since, but none of these appear to be associated with cytokinin perception. (For a review of these early approaches using mutants, the reader is referred to the review of Kakimoto, 1998.)

Identification and characterisation of cytokinin receptors
and their downstream elements

It was the use of activation tagging that eventually identified the first putative cytokinin receptor. For this, Kakimoto (1996) used a T-DNA tag driven by a tetrameric CAMV35S enhancer construct and transformed callus tissue derived from hypocotyl segments of *Arabidopsis thaliana*. Five mutant lines (designated *cki1-1* to *cki1-4* and *cki2*), were obtained, and callus derived from the four *cki1* mutants displayed cytokinin responses such as rapid proliferation of callus, greening, and shoot formation without any added cytokinin – i.e., a constitutive cytokinin response. Isolation of the *CKI1* gene revealed that it had homology to histidine-kinase mediated two-component signalling systems of prokaryotes. Further, over-expression of *CKI1* in *Arabidopsis* conferred a constitutive cytokinin phenotype without any added cytokinin.

In a second approach, Inoue et al. (2001) produced callus from hypocotyl segments of EMS mutants of *Arabidopsis thaliana* and screened these lines for lack of response to cytokinins. A mutant line, designated *cytokinin response 1 (cre1-1)*, was identified in which added kinetin was ineffective at inducing callus proliferation, greening and shoot formation when compared with wild-type callus. The mutant callus line was also shown to be insensitive to other cytokinins including *trans*-zeatin, isopentenyl adenine, benzyladenine and the phenylurea-type synthetic cytokinin, thidiazuron. A second *cre* mutant, *cre1-2*, was generated using T-DNA insertional mutagenesis, and in seedlings of both *cre1-1* and *cre1-2*, the

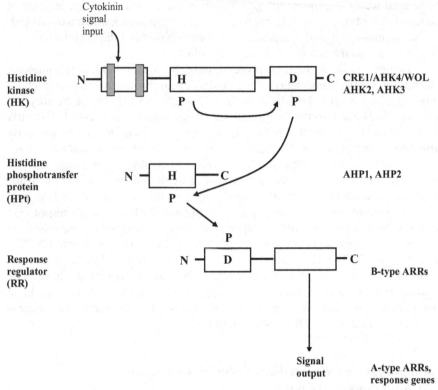

Figure 8.1. Overview of the His → Asp phospho-relay system that forms the basis of cytokinin signalling in *Arabidopsis* (see text for details; modified from Hwang et al., 2002).

addition of benzyladenine failed to inhibit root growth, although ethylene (as added ACC), auxin (added as IAA) and ABA were effective.

Using mapped-based cloning, the *CRE1* gene was isolated from a locus harbouring a putative histidine kinase of *Arabidopsis*, and shown to restore cytokinin sensitivity when transformed into callus derived from the *cre1-1* mutant. Sequencing of the *CRE1* gene revealed that it was homologous to other two-component signalling components. In addition to the basic two-component system outlined in our discussion of ETR, many bacteria and eukaryotes contain two component systems that typically have a transmembrane domain attached to the histidine kinase domain and a C-terminal receiver domain (a hybrid-type), which phosphorylates a soluble histidine-containing (phospho) transmitter (HPt). These HPt proteins then phosphorylate the receiver domain of a second response regulator (RR) (designated the His → Asp phospho-relay; see Figure 8.1) (for a review, see Hwang et al. 2002). Translation of the *CRE1* gene sequence reveals two transmembrane domains in the N-terminal, a histidine kinase (HK) domain and two receiver domains on the response regulator, only one of which contains the critical aspartate residue that is phosphorylated by the histidine kinase.

To show that CRE1 does indeed function as a cytokinin receptor, Inoue et al., (2001) utilised the two-component Δ*sln1* mutant of yeast. In the wild-type strain the membrane-localised HK protein, SLN1, signals to the phospho-transmitter (HPt) protein YPD1 to repress activity of the response regulator (RR) protein, SSK1, and so repress downstream MAP kinase signalling. In the mutant, the SLN1 protein is non-functional, and so SSK1 is free to initiate MAP kinase signalling and produce a lethal phenotype. However, a galactose-induced protein PTP2 can inhibit the MAPK pathway, so the mutant can be maintained viable by the addition of galactose, and Inoue et al. (2001) showed that transformation of the Δ*sln1* mutant with *CRE1* conferred a cytokinin-dependent, galactose-independent, viable phenotype to the transformants. They showed that the cytokinins, *trans*-zeatin, 2-isopentenyl, 6-benzyl aminopurine and thidiazuron were all effective. However, in Δ*ypd1* mutants (i.e., those with a defective soluble phospho-transmitter Hpt protein), no cytokinin-dependent relief of galactose-dependence could be demonstrated suggesting that CRE1 could only substitute for the receptor HK domain and not the downstream HPt protein. Finally, Inoue et al. (2001) showed that the original *cre1* mutant with a G467 → D substitution near the critical His residue could not rescue the Δ*sln1* mutant from galactose-dependence, nor could mutated CRE1 proteins in which the critical His residue was substituted in the HKI domain (H459 → Q) or the aspartate residue within the receiver domain (D973 → N).

Using computer searching techniques, Suzuki et al. (2001b) surveyed the published genome sequence of *Arabidopsis* to identify other possible cytokinin receptors in addition to the CKI1 of Kakimoto (1996). At least eleven two-component-like genes were found, of which five were the *ETR*-like and *ERS*-like ethylene receptors, plus *CKI1* and *CKI2* and one osmosensor, *AtH1* (to make a total of eight). The three remaining genes were *AtHK2, AtHK3* and *AtHK4* (identical to *CRE1*). These had been cloned earlier (see pers. comm. in Imamura et al., 1999). Using a similar approach as Inoue et al. (2001), Suzuki et al. (2001b) transformed *AtHK4* into the triple mutant of yeast, Δ*phk1/2/3* which lacks the HK proteins, Phk1, 2 and 3, and again showed that transformation of the *AtHK4* gene alone was not sufficient to correct the mutant phenotype – added cytokinin was also needed.

In another mutant of *Arabidopsis, wooden leg (wol)*, there is an absence of specific cell divisions in the root and lower hypocotyl during the late stages of embryogenesis (Scheres et al., 1995). Specifically, very few vascular initials develop because cell division ceases soon after the torpedo stage and only the protoxylem forms. In the seedling root, therefore, there is a narrow cylinder of vascular tissue comprising protoxylem cells, but with no phloem. The *WOL* gene has been mapped (Mahonen et al., 2000) and shown to be allelic to the *CRE/AtHK4* genes (see Hwang et al., 2002). However, the lesion in the *wol* mutant was a Thr301 → Ile substitution in a 270 residue proposed extracellular loop between the two transmembrane domains at the N-terminal of the protein. When the mutant *WOL* gene was transformed into the Δ*phk1/2/3* mutant, the normal phenotype was *not* restored even when cytokinin was added. Further, and most significantly, no binding of ^{3}H-isopentenyadenosine could be demonstrated in cells of *S. pombe* when transformed with the mutant *WOL* gene.

Together, these different studies show that the CRE1/AtHK4/WOL protein is a HK protein that can bind cytokinin such that binding of the hormone can initiate signalling to HPt proteins – this protein can therefore presently be considered as a cytokinin receptor. The next hurdle is to identify other potential components of cytokinin signalling pathways.

Response elements and modulating the cytokinin input

In the two-component regulators characterised in other species that are similar to CRE1/AtHK4/WOL, the membrane-associated HK protein signals to soluble histidine-containing phospho-transmitters (the HPts). In their assessment of the AtHK4 protein as a cytokinin receptor, Suzuki et al. (2001b) co-introduced two putative HPt proteins of *Arabidopsis*, AHP2 and AHP5, with AtHK4, into an *E. coli* mutant lacking its membrane-associated HK protein, RcsC (i.e., the AtHK4 homologue). They showed that the AHP proteins could, in a cytokinin-dependent manner, compete out the activation of a reporter gene construct, *cps:LacZ* by the endogenously occurring Hpt protein, suggesting that AtHK4 is signalling to AHP2 or AHP5 in response to cytokinins.

Concurrently with the identification of the cytokinin receptor CRE1/AtHK4/WOL, and the Hpt proteins AHP2 and AHP5, the genome of *Arabidopsis* was scrutinised for genes encoding proteins with identity to two-component response regulators (RRs). A summary of these early experiments is reviewed in Hwang et al. (2002) who show that the *RR* genes of *Arabidopsis* (the *ARRs*) could be divided into two major classes, the A- and B-types. The A-type genes contain a receiver domain, while the B-type genes contain a receiver domain together with a DNA (myb-like) binding domain. Several research groups have shown that the transcription of a number of the A-type *ARRs* (*ARR3, ARR4, ARR5, ARR6, ARR7, ARR8,* and *ARR9*) are induced by cytokinin in the presence of cyclohex-imide (Imamura et al., 1998; Brandstatter and Kieber, 1998; Urao et al., 1998; Imamura et al., 1999; D'Agostino et al., 2000), but transcription of the B-type *ARRs* is *not* induced by cytokinin (Imamura et al., 1999; Kiba et al., 1999). In terms of further dissection of signalling, Suzuki et al. (2001a) demonstrated, using the yeast two-hybrid assay, that the B-type RRs, ARR1 and ARR10, directly interacted with the HPt proteins, AHP1, AHP2 and AHP3, and that dephospho-rylation of AHP2 was dependent on the formation of a complex with the ARR proteins. However, as transcription of these activators was not up-regulated by endogenous cytokinin, the role of the B-type ARR proteins in cytokinin signalling was at this stage, therefore, uncertain (Kiba et al., 1999).

Hwang and Sheen (2001) then undertook a series of elegant experiments us-ing a reporter system based upon a 2.4 kb promoter sequence of *ARR6* (an A-type, cytokinin-responsive ARR) fused to the luciferase (LUC) reporter gene to give a construct designated *ARR6p:LUC*. When transformed into leaf mes-ophyll protoplasts of *Arabidopsis,* luciferase activity (and thus the induction of *ARR6* transcription), was induced only by the active cytokinins, *trans*-zeatin, 2-isopentenyladenine (2-IP) and 6-benzyladenine and not by IAA or ABA.

Having established the specificity of cytokinin, they proceeded to examine the role of CKI1, CRE1, the A-type, B-type ARRs and the AHPs in the cytokinin signalling pathway.

Initial experiments showed that if *CKI1* was co-infected with the *ARR6p:LUC* construct, activation of the *ARR6* promoter occurred in the absence of cytokinin, while applied cytokinin was required to induce *ARR6* transcription when cells were transformed with *CRE1/AHK4/WOL, AHK2,* and *AHK3.* Therefore, these experiments confirmed the earlier findings of Kakimoto (1996) which showed that CKI1 confers a constitutive cytokinin phenotype, while added cytokinin is required to activate the CRE1/AHK4/WOL, AHK2, and AHK3 proteins (Inoue et al., 2001; Suzuki et al., 2001b).

For the investigation of the role of soluble phospho-transmitter AHP proteins in cytokinin signalling, Hwang and Sheen (2001), using a *ARR6p:GUS* reporter system, showed that AHP1, AHP2 or AHP5 could not induce the transcription of *ARR6* with or without added *trans*-zeatin. They did show, using a AHP1-GFP fusion protein, that in the absence of added cytokinin, the fusion remained in the cytosol, but when cytokinin was added, the fusion protein was directed to the nucleus. This cytokinin-dependent targeting to the nucleus was also found for AHP2 but not for AHP5 (the significance of the observation of the binding of AHP5 to AHK4 remains to be explained). Suzuki et al. (2001b) showed that the AHP proteins could bind to the AHK4 protein and, in a separate study, that AHP1 physically interacts with the B-type ARR, ARR1, and that AHP2 interacts with the B-type ARR1, ARR2 and ARR10, but not the A-type ARR3 or ARR4.

With this information, it is reasonable to assume, therefore, that once phosphorylated, AHP1 and AHP2 translocate into the nucleus where they can interact with (at least) the B-type ARRs, ARR1, ARR2 and ARR10 (Suzuki et al., 2001a). The next issue to address, therefore, was how do the A-type and B-type ARR proteins interact?

Co-infection of the A-type (cytokinin-inducible ARR) *ARR4, ARR5, ARR6* and *ARR7* with the *ARR6p:LUC* construct, repressed the *trans*-zeatin induced expression of *ARR6.* However, *ARR1, ARR2* and *ARR10* (the B-type cytokinin-independent ARRs) induced *ARR6* expression without added cytokinin, and when cytokinin was added, a super stimulation of *ARR6* was observed (Hwang and Sheen, 2001). Therefore, the authors concluded that although not induced by cytokinins, the B-type ARRs, ARR1, ARR2 and ARR10 can induce the transcription of the A-type ARRs.

A model for cytokinin signalling in plants

After consideration of all the evidence for the putative cytokinin receptor and the downstream signalling components, Hwang and Sheen (2001) proposed a model to describe cytokinin signalling in (at least) *Arabidopsis.* Extracellular cytokinin is perceived by the two-component proteins CRE1/AHK4/WOL, AHK2 or AHK3, which then signal by phosphorylation of the histidine phospho-transmitters (Hpt) proteins, AHP1 and AHP2. In the phosphorylated state, AHP1 and AHP2

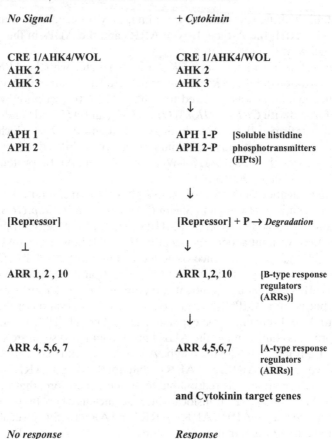

Figure 8.2. Conceptual representation of cytokinin signalling mediated via the HK proteins CRE1/AHK4/WOL in *Arabidopsis thaliana* (see text for details). ⊥ repression of the signalling function.

translocate to the nucleus and mediate a phosphorylation-based derepression of the transcription of the B-type response regulators, *ARR1, ARR2* and *ARR10*. The ARR1, ARR2 and ARR10 proteins then induce transcription of their target genes (the cytokinin-induced gene programme for that particular target tissue), including the A-type response regulators, *ARR4, ARR5, ARR6* and *ARR7* (see Figure 8.2 for a conceptual overview of cytokinin signalling in *Arabidopsis*). These proteins then serve as feed-back repressors of the programme of transcription so providing a mechanism to attenuate further cytokinin signalling. This model explains the observation that the B-type RR genes are themselves not induced by cytokinin – rather, the action of cytokinin is to relieve the repression of these response regulators.

This crucial role of the ARR1 (with ARR2 and ARR10) as the master regulators of the cytokinin response has been confirmed by Sakai et al. (2001). They showed that the level of ARR1 determines the sensitivity of the tissue to cytokinin; the higher the level of ARR1, the greater the level of response to cytokinin. Using a glucocorticoid-based gene expression system, Sakai et al. (2001) also demonstrated that transcription of *ARR1* could directly induce the transcription of *ARR6* even in the presence of cycloheximide. On the basis of these results, they concluded that the role of the cytokinins must be to relieve the suppression of ARR1 function (and the other B-type ARRs) so that these proteins are free to transactivate the expression of the A-type *RR* genes.

The genome of *Arabidopsis* has 10 members in the A-type ARR family, and 11 members in the B-type ARR family (Hwang et al., 2002; Tajima et al., 2004). Not all of A-type or B-type *ARR* genes have been assessed in detail. However, some studies do suggest that multi-members of the A-type and B-type *ARR* gene families provide sufficient scope to explain differential tissue-specific responses to the hormone and so determine the target status of any cytokinin-responsive tissue. For example, D'Agostino et al. (2000), using GUS reporter gene constructs found that *ARR5* is high in shoot and root meristems (rapidly dividing tissues), at the junction of the pedicel and the silique and in the central portion of mature roots. Sweere et al. (2001) have shown that *ARR4* is expressed in stems, leaves, flowers and roots, but in a protein blot analysis, ARR4 accumulation was not observed in the roots. For the B-type ARRs, Tajima et al. (2004) determined that, in *Arabidopsis thaliana,* the expression of certain members of this family (now divided into three subfamilies) is developmentally regulated with many expressed ubiquitously in all the tissues examined, while others have a more restricted distribution. The differential expression of these genes, and their potential post-transcriptional processing, can add a further level of modulation by which cytokinins may regulate specific physiological responses in a wide range of specific target tissues.

Gibberellin perception and the search for receptors

Unlike the putative receptors described for cytokinins, auxins and ethylene, there is as yet no cloned and sequenced gene encoding, unequivocally, a receptor for the gibberellin group of hormones. Nonetheless, a molecular basis of GA perception and response is emerging and, in common with that of auxin, ethylene and cytokinin, the role of the hormone appears to be to relieve a repression of specific programmes of gene expression, with distinctive developmental outcomes in specific target tissues.

Summary of the candidates for gibberellin binding proteins and receptors

A review of some of the key studies suggests that GA can be perceived extra-cellularly. By cross-linking the biologically active gibberellin GA_4 to Sepharose 6B beads such that these complexes could not traverse the cell membrane, Hooley et al. (1991) showed that this construct could induce α-amylase secretion from protoplasts of aleurones of the wild oat, *Avena fatua* L. Using a similar experimental system but a different approach, Hooley et al. (1992) made antibodies to monoclonal antibodies made to GA_4. These anti-idiotypic antibodies should recognise what GA_4 recognises, but be unable to cross the plasma membrane. That the antibodies antagonised the induction of α-amylase by GA_4-treated protoplasts of wild oat (*Avena fatua* L.) provided further evidence for GA perception external to the cell. This notion was supported by Gilroy and Jones (1994) who microinjected barley aleurone protoplasts with either the active gibberellin GA_3 or the inactive GA_8. Both GAs were ineffective at inducing α-amylase. Only when GA_3 was supplied in the external medium and made contact with the plasma membrane was induction of α-amylase achieved.

While these studies suggest that perception of GA is on the external surface of the cell, these approaches did not identify any protein candidates. However, Hooley et al. (1993) photoaffinity-labelled a 60 kDa protein and, again using isolated aleurone protoplasts of wild oat, showed that adding the biologically active GA_4 could compete with the labelling. Preliminary subfractionation studies revealed that the labelled peptide was present in the microsomal membrane-containing fraction, but was not a soluble protein. In later studies, Lovegrove et al. (1998) photoaffinity-labelled two polypeptides of 68 kDa and 18 kDa in plasma membrane preparations from cells of oat (*Avena sativa*) and from both wild-type and GA-insensitive mutants of sweet pea (*Lathyrus odoratus*) and *Arabidopsis thaliana*. The labelled peptides competed with the biologically active GA_4 and GA_1 while the biologically inactive GA_{34} did not. Importantly, work with the pea mutants showed that the semi-dominant semi-dwarf mutant displayed reduced binding.

With the demonstration of possible GA receptor proteins on the plasma membrane, the identification of downstream elements became important. Jones et al. (1998b) used Mas7, a cationic amphiphilic tetradecapeptide, to evaluate the role of heterotrimeric G proteins in the induction of α-amylase activity. They showed that Mas7 could induce α-amylase in aleurone protoplasts of wild oat in a similar manner to GA_1. Further, using an α-amylase promoter:GUS reporter construct they also showed that Mas7 could induce GUS activity, and that by adding GDP-β-S (an inhibitor of GDP/GTP exchange by G proteins), the GA_1 induction of the α-amylase:GUS promoter was inhibited. G proteins have been shown to operate on the cytoplasmic face of the plasma membrane in other eukaryotes, so these results are suggestive that heterotrimeric G proteins could be involved in signalling between perception of GA at the external membrane surface and the regulation of GA-induced gene expression in the cytosol. However, it is the dissection of events in the nucleus and the role of GA in relieving repression of GA-inducible programmes of gene expression that has attracted more recent research interest.

GA perception and signalling

The power of using mutants of *Arabidopsis* to determine hormone action has been applied usefully to GA signalling, with two broad groups identified: the GA-insensitive dwarfs and the constitutive GA-response mutants. The GA-insensitive dwarfs resemble GA-deficient mutants but are not rescued by added GA. In contrast, the GA constitutive response mutants all appear as if they have been exposed to GA (for example, they all have elongated stems) in the absence of any treatment with the hormone. These latter mutants show resistance to inhibitors of GA biosynthesis, so demonstrating that there is additionally a GA-independent activation of GA responses.

We do not wish to describe all of these mutants here, so the reader is referred to the review of Sun (2000). However, certain key studies are described as they contribute to the emerging view of the molecular mode of signalling of GA.

We begin with the study by Peng et al. (1997) who examined the *gai* mutant, described previously by Peng and Harberd (1993), and showed that it had reduced responsiveness to GA. They then cloned the *GAI* gene and a closely related *GRS* gene, and showed that the *gai* mutant differed from the wild-type *GAI* by the deletion of 17 amino acids from the N-terminal protein sequence. The GAI protein also possessed a nuclear localisation sequence, a *LXXLL* motif, and had some similarity with the *VHIID* domain family, suggesting that these proteins are transcriptional co-activators. Significantly, GAI was shown to be a repressor of GA responses, but the addition of an active GA released this repression. The mutant repressor *gai* is, however, resistant to the effects of added GA and repression is not relieved.

Using a slightly different approach, Silverstone et al. (1997) screened a series of recessive mutants for their ability to suppress a number of GA-associated growth defects, including stem elongation, flowering time and leaf abaxial trichome initiation that were apparent in the GA biosynthetic mutant, *ga1-3*. The *ga1-3* mutant had been shown by Sun and Kamiya (1994) to have a lesion in the enzyme, *ent*-kaurene synthetase A (copalyl diphosphate synthase; GA1), which catalyses the first committed step in the GA biosynthetic pathway, the conversion of geranyl-geranyldiphosphate to copapyl diphosphate. The endogenous levels of GA are, therefore, very low in the *ga1-1* mutant and the plants are severely dwarfed (Sun and Kamiya, 1994). Silverstone and colleagues reasoned that any mutant alleles that repressed this phenotype must be due to a GA perception and/or signalling that occurs despite the very low levels of endogenous hormone. One mutant, *rga* (for *repressor of ga1*), was identified, suggesting that the *RGA* gene could encode a negative regulator of GA signalling.

On cloning and sequencing the *RGA* gene, Silverstone et al. (1998) established that the RGA protein and the GAI protein (identified by Peng et al., 1997) show a high degree of identity. Both are members of the VHIID family, with nuclear localisation sequences, a serine/threonine rich domain, leucine heptad repeats and an LXXLL motif characteristic of a transcriptional regulator. The RGA protein was shown to be targeted to the nucleus and to share a common N-terminal DELLA sequence domain with GAI. These findings provided an impetus to research on GA transduction mechanisms.

The role of the DELLA protein in regulating the GA response

GAI and RGA belong to the plant-specific GRAS gene superfamily of regulatory proteins, but those members involved in GA signalling all belong to the DELLA subfamily, in which the proteins contain an acidic N-terminal with

Asp(D)-Glu(E)-Leu(L)-Leu(L)-Ala(A) as the first five amino acids. The *gai* mutants had been shown earlier to arise from a deletion of 17 amino acids in the DELLA domain. Significantly, Ikeda et al. (2001), working with the *SLR1* gene of rice (a member of the DELLA subfamily of GRAS), showed that a GA-insensitive dwarf could be created by the transformation of wild-type rice with a mutated version of the *SLR1* gene by the deletion of 17 amino acids of the N-terminus affecting the DELLA region. Wen and Chang (2002), working with another member of the DELLA family, *RGA-LIKE1 (RGL1)*, could create a GA-insensitive mutant displaying a dominant dwarf phenotype by the transformation of wild-type *Arabidopsis* with a CAMV35S promoter driven *RGL1* gene also mutated by the deletion of the 17 N-terminal amino acids. None of the mutant lines could be rescued by the GA_3 treatment, suggesting that the function of RGL1 was to act as a negative regulator of the GA response and that the DELLA domain is a critical mediator of this response.

The mechanism by which the DELLA motif confers repression of GA response has been investigated by Dill et al. (2001) who transformed wild-type *Arabidopsis* with an *RGA* gene mutated by the deletion of 17 N-terminal amino acids within the DELLA domain (this creates the *gai-1* mutant isolated by Peng et al., 1997). The transformants displayed a GA-unresponsive severe dwarf phenotype – i.e., the mutant could not be rescued by the addition of GA. As well, the mutant protein was resistant to degradation by GA, while the wild-type protein was degraded. Dill et al. (2001) therefore proposed that mutation in the DELLA domain stabilised the RGA protein and so, regardless of the endogenous concentration of GA, the protein became a constitutive repressor of GA signalling. Further, GA might play a role in regulating the degradation of these repressor proteins. Silverstone et al. (2001) have noted that levels of the RGA protein are indeed lower in response to added GA.

The significance of GA and DELLA protein degradation (and subsequent GA signalling) has been investigated directly with the SLN1 protein in barley, a DELLA protein that is destabilised by GA treatment. Fu et al. (2002) showed that proteasome-mediated protein degradation is necessary for GA-mediated destabilisation of SLN1 and for the classic GA responses in barley such as α-amylase induction in the aleurone.

This proposed role for the proteasome and ubiquitin-mediated turnover of the DELLA protein repressors has been studied further in an elegant series of experiments by Sasaki et al. (2003) using a GA-insensitive dwarf mutant of rice, *gid2,* in which mutations in this ubiquitin-mediated pathway conferred GA insensitivity. They showed that the GID2 protein was a putative F-box protein, and this protein could bind to another, the rice *skp1* homolog, and thus create part of the SCF complex, one of the protein complexes in the ubiquitin-directed proteasome-mediated degradation pathway. They followed the accumulation of the SLR1 protein, a DELLA protein GA signalling repressor, the disappearance of which had been shown previously to be important for GA signalling (Itoh et al. 2002), and observed that SLR1 from rice accumulated in the *gid2* genetic background when compared with the wild-type, and that added GA did not degrade the protein in this *gid2* background. However, the protein was rapidly degraded

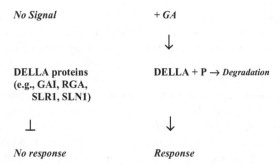

Figure 8.3. Conceptual representation of gibberellin signalling mediated via functional DELLA proteins (see text for details). ⊥ repression of the signalling function.

after the addition of GA in the wild-type background. Further, the SLR1 protein was found to accumulate as two proteins in the *gid2* background; and through the use of phosphatase treatment, SLR1 was shown to be phosphorylated in the *gid2* background. Reasoning that GA may be involved in this phosphorylation, the authors treated both wild-type and *gid2* plants with uniconazol (an inhibitor of GA biosynthesis) and showed that SLR1 was phosphorylated in the wild-type background and then disappeared after GA treatment. In contrast, more phosphorylated SLR1 protein was found in the *gid2* mutant background, and this increased in response to GA.

Sasaki et al. (2003) concluded, therefore, that GA promotes the phosphorylation of SLR1 which then marks it for ubiquitin-mediated protein degradation. With the removal of the repressor protein, the transcriptional activators of GA-induced gene expression are thus free to mediate gene expression. By extension, the observed mutations in these repressor proteins must interrupt this GA-mediated turnover, hence these mutations can manifest as GA-unresponsive with a dwarf phenotype irrespective of the internal concentration of GA. As many of these mutations can be created with the deletion of the DELLA motif, phosphorylation of this motif has been proposed as the initiator of GA-mediated degradation. See Figure 8.3 for a conceptual representation of GA signalling in plants.

Elucidation of the key role of the DELLA proteins in GA signalling is now being extended to many physiological processes in plants (e.g., Cheng et al., 2004), but nothing is yet known of how GA induces phosphorylation of this protein group, so these steps in the pathway are yet to be determined. Many studies have shown that these proteins are nuclear-localised (e.g., Silverstone et al., 1998; Gubler et al., 2002; Itoh et al., 2002; Wen and Chang, 2002) and that GA degrades the protein in the nucleus (e.g., Gubler et al., 2002; Itoh et al., 2002). Some molecular targets for derepression have been identified. For example, in barley aleurone cells, GA has been shown to induce the degradation of the negative regulator SLN1 while up-regulating the transcription of the GA-regulated transcriptional activator of α-amylase gene expression *GAMYB* (Gubler et al., 2002). These workers also demonstrated that GAMYB protein accumulated after the GA-induced decline of the SLN1 protein.

The GA-induced relief of repression of GA-mediated gene expression has parallels to the IAA-induced, proteasome-dependent degradation of the Aux/IAA repressor proteins. In this regard, therefore, there may be significance in a recent study showing that auxin can promote root growth in *Arabidopsis*, probably by modulating the gibberellin response (Fu and Harberd, 2003). In this study, added GA enhanced root growth in the *ga1-3* mutant of *Arabidopsis* which had shorter roots than those of the wild-type. As discussed earlier, the *GA1* gene encodes the first enzyme in the GA biosynthetic pathway, and so *ga1-3* mutants are GA-deficient, but when GA is supplied, root growth is resumed. Also, the DELLA proteins GAI and RGA were shown to cooperate to mediate GA-induced root growth. The role for auxin was established when it was shown that GA could not restore growth of roots of *ga1-1* mutants which had been decapitated (and therefore lacked the source of auxin from the shoot apex), but if auxin was added to the cut surface (and then transported), then GA could again restore root growth. Fu and Harberd (2003) then examined the disappearance of an RGA promoter driven RGA:GFP reporter gene construct. Normally, added GA mediates the disappearance of RGA, but if the source of auxin from the shoot apex was also removed, the added GA failed to induce the disappearance of the RGA:GFP reporter construct. However, application of IAA again led to the disappearance of the RGA protein in response to added GA. Thus auxin, transported in a polar fashion from the shoot apex, was necessary for the GA-induced reduction in RGA concentration, and for the relief of the suppressed root growth. The role of transported auxin was then addressed more directly using 1-*N*-naphthylphthalamic acid (NPA), an inhibitor of auxin efflux and the *AtPIN1* gene, that codes for an auxin efflux protein. As expected, NPA inhibited GA-restored growth in the *ga1-1* mutant, and also decreased the GA-induced disappearance of the GFP:RGA protein. Further, transgenic plants in which *AtPIN1* expression was knocked out using RNAi constructs also showed that GA treatment did not mediate the disappearance of the GFP:RGA protein. Together, these experiments underline the importance of polar transport of auxin to the regulation of RGA disappearance and subsequent root growth. Finally, Fu and Harberd (1993) investigated the mechanism by which this control of RGA protein stabilization might occur. Using the *axr1-12* mutation that has an attenuated proteasome mediated response, they found that the GFP-GA protein was resistant to degradation in this mutant background when compared with the *AXR1* background. These results suggest that there are tantalising similarities between the role of the proteasome in removing the repressors of IAA (the Aux/IAA proteins) or GA (the DELLA domain proteins) responses. Probably these pathways may be shared (see Chapter 9 for a more detailed overview of hormonal cross-talk).

Perception of the brassinosteroids

The brassinosteroids are becoming of increasing interest as molecular regulators of plant development, so we examine how brassinosteroid perception might be interpreted from the target cell viewpoint. We continue the theme that the central signalling role for a hormone is to relieve a repression in a particular programme of gene expressions.

After the identification of the brassinolide insensitive mutant *bri1* by Clouse et al. (1996; see Chapter 2), Li and Chory (1997) isolated 18 further mutants of *Arabidopsis* that were all allelic to *bri1*. To do this, they first screened 80,000 EMS-generated M_2 mutants for phenotypes that resembled the *det2* or *cpd* brassinolide biosynthesis mutants described by Li et al. (1996) and Szekeres et al. (1996). From 200 found, only 18 could not be rescued by the addition of 1 μM brassinolide, and only one (*bin1-1*) produced viable seed, the other 17 lines being male-sterile. Subsequent genetic crosses determined that *bin1-1* was allelic to *bri1* and so the Li and Chory mutant series was re-designated *bri1-101* to *bri1-118*. On sequencing, *BRI1* was found to be a member of the receptor-like kinase family in plants, with 25 extracellular leucine-rich regions (LRRS), and a more detailed description of the characterisation of BRI1 is set out in Box 8.1.

Box 8.1 Characterisation of BRI1 as the brassinolide receptor

Sequencing of BRI1 determined that the protein is a member of the receptor-like kinase (RLK) family in plants, with 25 extracellular leucine rich repeats (LRRs). While the function of the majority of these RLK-LRR proteins is un-known, a unique feature of BRI1 amongst the group is that it has a 70 amino acid island between the 21st and 22nd extracellular LRRs. Further, homol-ogy of BRI1 was highest to CLAVATA and ERECTA from *Arabidopsis*, two receptors important in regulating cell differentiation and signalling (see Chapter 3). Li and Chory (1997) determined that of the five mutant alleles examined, four, *bri1-101, bri1-104, bri1-107* and *bri1-115* had single bp substitutions in the kinase domain, but the fifth, *bri1-113*, had a miss-sense mutation to give $Gly_{611} \rightarrow Glu$ in the unique 70-amino-acid extracellular domain. Together, these results suggested that mutations in the unique extracellular domain or in the kinase domain repressed the response to added brassinolide. Some confir-mation of the function of BRI1 came from the experiments by He et al. (2000), who fused the extracellular domain of BRI1 (specifically the extracellular and transmembrane domains and 65 residues of the intracellular domain) to the cytoplasmic serine/threonine kinase domain of XA21, a related RLK that has been shown to confer resistance of rice cells to *Pseudomonas oryzae* pv *oryzae* (Xoo). Using this system, the addition of brassinolide conferred resistance to transformed rice cells against the Xoo pathovar (as determined by the induc-tion of the hypersensitive response characterised by a brief oxidative burst and subsequent cell death). In control experiments, the authors created a mu-tation in one residue ($Gly_{611} \rightarrow Glu$) of the 70-amino-acid island between the 21st and 22nd LRR (thus re-creating the *bri1-113* mutant allele) and showed that this mutant did not induce the HR response, nor did a construct with a mutation in the kinase domain ($Lys_{737} \rightarrow Glu$) of the XA21 protein. These experiments demonstrated that brassinolide was perceived by the extracellu-lar domain of BRI1, and that this binding event could be transduced by the cytoplasmically localised serine/threonine kinase domain.

Friedrichsen et al. (2000) carried out expression studies on *BRI1,* using *BRI1-GFP* constructs, and determined that the gene was expressed ubiquitously in the meristems, roots and shoots of young seedlings of *Arabidopsis* (i.e., BR target tissues), but expression was less in older tissues. Using confocal microscopy, they showed that BRI1 is localised to the plasma membrane of cells in which expression was observed. In terms of biochemical analysis, they also showed that BRI1 had autokinase activity with both serine and threonine residues phosphorylated. Oh et al. (2000) identified a minimum of 12 sites of autophosphorylation in the cytoplasmic domain of the protein, and using a mutated BRI1 protein with an inactive kinase as substrate, they showed that BRI1 could not transphosphorylate suggesting that autophosphorylation is intramolecular. To further substantiate the evidence that BRI1 is a brassinolide binding protein, Wang et al. (2001) overexpressed a *BRI1-GFP* construct and showed that the binding of brassinolide, and the intensity of response was related to the level of BRI1 protein present. As part of these experiments, they calculated a Ki value for brassinolide of 10.8 ± 3.2 nM for the wild-type protein. Finally, using immune-precipitation of extracts from BL-treated plants and with or without alkaline phosphatase, they showed that BRI1 was indeed phosphorylated *in vivo* in response to brassinolide treatment.

The accumulated evidence described in Box 8.1 suggests that BRI1 does bind brassinolide at the extracellular face of the plasma membrane, and that this binding event is transduced to activate autophosphorylation of BRI1. Further, the results of He et al. (2000) using the XA21 protein fusion, indicated that autophosphorylation could induce further downstream signalling. With this result, the possible downstream signalling components were then keenly sought.

As the opening series of experiments, Li et al. (2001) identified a semi-dwarf mutant, *bin2*, that was brassinolide-insensitive and not allelic to *bri1(bin1)*. On cloning the *BIN2* gene, Li and Nam (2002) established that it had identity to cytoplasmic serine/threonine kinases of which there were 10 in the gene family of *Arabidopsis* alone. BIN2 was also closely related to mammalian GSK3β kinases and the SHAGGY kinases in *Drosophila* which are important in numerous signalling pathways including cell fate determination and tissue patterning. Li and Nam (2002) expressed BIN2 in *E. coli* and demonstrated that the protein had kinase activity. However, of perhaps more significance was the observation that if *BIN2* was overexpressed in a weak *bri1* mutant (thus the phenotype was akin to wild-type), it led to a stronger *bri1* phenotype – that is, the plants became more unresponsive to added BR. Conversely, if *BIN2* expression was reduced in a weak *bri1* phenotype, then the plants became more responsive to BR. Taken together, these results suggest that BIN2 is a negative regulator of BRI1 signalling. In other eukaryotes, GSK3 and SHAGGY are constitutively active kinases that negatively regulate a variety of substrates by phosphorylation, and further, both GSK3 and SHAGGY are themselves negatively regulated by protein posphorylation. With this in mind, Li and Nam (2002), using two-hybrid assays, examined the possibility that BRI1 and BIN2 interacted directly, as a prelude to determining whether BRI1 phosphorylated BIN2. No evidence of any interaction was reported.

However, the downstream substrates of BIN2 have been identified. Yin et al. (2002) isolated the *bes1* mutant and showed that it displayed a constitutive BR response phenotype that included long and bending petioles, curly leaves, and accelerated senescence. To do this, 40,000 EMS-mutagenised homozygotes of *bri1-119*, a weak *bri1* allele, were screened for suppression of the phenotype, and one line, designated *bes1* (for *bri1*-EMS-suppressor 1), was identified. The *bes1* mutant was constitutively active and did not need a functional BRI1 protein, so its phenotype was independent of brassinosteroids. The *bes1/bri1-119* crosses displayed the same phenotype as *bes1* plants – i.e., they were similar to plants overproducing brassinosteroids or overexpressing *BRI1*. These observations indicated that *bes1* suppresses the *bri1-119* mutation and leads to the BL-responsive phenotype, and so BES1 is a positive regulator of the BR response. In addition, Yin et al. (2002) determined that the strongest expression of *BES1* was observed in leaf petioles, hypocotyls and vascular tissue, in a pattern that is similar to *BRI1* expression. Using a *BES-GFP* fusion, they found BES1 to be located to the nucleus in response to brassinolide treatment, while the protein was phosphorylated and destabilised by BIN2.

The lesion in the *bes1* mutant ($Pro_{233} \rightarrow Leu$) was identical to another downstream BR mutant *bzr1* characterised by He et al. (2002). These authors also showed that the BZR1 protein, a member of the same family as BES1 in *Arabidopsis*, when phosphorylated, was degraded by a proteasome-mediated pathway, and thus the likely function of the brassinosteroids was to inhibit the phosphorylation of these proteins (and thus enhance their accumulation). Previous work had shown that the *brassinazole resistant 1-1D* mutant (*bzr1-1D*) could suppress both the *bri1* mutant and the BR biosynthetic mutant, *det2,* suggesting that BZR1 acts downstream of BRI1, and that BZR1 mediates both the BR-induced growth response and feed-back inhibition of BR biosynthesis (Wang et al., 2002). Both BR treatment and the *bzr1-1D* mutation increased BZR1 accumulation. Overexpression of BZR1 suppressed a weak allele of *bri1* suggesting that BZR1, in common with BES1, is a positive regulator of the BR response in *Arabidopsis*.

BR enhancement of BZR1 accumulation in nuclei is particularly high in rapidly elongating cells. As BIN2 directly interacts with BZR1 and negatively regulates BZR1 accumulation *in vitro,* Wang and He (2004) suggest that BIN2 phosphorylates BZR1 and BES1 and thereby targets the proteins for proteasome-directed degradation. The phosphorylation of BZR1 and BES1 by BIN2 has been confirmed by Zhao et al. (2002a). Therefore, the role of BRI1 signalling must be to induce the phosphorylation of BIN2 and thus its subsequent inactivation and degradation. With BIN2 inactivated, the (unphosphorylated) positive activators BZR1 and BES1 can accumulate in the nucleus, thereby permitting the BR-regulated gene response (see Figure 8.4 for a diagrammatic overview). The model can be developed further in that the accumulation of BZR1 and BES1 also feed back and inhibit the biosynthesis of brassinosteroids, so there is less

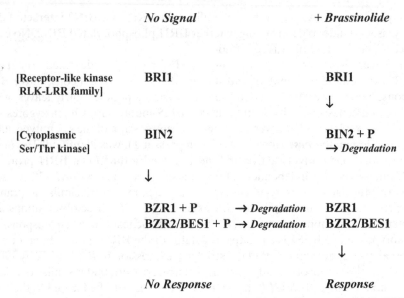

Figure 8.4. Conceptual representation of brassinosteroid signalling mediated via BRI1 in *Arabidopsis thaliana* (see text for details).

brassinolide-induced signalling through BRI1. BIN2 is, therefore, no longer repressed and so starts to inactivate BZR1 and BES1 via phosphorylation thus acting as a feed-back brake on the BR response.

A molecular model of plant hormone action and the target cell concept

In the previous two chapters, we considered what is now known of the molecular mechanisms by which some of the major plant signals are perceived. Our coverage has not been exhaustive, but enough examples are provided for the reader to derive some notion of the commonality of these perception and transduction mechanisms. This commonality, we believe will direct research towards an extension of the target cell concept in the future.

We consider that the most striking feature for the successful role of a signalling cue is to relieve some form of pre-existing repression of a programme of gene expression. The manifestation of that programme of gene expression can still be referred to as being switched on by the signal, but the dissection of the events that precede this show that the programme is actually held in a state of suppression, and only on perception and transduction of the correct (hormonal) signal can the specific programme of gene expression become free to unfold.

This analysis of events at the molecular level emphasises that the competence to respond to a developmental cue resides within the cell itself. For auxin, ethylene and cytokinins, the developmentally regulated array of ARF, EIN3, and the B-type ARR transcription factors determines the specificity of response when these cells are exposed to inputs of auxin, ethylene, or cytokinin, respectively. There is ample evidence that, as examples, the *ARF* (auxin), *EIN3* (ethylene) and B-type *ARRs* (cytokinin) genes are developmentally regulated and, significantly, the *ARF* genes or the B-type *ARR* genes are not induced by auxin or cytokinin directly. However, each hormone may interact with other developmental cues to regulate the abundance of these primary transcription factors.

Why then, should such a repression and derepression mechanism operate in plants? The characterisation of the mode of ethylene action was the first to illustrate that the binding of the hormone to its receptor, ETR1, served to relieve the repression caused by the presence of the ETR protein – that is, that ETR1 was constitutively a negative regulator of the ethylene response. At that time, it was considered advantageous for a sessile plant to possess a very rapid switch-on programme to respond to a particular environmental challenge, and for biotic and abiotic stresses, this notion of 'rapid response' may be entirely relevant. Now, hormones other than ethylene have been shown to influence the state of repression of potential expression patterns, including those that are part of differentiation programmes of organ ontogeny.

The answer may lie in a consideration of the fundamental nature of the controls that operate during plant development. While vascular tissues are clearly important in long-distance transport processes, any one target cell must be influenced profoundly by the type of cell that is its neighbour and by the array of low molecular compounds to which it is exposed. Depending on circumstances, most of these molecules will possess a signalling capacity at one time or another. For any one cell, the concentrations and/or composition of such chemical signals will differ – so the influence of threshold levels of these morphogens and the steepness of their gradients during plant development become significant. A molecular mechanism for the establishment of an auxin gradient was described recently during the development of polarity in embryos of *Arabidopsis* (Friml et al., 2003). Here, the distribution of the auxin efflux PIN proteins was shown to determine the formation of the gradient via the formation first of a physical auxin diffusion gradient. One can envisage, therefore, a mechanism by which even small changes in the cellular concentration of auxin can uncover subtle changes in the composition of ARF proteins and so initiate programmes in gene expressions that determine the developmental status of each cell during development. However, any worthwhile insight into understanding the significance of repression/derepression must also take into account the fact that every plant cell is a target for any number of signalling inputs and must have a mechanism for ignoring those that are essentially too low to constitute more than background noise or enhancing others by synergism with another signal; such partnerships are the basis of hormonal cross-talk.

The notion of hormonal cross-talk is already well established in plants, but in the context of our discussions, a significant advance has been the realisation that different signals can participate in and share the same downstream elements. This final feature of our discussion on target cells is covered in Chapter 9.

9

The Phenomenon of
Hormonal Cross-Talk

In the last two chapters, discussion has been concerned with the evidence for specific hormone receptors and the downstream signalling events in cells that form part of the transduction chain initiated by the binding of a ligand (the hormone) to its respective receptor. Each major plant hormone has been considered and classified as a separate operational entity, but it is clear that while the same hormone can have different effects in different tissues, a similar response in the same tissue can also be brought about by more than one hormone, the interactions involved being highly dependent upon the genetic background of the tissue in question. With the unravelling of intracellular signalling downstream of hormone perception, it is now becoming clear that more than one signal can utilise a particular transduction pathway. In this final chapter, we refer to examples of such apparently duplicated hormonal responses and how this cross-talk in perception and signalling has been revealed through the use of specific phenotypically expressed mutants. The list is not exhaustive but it serves to illustrate the level of flexibility that a cell can sustain, combined with the basic concept of every cell as an individual target cell.

It has become evident that plants are quite versatile in the cross-talk of their molecular communication language, as represented by situations where one hormone can substitute in function for another. For example, in the ethylene-enhanced elongation of the internodes of certain water plants (*Callitriche platycarpa*) the growth response can be further increased by the simultaneous addition of either GA or IAA to give what is seen as a 'super-growth' cell extension (Musgrave et al., 1972; Osborne, 1984). In another example, GA, auxin and ethylene all promote elongation of the hypocotyls of *Arabidopsis thaliana* in the light, where GA is the primary signal that is modulated by addition of IAA or ethylene (Saibo et al., 2003).

During their work on the ethylene signalling *ein2* mutant of *Arabidopsis*, Alonso et al. (1999) were aware that this mutant had been detected by others previously in, for example, screens for resistance to auxin transport inhibitors (Fujita and Syono, 1996), and to lack certain responses to cytokinins (Su and Howell, 1992). None of the other ethylene signalling loci (e.g., *etr1, ctr1*) were identified in such screens. With such information, Alonso et al. (1999) speculated that EIN2 might mediate cross-talk between several different hormonal signalling pathways. They were aware that ethylene and methyljasmonate must be added concomitantly for the induction of the pathogen-responsive gene *PDF1.2* in *Arabidopsis* (Penninckx et al., 1998), and they deduced, then ascertained, that applied jasmonate could not induce the expression of *PDF1.2* in an *ein2-5* mutant background. However, if the hydrophilic C-terminal domain of the EIN2 protein was expressed in the *ein2-5* genetic mutants to give *ein2-5*:CEND plants, then added jasmonate alone could induce expression of the *PDF1.2* gene. Experiments of this type, in which specific genetic backgrounds are modified, not only confirm a commonality in the ethylene and jasmonate response pathways but illustrate the significant potential of cross-talk between the two signalling molecules.

Yet another example of ethylene in a cross-talk relationship is seen in experiments with ABA in seed dormancy. Beaudoin et al. (2000) generated EMS mutants of the ABA-insensitive *abi1-1* to screen for either enhanced or suppressed ABA insensitivity in *Arabidopsis*. The *abi1-1* mutant displayed reduced seed dormancy but another mutant generated that was allelic to *ctr1* (designated *abi1-1 ctr1-10*) was identified that was even less responsive to ABA in terms of maintaining dormancy (i.e., higher concentrations of ABA were needed to reduce germination when compared with the *abi1-1* mutants). In contrast to these plants, a further mutant was identified that was allelic to *ein2* (designated *abi1-1 ein2-45*) in which the seeds required much less ABA to suppress germination, when compared with the *abi1-1* lines. The role of ethylene in ABA suppression of germination was examined further using two of the earliest ethylene signalling mutants identified, *ctr1-1* (Kieber et al., 1993) and *ein2-1* (Guzman and Ecker, 1990). In germination assays, the *ctr1-1* plants were less responsive to ABA (when compared with wild-type) whereas the *ein2-1* mutants were highly sensitive to ABA (i.e., less added ABA was required to repress germination). Taken together, these results indicate a level of cross-talk between ethylene and ABA at an interconnecting junction in their otherwise separate signalling pathways.

Molecular evidence for a direct link between two signals in a common signal transduction pathway has come from the work of Xie et al. (2003). They sequenced the gene lesion in the JA-insensitive *coi1* mutant which was identified earlier as coronatine insensitive (Feys et al., 1994). Xie and colleagues determined that the COI1 protein had 16 leucine rich repeats and an F-box, with similarity to the F-box protein TIR1 that is part of the SCF complex that mediates the ubiquitination pathway operating in auxin signalling. This result suggests that JA and IAA signalling may function via a similar proteasome-dependent process, though not identical pathway, since the *coi1* mutant shows no changes in phenotype when treated with auxin other than those already observed in the wild-type. More direct evidence that the proteasome-dependent signalling pathway could be shared by IAA and JA is seen from the work of Tiryaki and Staswick (2002) who, while screening for mutants that were insensitive to MeJA, isolated a mutant, *axr1-24*, that was allelic to *axr1-3*. The AXR1 protein is another component of the proteasome-dependent signalling process in auxin signalling that, with ECR1, forms a heterodimer that activates the ubiquitin-like RUB protein. The performance of the *axr1-24* mutant, therefore, suggests that both IAA and JA signalling could be mediated through this similar mechanism.

Evidence for a cross-talk between JA, salicylic acid (SA), GA and the effects of mechanical damage comes from the assessment of trichome formation in epidermal cells of leaves of *Arabidopsis* (Traw and Bergelson, 2003). While damage and JA both increased the abundancy of trichomes, as did JA and GA when supplied together, no increase occurred with JA in the jasmonate pathway mutant *jar 1-1*. SA reduced trichome production in the presence or absence of JA, even in the *nim1-1* salicylate-dependent mutant, indicating a negative cross-talk between the two.

In an ABA-auxin-cytokinin example of cross-talk, Lu and Fedoroff (2000) showed that the hyponastic leaves mutation (*hyl 1*) of *Arabidopsis* was associated with a reduced responsiveness to both auxin and cytokinin, but an increased responsiveness to ABA. The auxin transport inhibitor (TIBA) normalized the mutant to a certain extent, but NPA (another auxin transport inhibitor) had the reverse effect causing the *hyl 1* mutant to become even more hyponastic. The *HYL* gene was shown to be ABA-regulated and encoding a nuclear dsRNA binding protein which was deduced as possessing activity in the cell either at the transcriptional or post-transcriptional level with a possible transgene-induced gene silencing function. This is a scene of considerable cross-talk complexity.

In whole plant and cell growth studies, many of the effects of auxin can be induced by brassinolide (BL). This suggests immediately that there must be commonalities in their signalling pathways. The question of whether such cross-talk is sustained during cell proliferation in cell suspension cultures has been addressed by Miyazawa et al. (2003). Using the BY-2 cultured cell line of tobacco and CYM and histone H4 DNA fragments as probes for markers of M- and S-phase expression, they demonstrated that BL-promoted cell division was linked to the accumulation of cell-cycle–related gene products although the mechanism supporting such divisions was distinct from that regulated by the balance between auxin and cytokinin levels. The major difference appears to lie in the lack of plastid or mitochondrial DNA synthesis in the BL-treated cultures during the initial phase of cell proliferation. This contrasts with the preferential organellar DNA synthesis activated by auxin, which is necessary for the continued division of BY-2 cells in auxin-cytokinin regulated cultures.

These complex interactions between hormonal or signal molecules, especially in different genetic backgrounds, indicate the subtle interplay and cross-talk that operates continuously throughout the lifetime of every cell.

One of the more striking examples of cross-talk arises from the studies of Montoya et al. (2002) and Scheer and Ryan (2002), who together have shown that in tomato, brassinosteroids and systemin are perceived by the same receptor protein. The isolation of the systemin receptor SR160 by Scheer and Ryan (2002) using direct biochemical approaches based upon using labelled systemin as a bait protein was described previously in Chapter 3. Montoya et al. (2002) identified a tomato mutant that was similar in phenotype to the brassinosteroid dwarf mutants possessing dark-green curled leaves, and roots that were only partially inhibited by added brassinolide (BL). This mutant was designated as *altered brassinosteroid1 (abs1)*. Montoya et al. (2002) reasoned that *abs1* must be distinct from another tomato brassinosteroid mutant, *curl3 (cu3)* that also had an extreme dwarfed phenotype, including the dark-green and curled leaves, but was completely insensitive to added BL (Koka et al., 2000). However, double mutant analysis established that *abs1* was in fact a weak, recessive allele at the *cu3* locus. To study dwarf mutants, Montoya et al. (2002) examined the brassinosteroid content in both *cu3* and *cu3^{abs1}* and determined that they and wild-type tomato plants contained no brassinolide, but, instead, the mutants had an enhanced level of castasterone, indicating that this related compound can substitute for brassinolide, and could act as the active brassinosteroid in tomato tissues. These results confirmed that dwarfism in the two mutants was not caused by a deficiency in BR (as castasterone) content, and that both were BR signalling mutants. To examine BR signalling further, the homologue of the *Arabidopsis (At)BRI1* gene in tomato, *tBRI1*, was isolated and sequenced. Sequencing showed that *cu3* was a nonsense (Gly$_{749}$ → Z) mutant and *Cu3^{abs1}* was a miss-sense mutant (His$_{101}$ → Tyr) of tBRI1. Given that tBRI1 is nearly identical to the systemin receptor SR160, the possibility of shared perception and signalling between the brassinosteroids and systemin presents itself. Montoya et al. (2002) did not show whether systemin actually binds to tBRI1, but Scheer and Ryan (2002) tried to determine whether added BL could compete with systemin-induced alkalisation of the medium that occurs in tomato cells in culture (see Chapter 3). There was no competition up to a concentration of 1 μM, but the possibility cannot be ruled out that BL may require additional peptides or proteins *in vivo* to mediate any effects through the SR160 receptor.

In this overview, we have included only a very few examples of different hormonal cross-talk primarily to illustrate the complexity of the interpretation of the target cell state in plants. Evidence for cross-talk of plant signals now extends to auxin and the brassinosteroids in *Arabidopsis* (Nakamura et al., 2003) and the requirement of auxin for the biosynthesis of active gibberellins during inflorescence development in barley (Wolbang et al., 2004). There are many more examples known and many more will be discovered in the future. The evidence that ethylene, auxin and GA inputs can all be attributed to effects on DELLA protein function (Achard et al., 2003) further illustrates that such commonalities will surely extend to many more interacting plant signals.

Concluding remarks

We have shown how the concentration of hormones and receptors can regulate cell, tissue and whole plant responses and how the levels of non-receptors but interacting proteins can modify the effectiveness of receptors. What is of significance for the cell's sensitivity (or ability of a cell to respond to a signal) is the extent to which the receptor should approach saturation at any one time. If the level of receptor is low, then only a few signal molecules are required to evoke a hormonal response (or to repress or derepress a specific gene), but if the receptors are freely available then many more molecules of the effector signal are necessary for any given cell to recognize the molecular messenger above the minimal threshold level for response.

But receptors are not the only controllers of cross-talks within a cell. Any regulation of a hormonal or signal biosynthesis gene will generate perturbations that have repercussions at the level of multiple gene expressions. This is particularly evident from our knowledge of auxin–ethylene interactions, where the level of auxin can determine the up- or down-regulation of ethylene biosynthesis while the levels of ethylene produced will, in turn, control the transport of auxin molecules from cell to cell, the polar movement through tissues and rates of auxin conjugation to non-active states, or auxin degradative loss by oxidation or decarboxylation.

If we add to the interactive complexities of any one cell in contact with its neighbours, the constant and progressive developmental ageing sequence that follows as the cell becomes displaced from the meristem, the positional information alone provides abundant evidence that no two cells can ever be identical. At the first cell division in any lineage, the wall of the new cell plate is always developmentally distinct from the walls of the mother cell, and all are in a changing structural and compositional progression throughout enlargement – the wall components are in a state of dynamic modification throughout the lifetime of the cell (Knox, 1997).

Compared with the cells of animals, which lack cell walls, the plant cell has a whole lexicon of reserve information within the molecular organization of its cell wall. Emphasis has been given in Chapters 2 and 3 as to how wall assembly and the organised dissolution of the wall can release informational mixed saccharide and pectolytic fragments that have a powerful regulatory influence upon cell behaviour, in both short- and long-distance actions. The field of pectinomics in plant performance must be seen as tightly integrating with cross-talks determined at the level of the gene and the regulatory proteins that genes generate. The dynamics of cell wall turnover provide an almost limitless diversity of opportunity for signalling molecules derived from the complex assemblies of polysaccharides and proteins, and we predict many exciting new discoveries will result from their further study.

The case for every plant cell being an individual target for the constant and constantly changing levels of hormonal or environmental signals that it receives is well made but, in fact, it is still quite poorly understood. Much has been achieved in unravelling the mechanisms involved in gene control, but understanding how the cell perceives and responds to the many interconnecting signal cascades has far to go. This is because although we tacitly assume that every plant cell is totipotent (though this may not actually be so) and accept that every cell is controlled by a similar parental genome, we are also well aware that different sets of proteins, their levels, their rates of turnover and their cellular locations are certainly not the same.

Whereas the systematic approaches to characterising genomes are now well established for most living organisms, and this includes plant developmental studies, the understanding of the proteome – the complete set of proteins expressed by the genome – lags behind. Unlike genes, which all use the same code, each protein has unique properties, both chemical and physical. Many thousands of proteins are present in each cell and each is estimated to be present at molecular abundances ranging from fifty to over a million.

As well as to pectinomics, it is to proteome analysis and to proteomic technologies and the relationship of the proteome to the genome that the plant scientist must now resort to reach a fuller understanding of how the plant regulates its society of cells. If this can be attempted with yeast (Huh et al., 2003), then it is possible we can do it for *Arabidopsis*. But we cannot do this without an equally precise knowledge of in which of the many different cell types the informational signals arise, at which stages in the cell cycle they are generated, from which location in the cell the signal recognition takes place, and the status of the protein-determined signal cascade that then operates each gene.

The use of mutants, transformed plants, 'knock out' genes, and anti-sense transcriptions has provided a wealth of information for developmental studies during the past two decades. But it is now to the global interpretations of individual cells that the future experimentation must turn and to the complex role played by the dynamic and ever-changing states of proteins.

Because we still believe each living plant cell to be totipotent, then positional multi-protein cell monitoring by advanced methodologies that include highly sensitive green fluorescent protein tagging, precise subcellular *in situ* hybridisations and quantitative immuno-blotting must yield a new and spectacular world of cell-to-cell protein cross-talk directing each gene performance in the organelle-encapsulated nuclear and satellite mitochondrial and chloroplast genomes.

In the protein landscape, it is evident that we must now evaluate the language of the proteins that collectively control the performance of the genes rather than considering only the genes as holding dictatorship control as the one-gene, one-protein dictum was originally presented and interpreted.

We conclude that the complex interacting components of hormones, signals and the availability of specific receptors and attendant proteins, linked to the transduction events required to instruct the plant genome, will take many more years to fully elucidate. But for certain, each target cell will be seen to develop its own individual molecularly tuned informational network that controls precisely the way it shall perform.

References

Abel, S. and Theologis, A. (1996) Early genes and auxin action. *Plant Physiology* **111**, 9–17.

Abel, S., Oeller, P.W. and Theologis, A. (1994) Early auxin-induced genes encode short-lived nuclear proteins. *Proceedings of the National Academy of Sciences, USA* **91**, 326–330.

Abernethy, G.A., Fountain, D.W. and McManus, M.T. (1998) Observations of the leaf anatomy of *Festuca novae-zelandiae* (Hack.) Cockayne and biochemical responses to a water deficit. *New Zealand Journal of Botany* **36**, 113–123.

Achard, P., Vriezen, W.H., Van Der Straeten, D. and Harberd, N. (2003) Ethylene regulates *Arabidopsis* development via the modulation of DELLA protein growth repressor function. *Plant Cell* **15**, 2816–2825.

Adams, D.O. and Yang, S.F. (1977) Methionine metabolism in apple tissue: Implication of *S*-adenosylmethionine as an intermediate in the conversion of methionine to ethylene. *Plant Physiology* **60**, 892–896.

Adams, D.O. and Yang, S.F. (1979) Ethylene biosynthesis: Identification of 1-aminocyclopropane-1-carboxylic acid as an intermediate in the conversion of methionine to ethylene. *Proceedings of the National Academy of Sciences, USA* **76**, 170–174.

Addicott, F.T., Cairns, H.R., Cornforth, J.W., Lyon, J.L., Milborrow, B.V., Ohkuma, K., Ryback, G., Smith, G., Thiessen, W.E. and Wareing, P.F. (1968) Abscisic acid: A proposal for the redesignation of abscisin II (dormin). In: *Biochemistry and Physiology of Plant Growth Substances*, Wightman, F. and Setterfield, G. (eds). Runge Press, Ottawa, pp. 1527–1529.

Ainley, W.M., Walker, J.C., Nagao, R.T. and Key, J.L. (1988) Sequencing and characterization of two auxin-regulated genes from soybean. *Journal of Biological Chemistry* **263**, 10658–10666.

Akiyoshi, D.E., Klee, H., Amasino, R.M., Nester, E.W. and Gordon, P.M. (1984) T-DNA of *Agrobacterium tumefaciens* encodes an enzyme of cytokinin biosynthesis. *Proceedings of the National Academy of Sciences, USA* **81**, 5994–5998.

Albersheim, P. and Valent, B.S. (1978) Host–pathogen interactions in plants. Plants when exposed to oligosaccharins of fungal origin defend themselves by accumulating antibiotics. *Journal of Cell Biology* **78**, 627–643.

Aldington, S. and Fry, S.C. (1993) Oligosaccharins. *Advances in Botanical Research* **19**, 2–101.

Aldridge, D.C., Galt, S., Giles, D. and Turner, W.B. (1971) Metabolites of *Lasiodiplodia theobromae*. *Journal of the Chemical Society. (C), Organic Chemistry* 1623–1627.

Allan, A.C., Fricker, M.D., Ward, J.L., Beale, M.H. and Trewavas, A.J. (1994) Two transduction pathways mediate rapid effects of abscisic acid in *Commelina commonis* guard cells. *Plant Cell* **6**, 1319–1328.

Allen, G.J., Kuchitsu, K., Chu, S.P., Murata, Y., and Schroeder, J.L. (1999) *Arabidopsis abi1-1* and *abi2-1* phosphatase mutations reduce abscisic acid induced cytosolic calcium rises in guard cells. *Plant Cell* **11**, 1785–1798.

Allen, G.J., Chu, S.P., Schumacher, K., Shimazaki, C., Vafeados, D., Kemper, A., Hawke, S.D., Tallman, G., Tsien, R.Y., Harper, J.F., Chory, J. and Schroeder, J.I. (2000) Alteration of stimulus-specific guard cell calcium oscillations and stomatal closing in *Arabidopsis det3* mutant. *Science* **289**, 2338–2342.

Alonso, J.M., Hirayama, T., Roman, G., Nourizadeh, S. and Ecker, J.R. (1999) EIN2, a bifunctional transducer of ethylene and stress response in *Arabidopsis. Science* **284**, 2148–2152.

Altamura, M.M., Zagli, D., Salvi, G., de Lorenzo, G. and Bellincampi, D. (1998) Oligogalacturonides stimulate pericycle cell wall thickening and cell divisions leading to stoma formation in tobacco leaf explants. *Planta* **204**, 429–436.

Anderson, B.E., Ward, J.M. and Schroeder, J.I. (1994) Evidence for an extracellular reception site for abscisic acid in *Commelina* guard cells. *Plant Physiology* **104**, 1177–1183.

Aoyagi, S., Sugiyama, M. and Fukuda, H. (1998) *BEN1* and *ZEN1* cDNAs encoding S1-type DNases that are associated with programmed cell death in plants. *FEBS Letters* **429**, 134–138.

Arimura, G., Ozawa, R., Shimoda, T., Nishioka, T., Boland, W. and Takabayashi, J. (2000) Herbivory-induced volatiles elicit defence genes in lima bean leaves. *Nature* **406**, 512–514.

Ashikari, M., Wu, J., Yano, M., Sasaki, T. and Yoshimura, A. (1999) Rice gibberellin-insensitive dwarf mutant gene *Dwarf 1* encodes the α-subunit GTP-binding protein. *Proceedings of the National Academy of Sciences, USA* **96**, 10284–10289.

Assman, S.M. (1994) Ins and outs of guard cell ABA receptors. *Plant Cell* **6**, 1187–1190.

Augur, C., Yu, L., Sakai, K., Ogawa, T., Sinai, P., Darvill, A. and Albersheim, P. (1992) Further studies on the ability of xyloglucan oligosaccharides to inhibit auxin-stimulated growth. *Plant Physiology* **99**, 180–185.

Augur, C., Benhamou, N., Darvill, A. and Albersheim, P. (1993) Purification, characterization and cell wall localization of an α-fucosidase that inactivates a xyloglucan oligosaccharin. *Plant Journal* **3**, 415–426.

Avers, C.J. (1963) Fine structure studies of *Phleum* root meristem cells. 11. Mitotic asymmetry and cellular differentiation. *American Journal of Botany* **50**, 140–148.

Ballas, N., Wong, L.-M. and Theologis, A. (1993) Identification of the auxin-responsive element, AuxRE, in the primary indoleacetic acid-inducible gene, PS-IAA4/5, of pea *(Pisum sativum). Journal of Molecular Biology* **233**, 580–596.

Barbier-Brygoo, H., Ephritikhine, G., Klambt, D., Ghislain, M. and Guern, J. (1989) Functional evidence for an auxin receptor at the plasmalemma of tobacco mesophyll protoplasts. *Proceedings of the National Academy of Sciences, USA* **86**, 891–895.

Barbier-Brygoo, H., Ephritikhine, G., Klämbt, D., Maurel, C., Palme, K., Schell, J. and Geurn, J. (1991) Perception of the auxin signal at the plasma membrane of tobacco mesophyll protoplasts. *Plant Journal.* **1**, 83–94.

Barker, C. (2000) Systemic acquired resistance. In: *Molecular Plant Pathology*, Dickinson, M. and Benyon, J. (eds.), Annual Plant Reviews, Vol 4. Academic Press, Sheffield, pp. 198–217.

Barlow, P.W. (1995) Gravity perception in plants: A multiplicity of systems derived by evolution? *Plant Cell and Environment* **18**, 951–962.

Barry, G.F., Rogers, S.G., Fraley, R.T. and Brand, L. (1984) Identification of a cloned cytokinin biosynthetic gene. *Proceedings of the National Academy of Sciences, USA* **97**, 14778–14783.

Bartel, B. (1997) Auxin Biosynthesis. *Annual Review of Plant Physiology and Plant Molecular Biology* **48**, 51–66.

Bartel, B. and Fink, G.R. (1995) ILR1, an amidohydrolase that releases active indole-3-acetic acid from conjugates. *Science* **268**, 1745–1748.

Bartel, B., LeClere, S., Magidin, M. and Zolman, B.K. (2001) Inputs to the active indole-3-acetic acid pool: *De novo* synthesis, conjugate hydrolysis and indole-3-butyric acid B-oxidation. *Journal of Plant Growth Regulation* **20**, 198–216.

Bartels, D. and Salamini, F. (2001) Desiccation tolerance in the resurrection plant *Craterostigma plantgineum.* A contribution to the study of drought tolerance at a molecular level. *Plant Physiology* **127**, 1346–1353.

Bartling, D., Seedorf, M., Mithofer, A. and Weiler, E.W. (1992) Cloning and expression of an *Arabidopsis* nitrilase which can convert indole-3-acetonitrile to the plant hormone, indole-3-acetic acid. *European Journal of Biochemistry* **205**, 417–424.

Barton, M.K. and Poethig, S. (1993) Formation of the shoot apical mersitem in *Arabidopsis thaliana:* An analysis of development in the wild-type and in the shoot meristemless mutant. *Development* **119**, 823–831.

Bassett, C.L., Artlip, T.S. and Callahan, A.M. (2002) Characterization of the peach homologue of the ethylene receptor, *PpETR1*, reveals some unusual features regarding transcript processing. *Planta* **215**, 679–688.

Bauly, J.M., Sealy, I.M., Macdonald, H., Brearley, J., Dröge, S., Hillmer, S., Robinson, D.G., Venis, M.A., Blatt, M.R., Lazarus, C.M. and Napier, R.M. (2000) Overexpression of auxin-binding protein enhances the sensitivity of guard cells to auxin. *Plant Physiology* **124**, 1229–1238.

Baydoun, E. A-H. and Fry, S.C. (1985) The immobility of pectic substances in injured tomato leaves and its bearing on the identity of the wound hormones. *Planta* **165**, 269–276.

Beaudoin, N., Serizet, C., Gosti, F. and Giraudat, J. (2000) Interactions between abscisic acid and ethylene signalling cascades. *The Plant Cell* **12**, 1103–1116.

Beeckman, T., Burssens, S. and Inze, D. (2001) The peri-cell-cycle in *Arabidopsis. Journal of Experimental Botany* **52**, 403–411.

Belanger, K.D., Wyman, A.J., Sudol, M.N., Singla-Pareek, S.L. and Quatrano, R.S. (2003) A signal peptide secretion screen in *Fucus distichus* embryos reveals expression of glucanase, EGF domain-containing, and LRR receptor kinase-like polypeptides during asymmetric growth. *Planta* **217**, 931–950.

Bellincampi, D., Cardarelli, M., Zaghi, D., Serino, G., Salvi, G., Gatz, C., Cervone, F., Altamura, M.M., Costantino, P. and De Lorenzo, G.D. (1996) Oligogalacturonides prevent rhizogenesis in *rolB*-transformed tobacco explants by inhibiting auxin-induced expression of the *rolB* gene. *Plant Cell* **8**, 477–487.

Bellincampi, D., Salvi, S., De Lorenzo, G., Cervone, F., Marfa, V., Eberhard, S., Darvill, A. and Albersheim, P. (1993) Oligogalacturonides inhibit the formation of roots on tobacco explants. *Plant Journal* **4**, 207–213.

Bengochea, T., Acaster, M.A., Dodds, J.H., Evans, D.E., Jerie, P.H. and Hall, M.A. (1980) Studies on ethylene binding by cell-free preparations from cotyledons of *Phaseolus vulgaris*. II. Effects of structural analogues of ethylene and of inhibitors. *Planta* **148**, 407–411.

Benjamins, R., Quint, A., Weijers, D., Hooykaas, P. and Offringa, R. (2001) The pinoid protein kinase regulates organ development in *Arabidopsis* by enhancing polar auxin transport. *Development* **128**, 4057–4067.

Bennett, M.J., Marchant, A., Green, H.G., May, S.T., Ward, S.P., Millner, P.A., Walker, A.R., Schulz, B. and Feldman, K.A. (1996) *Arabidopsis AUX1* gene: a permease-like regulator of root gravitropism. *Science* **273**, 948–950.

Berger, F., Taylor, A. and Brownlee, C. (1994) Cell fate determination by the cell wall in early *Fucus* development. *Science* **263**, 1421–1423.

Berger, S., Menudier, A., Julien, R. and Karamanos, Y. (1996) Regulation of de-N-glycosylation enzymes in germinating radish seeds. *Plant Physiology* **112**, 259–264.

Bernier, G. (1988) The control of floral evocation and morphogenesis. *Annual Review of Plant Physiology and Plant Molecular Biology* **39**, 175–219.

Bethke, P.C., Badger, M.R. and Jones, R.L. (2004) Apoplastic synthesis of nitric oxide by plant tissues. *Plant Cell* **16**, 332–341.

Bethke, P.C., Lonsdale, J.E., Fath, A. and Jones, R.L. (1999) Hormonally regulated programmed cell death in barley aleurone cells. *Plant Cell* **11**, 1033–1046.

Beven, A., Guan, Y., Peart, J., Cooper, C. and Shaw, P. (1991) Monoclonal antibodies to plant nuclear matrix reveal intermediate filament related components within the nucleus. *Journal of Cell Science* **98**, 293–302.

Beveridge, C.A. (2000) The ups and downs of signalling between root and shoot. *New Phytologist* **147**, 413–416.

Beveridge, C.A., Murfet, I.C., Kerhoas, L., Sotta, B., Miginiac, E. and Rameau, C. (1997a) The shoot controls zeatin riboside export from pea roots. Evidence from the branching mutant *rms4*. *Plant Journal* **11**, 339–345.

Beveridge, C.A., Symons, G.M., Murfet, I.C., Ross, J.J. and Rameau, C. (1997b) The *rms1* mutant of pea has elevated indole-3-acetic acid levels and reduced root sap zeatin riboside content but increased branching controlled by graft transmissible signal(s). *Plant Physiology* **115**, 1251–1258.

Biale, J.B., Young, R.E. and Olmstead, A.J. (1954) Fruit respiration and ethylene production. *Plant Physiology* **29**, 168–174.

Bierhorst, D.W. (1977) On the stem apex, leaf initiation and early leaf ontogeny in filicalean ferns. *Amercian Journal of Botany* **64**, 125–152.

Binns, A.N. (1994) Cytokinin accumulation and action: Biochemical, genetic and molecular approaches. *Annual Review of Plant Physiology and Plant Molecular Biology* **45**, 173–196.

Bleecker, A.B., Estelle, M.A., Somerville, C. and Kende, H. (1988) Insensitivity to ethylene conferred by a dominant mutant in *Arabidopsis thaliana*. *Science* **241**, 1086–1089.

Bonfante, P., Genre, A., Faccio, A., Martini, I., Schauser, L., Stougaard, J., Webb, J. and Parniske, M. (2000) The *Lotus japonicus Lj Sym-4* gene is required for the successful symbiotic infection of root epidermal cells. *Molecular Plant-Microbe Interactions* **13**, 1109–1120.

Bonghi, C., Rascio, N., Ramina, A. and Casadoro, G. (1992) Cellulase and polygalacturonase involvement in the abscission of leaf and fruit explants of peach. *Plant Molecular Biology* **20**, 839–848.

Boss, P.K. and Thomas, M.R. (2002) Association of dwarfism and floral induction with a grape "green revolution" mutation. *Nature* **416**, 847–850.

Boubriak, I., Naumenko, N., Lyne, L. and Osborne, D.J. (2000) Loss of viability in rye embryos at different levels of hydration: senescence with apoptopic nucleosome cleavage or death with random DNA fragmentation. In: *Seed Biology: Advances and Applications*, Black, M.J., Bradford, K.J. and Vazquez-Ramos. J. (eds). CAB International, Oxford, pp. 205–214.

Bradford, K.J. and Yang, S.F. (1980) Xylem transport of 1-aminocyclopropane-1-carboxylic cid, an ethylene precursor, in waterlogged tomato plants. *Plant Physiology* **65**, 322–326.

Branca, C., De Lorenzo, G. and Cervone, F. (1988) Competitive inhibition of the auxin-induced elongation by α-$_D$-oligogalacturonides in pea stem segments. *Physiologia Plantarum* **72**, 499–504.

Brandstatter, I. and Kieber, J.J. (1998) Two genes with similarity to bacterial response regulators are rapidly and specifically induced by cytokinin in *Arabidopsis*. *Plant Cell* **10**, 1009–1020.

Brault, M., Caiveau, O., Pédron, J., Maldiney, R., Sotta, B. and Miginiac, E. (1999) Detection of membrane-bound cytokinin-binding proteins in *Arabidopsis thaliana* cells. *European Journal Biochemistry* **260**, 512–519.

Brinegar, A.C. and Fox, J.E. (1985) Resolution of the subunit composition of a cytokinin-binding protein from wheat embryos. *Biological Plantarum* **27**, 100–104.

Brinegar, A.C., Stevens, A. and Fox, J.E. (1985) Biosynthesis and degradation of a wheat triticum-durum embryo cytokinin-binding protein during embryogenesis and germination. *Plant Physiology* **79**, 706–710.

Brown, J.C. and Jones, A.M. (1994) Mapping the auxin-binding site of auxin-binding protein 1. *Journal of Biological Chemistry* **269**, 21136–21140.

Brownlee, C. and Wood, J.W. (1986) A gradient of cytoplasmic free calcium in growing rhizoid cells of *Fucus serratus*. *Nature* **320**, 624–626.

Brugiere, N., Rothstein, S.J. and Cui, Y. (2000) Molecular mechanisms of self-recognition in *Brassica* self-incompatability. *Trends in Plant Science* **5**, 432–438.

Bui, A.Q. and O'Neill, S.D. (1998) Three 1-aminocyclopropane-1-carboxylate synthase genes regulated by primary and secondary pollination signals in orchid flowers. *Plant Physiology* **116**, 419–428.

Buitink, J., Vu, B.L., Satour, P. and Leprince, O. (2003) The re-establishment of desiccation tolerance in germinated radicles of *Medicago truncatula* Geertn. seeds. *Seed Science Research* **13**, 273–286.

Burg, S.P. and Burg, E.A. (1962a) Role of ethylene in fruit ripening. *Plant Physiology* **37**, 179–189.

Burg, S.P. and Burg, E.A. (1962b) Post-harvest ripening of avocados. *Nature* **194**, 398–399.

Burlat, V., Kwon, M., Davin, L.B. and Lewis, N.G. (2001) Dirigent protein and dirigent sites in lignifying tissues. *Phytochemistry* **57**, 883–897.

Burnett, E.C., Desikan, R., Moser, R.C. and Neill, S.J. (2000) ABA activation of an MBP kinase in *Pisum sativum* epidermal peels correlates with stomatal responses to ABA. *Journal of Experimental Botany* **51**, 197–205.

Burroughs, L.F. (1957) 1-Aminocyclopropane-1-carboxylic acid: A new amino acid in perry pears and cider apples. *Nature* **179**, 360–361.

Bush, D.S. (1996) Effects of gibberellic acid and environmental factors on cytosolic calcium in wheat aleurone cells. *Planta* **199**, 89–99.

Bush, M.S. and McCann, M.C. (1999) Pectic epitopes are differently distributed on the cell walls of potato (*Solanum tuberosum*) tubers. *Physiologia Plantarum* **107**, 201–213.

Bush, M.S., Marry, M., Huxham, M.I., Jarvis, M.C. and McCann, M.C. (2001) Developmental regulation of pectic epitopes during potato tuberization. *Planta* **213**, 869–880.

Byard, E.H. and Lange, B.M.H. (1991) Tubulin and microtubules. *Essays in Biochemistry* **26**, 13–25.

Campbell, A.D. and Labavitch, J.M. (1991a) Induction and regulation of ethylene biosynthesis by pectic oligomers in cultured pear cells. *Plant Physiology* **97**, 699–705.

Campbell, A.D. and Labavitch, J.M. (1991b) Induction and regulation of ethylene biosynthesis and ripening by pectic oligomers in tomato pericarp discs. *Plant Physiology* **97**, 706–713.

Cancel, J.D. and Larsen, P.B. (2002) Loss-of-function mutations in the ethylene receptor *ETR1* cause enhanced sensitivity and exaggerated response to ethylene in *Arabidopsis*. *Plant Physiology* **129**, 1557–1567.

Carle, S.A, Bates, G.W. and Shannon, T.A. (1998) Hormonal control of gene expression during reactivation of the cell cycle in tobacco mesophyll protoplasts. *Journal of Plant Growth Regulation* **17**, 221–230.

Casimiro, I., Beeckman, T., Graham, N., Bhalerao, R., Zhang, H., Caseros, P., Sandberg, G. and Bennett, M.J. (2003) Dissecting *Arabidopsis* lateral root development. *Trends in Plant Science* **8**, 165–171.

Cassab, G.I., Lin, J.-J., Lin, L.S. and Varner, J.E. (1988) Ethylene effect on extension and peroxidase distribution in the subapical region of pea epicotyls. *Plant Physiology*. **88**, 522–524.

Chailakhyan, M.H. (1936) On the mechanism of photoperiodic interaction. *Comptes Rendus (Doklady) Academie des Sciences, USSR* **10**, 89–93.

Chang, C., Kwok, S.F., Bleecker, A. B. and Meyerowitz, E.B. (1993) *Arabidopsis* ethylene response gene ETR1-similarity of product to two-component regulators. *Science* **262**, 539–544.

Chang, C. and Meyerowitz, E.M. (1995) The ethylene hormone response in *Arabidopsis* – An eukaryotic two-component signaling system. *Proceedings of the National Academy of Sciences, USA* **92**, 4129–4133.

Cheah, K.S.E. and Osborne, D.J. (1978) DNA lesions occur with loss of viability in embryos of ageing rye seed. *Nature* **272**, 593–599.

Chen, J.-G., Shimomura, S., Sitbon, F., Sandberg, G. and Jones, A.M. (2001a) The role of auxin-binding protein 1 in the expansion of tobacco leaf cells. *Plant Journal* **28**, 607–617.

Chen, J.G., Ullah, H., Young, J.C., Sussman, M.R. and Jones, A.M. (2001b) ABP1 is required for organized cell elongation and division in *Arabidopsis* embryogenesis. *Genes and Development* **15**, 902–911.

Chen, R., Hilson, P., Sedbrook, J., Rosen, E., Casper, T. and Masson, P.H. (1998) The *Arabidopsis thaliana AGRAVITROPIC1* gene encodes a component of the polar-auxin-transport efflux carrier. *Proceedings of the National Academy of Sciences, USA* **95**, 15112–15117.

Chen, Y.F., Randlett, M.D., Findell, J.L. and Schaller, G.E. (2002) Localization of the ethylene receptor ETR1 to the endoplasmic reticulum of *Arabidopsis*. *Journal of Biological Chemistry*, **277**, 19861–19866.

Cheng, H., Qin, L., Lee, S., Fu, X., Richards, D.E., Cao, D., Luo, D., Harberd, N.P. and Peng, J. (2004) Gibberellin regulates *Arabidopsis* floral development via suppression of DELLA protein function. *Development* **131**, 1055–1064.

Cheung, A.Y. and Wu, H.M. (1999) Arabinogalactan proteins in plant sexual reproduction. *Protoplasma* **208**, 87–98.

Chibnall, A.C. (1939) *Protein Metabolism in the Plant.* Yale University Press, New Haven; H. Milford, Oxford University Press, U.K.

Chlyah, H. (1974a) Inter-tissue correlations in organ fragments: Organogenetic capacity of tissues excised from stem segments of *Torania fournieri* Lind. cultured separately *in vitro. Plant Physiology* **54**, 341–348.

Chlyah, H. (1974b) Formation and propagation of cell division-centers in the epidermal layer of internodal segments of *Torrenia fournier* grown *in vitro.* Simultaneous surface observations of all epidermal cells. *Canadian Journal of Botany* **52**, 867–872.

Chlyah, H. (1978) Intercellular correlations: Relation between DNA synthesis and cell division in early stages of *in vitro* bud neoformation. *Plant Physiology.* **62**, 482–485.

Cho, H.-T. and Cosgrove, D.J. (2000) Altered expression of expansin modulates leaf growth and pedicel abscission in *Arabidopsis thaliana. Proceedings of the National Academy of Sciences, USA* **97**, 9783–9788.

Choe, S., Dilkes, B.P., Fujioka, S., Takasuto, S., Sakurai, A. and Feldmann, K.A. (1998) The *DWF4* gene of *Arabidopsis* encodes a cytochrome P450 that mediates multiple 22α-hydroxylation steps in brassinosteroid biosynthesis. *Plant Cell* **10**, 231–244.

Choe, S., Tanaka, A., Noguchi, T., Fujioka, S., Takatsuto, S., Ross, A.S., Tax, F.E., Yoshida, S. and Feldmann, K.A. (2000) Lesions in the sterol Δ^7 reductase gene of *Arabidopsis* cause dwarfism due to a block in brassinosteroid biosynthesis. *Plant Journal* **21**, 431–443.

Christensen, S.K., Dagenais, N., Chory, J. and Weigel, D. (2000) Regulation of auxin response by the protein kinase PINIOD. *Cell* **100**, 469–478.

Ciardi, J.A., Tieman, D.M., Lund, S.T., Jones, J.B., Stall, R.E. and Klee, H.J. (2000) Response to *Xanthomoanas campestris* pv. *vesicatoria* in tomato involves regulation of ethylene receptor gene expression. *Plant Physiology* **123**, 81–92.

Clark, A.M., Verbeke, J.A. and Bohnert, H.J. (1992) Epidermis-specific gene expression in *Pachyphytum. Plant Cell* **4**, 1189–1198.

Clark, K.L., Larsen, P.B., Wang, X. and Chang, C. (1998) Association of the *Arabidopsis* CTR1 raf-like kinase with the ETR and ERS ethylene receptors. *Proceedings of the National Academy of Sciences, USA* **95**, 5401–5406.

Clark, S.E., Running, M.P. and Meyerowitz, E.M. (1993) CLAVATA1, a regulator of meristem and flower development in *Arabidopsis. Development* **119**, 397–418.

Clark, S.E., Running, M.P., and Meyerowitz, E.M. (1995) CLAVATA3 is a specific regulator of shoot and floral meristem development affecting the same processes as CLAVATA1. *Development* **121**, 2057–2067.

Clark, S.E., Williams, R.W. and Meyerowitz, E.M. (1997) The CLAVATA1 gene encodes a putative receptor kinase that controls shoot and floral meristem size in *Arabidopsis. Cell* **89**, 575–585.

Clements, J.C. and Atkins, C.A. (2001) Characterization of a non-abscission mutant in *Lupinus angustifolius* L.: Physiological aspects. *Annals of Botany* **88**, 629–635.

Close, T.J. (1996) Dehydrins: Emergence of a biochemical role of a family of plant dehydration proteins. *Physiologia Plantarum* **97**, 795–803.

Clouse, S.D., Zurek, D.M., McMorris, T.C. and Baker, M.E. (1992) Effect of brassinolide on gene expression in elongating soybean epicotyls. *Plant Physiology* **100**, 1377–1383.

Clouse, S.D., Hall, A.F., Langford, M., McMorris, T.C. and Baker, M.E. (1993) Physiological and molecular effects of brassinosteroids on *Arabidopsis thaliana. Journal of Plant Growth Regulation* **12**, 61–66.

Clouse, S.D., Langford, M. and McMorris, T.C. (1996) A brassinosteroid-insensitive mutant in *Arabidopsis thaliana* exhibits multiple defects in growth and development. *Plant Physiology* **111**, 671–678.

Clowes, F.A.L. (1978) Origin of the quiescent centre in *Zea mays*. *New Phytologist.* **80**, 409–419.

Cohen, J.D. and Bandurski, R.S. (1982) Chemistry and physiology of the bound auxins. *Annual Review of Plant Physiology* **33**, 403–430.

Compaan, B., Tang, W.C., Bisseling, T. and Franssen, H. (2001) ENOD40 expression in the pericycle precedes cortical cell division in *Rhizobium*-legume interaction and the highly conserved internal region of the gene does not encode a peptide. *Plant and Soil* **230**, 1–8.

Cooper, W.C. and Henry, W.H. (1971) Abscission chemicals in relation to citrus fruit harvest. *Journal of Agricultural and Food Chemistry* **19**, 559–563.

Cornford, C.A., Black, M., Chapman, J.M. and Baulcombe, D.C. (1986) Expression of α-amylase and other gibberellin-regulated genes in aleurone tissue of developing wheat grains. *Planta* **169**, 420–428.

Cornforth, J.W., Milborrow, B.V., Ryback, G. and Wareing, P.F. (1965) Chemistry and physiology of 'dormins' in sycamore. Identity of sycamore 'dormin' with abscisin II. *Nature* **205**, 1269–1270.

Coursol, S., Fan, L.-M., Le Stunff, H., Spiegel, S., Gilroy, S. and Assmann, S.M. (2003) Sphingolipid signalling in *Arabidopsis* guard cells involves heterotrimeric G proteins. *Nature* **423**, 651–654.

Cousson, A. and Vavasseur, A. (1998) Putative involvement of cytosolic Ca^{2+} and GTP-binding proteins in cyclic-GMP-mediated induction of stomatal opening by auxin in *Commelina communis* L. *Planta* **206**, 308–314.

Crabalona, L. (1967) Sur la présence de jasmonate de méthyle lévogyre [(pentène-2yl)-2 oxo-3 cyclopentylacétate de méthyle, cis] dans l'huile essentielle de romarin de Tunisie. *Comptes Rendus Hebdomadaires des Seances de l'Academie des Sciences. Serie C.* **264**, 2074–2076.

Creelman, R.A. and Mullet, J.E. (1995) Jasmonic acid distribution and action in plants: Regulation during development and response to biotic and abiotic stress. *Proceedings of the National Academy of Sciences, USA* **92**, 4114–4119.

Crick, F. (1970) Diffusion in embryogenesis. *Nature* **225**, 420–422.

Cusick, F. (1966) On phylogenetic and ontogenetic fusion. In: *Trends in Plant Morphogenesis*, Cutter, E.G. (ed.), Longmans, Green and Co. Ltd. London, pp. 170–183.

D'Agostino, I.B., Deruère, J. and Kieber, J.J. (2000) Characterization of the response of the *Arabidopsis* response regulator gene family to cytokinin. *Plant Physiology* **124**, 1706–1717.

Dan, H., Imaseki, H., Wasteneys, G.O. and Kazama, H. (2003) Ethylene stimulates endoreduplication but inhibits cytokinesis in cucumber hypocotyls epidermis. *Plant Physiology* **133**, 1726–1731.

Darvill, A., Augur, C., Bergmann, C., Carlson, R.W., Cheong, J.-J., Eberhard, S., Hahn, M.G., Ló, V.-M., Marfa, V., Meyer, B., Mohnen, D., O'Neill, M.A., Spiro, M.D., van Halbeek, H., York, W.S. and Albersheim, P. (1992) Oligosaccharins – oligosaccharides that regulate growth, development and defence responses in plants. *Glycobiology* **2**, 181–198.

Darwin, F. and Pertz, D.F.M. (1911) On a new method of estimating the aperture of stomata. *Philosophical Transactions of the Royal Society (London)* **B84**, 136–154.

Davies, R.T., Goetz, D.H., Lasswell, J., Anderson, M.N. and Bartel, B. (1999) *IAR3* encodes an auxin conjugate hydrolase from *Arabidopsis*. *Plant Cell* **11**, 365–376.

Day, C.D., Galgoci, B.F.C. and Irish, V.P. (1995) Genetic ablation of petal and stamen primordia to elucidate cell interactions during floral development. *Development* **121**, 2887–2895.

Del Campillo, E. and Bennett, A.B. (1996) Pedical breakstrength and cellulase gene expression during tomato flower abscission. *Plant Physiology* **111**, 813–820.

Del Pozo, J.C., Timpte, C., Tan, S., Callis, J. and Estelle, M. (1998) The ubiquitin-related protein RUB1 and auxin responses in *Arabidopsis. Science* **280**, 1760–1763.

Del Pozo, J.C., Dharmasiri, S., Hellmann, H., Walker, L., Gray, W.M. and Estelle, M. (2002) AXR1-ECR1-dependent conjugation of RUB1 to the *Arabidopsis* cullin AtCUL1 is required for auxin responses. *Plant Cell* **14**, 421–433.

Delledonne, M., Xia, Y., Dixon, R.A. and Lamb, C. (1998) Nitric oxide functions as a signal in plant disease resistance. *Nature* **394**, 585–588.

Demole, E., Lederer, E. and Mercier, D. (1962) Isolement et détermination de la structure du jasmonate de méthyle constituant odorant caractéristique de l'essence de jasmin. *Helvetica Chimica Acta* **45**, 675–685.

Demura, T. and Fukuda, H. (1994) Novel vascular cell-specific genes whose expression is regulated temporally and spatially during vascular system development. *Plant Cell* **6**, 967–981.

Devoto, A. and Turner, J.G. (2003) Regulation of jasmonate-mediated plant responses in *Arabidopsis. Annals of Botany* **92**, 329–337.

Diekmann, W., Venis, M.A. and Robinson, D.G. (1995) Auxins induce clustering of the auxin binding protein at the surface of maize coleoptile protoplasts. *Proceedings of the National Academy of Sciences, USA* **92**, 3425–3429.

Dill, A., Jung, H.-S. and Sun, T.-P. (2001) The DELLA motif is essential for gibberellin-induced degradation of RGA. *Proceedings of the National Academy of Sciences, USA* **98**, 14162–14167.

Ding, C.-K. and Wang, C.Y. (2003) The dual effects of methyl salicylate on ripening and expression of ethylene biosynthetic genes in tomato fruit. *Plant Science* **164**, 589–596.

Dingwall, C. (1991) Transport across the nuclear envelope: Enigmas and explanations. *BioEssays* **13**, 213–218.

Dixit, R. and Nasrallah, J.B. (2001) Recognizing self in the self-incompatability response. *Plant Physiology* **125**, 105–108.

Doan, D.N.P., Linnestad, C. and Olsen, O.-A. (1996) Isolation of molecular markers from the barley endosperm coenocyte and the surrounding nucellus cell layers. *Plant Molecular Biology* **31**, 877–886.

Doares, S.H., Syrovets, T., Weiler, E.W. and Ryan, C.A. (1995) Oligogalacturonides and chitosan activate plant defensive genes through the octadecanoid pathway. *Proceedings of the National Academy of Sciences, USA* **92**, 4095–4098.

Dolan, L. (1996) Pattern in root epidermis: An interplay of diffusible signals and cellular geometry. *Annals of Botany* **77**, 547–553.

Dolan, L., Linstead, P. and Roberts, K. (1997) Developmental regulation of pectic polysaccharides in the root meristem of *Arabidopsis. Journal of Experimental Botany* **48**, 713–720.

Dolmetsch, R.E., Xu, K. and Lewis, R.S. (1998) Calcium oscillations increase the efficiency and specificity of gene expression. *Nature* **392**, 933–936.

Donovan, N., Peart, J., Roberts, K., Knox, J.P., Wang, M. and Neill, S.J. (1993) Production and characterisation of monoclonal antibodies against guard cell protoplasts of *Pisum sativum. Journal of Experimental Botany* **44**, Supplement, P1.16.

Draper, J. (1997) Salicylate, superoxide synthesis and cell suicide in plant defence. *Trends in Plant Science* **2**, 162–165.

Dubrovsky, J.G., Doerner, P.W., Colon-Carmona, A. and Rost, T.L. (2000) Pericycle cell proliferation and lateral root initiation in *Arabidopsis. Plant Physiology* **124**, 1648–1654.

Durner, J. and Klessig, D.F. (1999) Nitric oxide as a signal in plants. *Current Opinion in Plant Biology* **2**, 369–374.

Durner, J., Wendehenne, D. and Klessig, D.F. (1998) Defense gene induction in tobacco by nitric oxide, cyclic GMP and cyclic ADP-ribose. *Proceedings of the National Academy of Sciences, USA* **95**, 10328–10333.

Dwek, R.A. (1995) Glycobiology: More functions for oligosaccharides. *Science* **269**, 1234–1235.

Eagles, C.F. and Wareing, P.F. (1963) Dormancy regulation in woody plants. Experimental induction of dormancy in *Betula pubescens*. *Nature* **199**, 874–875.

Eberhard, S., Doubrava, N., Marfa, V., Mohnen, D., Southwick, A., Darvill, A. and Albersheim, P. (1989) Pectic cell wall fragments regulate tobacco thin-cell-layer explant morphogenesis. *Plant Cell* **1**, 747–755.

Elder, R. H. and Osborne, D. J. (1993) Function of DNA synthesis and DNA repair in the survival of embryos during early germination and in dormancy. *Seed Science Research* **3**, 43–53.

Enari, M., Sakahira, H., Yokoyama, H., Okawa, K., Iwamatsu, A. and Nagata, S. (1998) A caspase-activated DNase that degrades DNA during apoptosis, and its inhibitor ICAD. *Nature* **391**, 43–50.

Engelmann, W., Sommerkamp, A., Veit, S. and Hans, J. (1997) Methyl jasmonate affects the circadian petal movement of Kalanchloe flowers. *Biological Rhythms Research* **28**, 377–390.

English, P.J., Lycett, G.W., Roberts, J.A. and Jackson, M.B. (1995) Increased 1-aminocyclopropane-1-carboxylic acid oxidase activity in shoots of flooded tomato plants raises ethylene production to physiologically active levels. *Plant Physiology* **109**, 1435–1440.

Ephritikhine, G., Barbier-Brygoo, H., Muller, J.F. and Guern, J. (1987) Auxin effect on the transmembranes potential difference of wild-type and mutant tobacco protoplasts exhibiting a differential sensitivity to auxin. *Plant Physiology* **83**, 801–804.

Esau, K. (1965) *Plant Anatomy*, 2nd Edition. John Wiley & Sons Inc., New York.

Evans, D.E., Dodds, J.H., Lloyd, P.C., ap Gwynn, I. and Hall, M.A. (1982) A study of the subcellular localisation of an ethylene binding site in developing cotyledons of *Phaseolus vulgaris* by high resolution autoradiography. *Planta* **154**, 48–52.

Evans, M., Black, M. and Chapman, J. (1975) Induction of hormone sensitivity by dehydration is one possible role for drying in cereal seeds. *Nature* **258**, 144–145.

Fahn, A. (1990) *Plant Anatomy*, 4th Edition, Pergamon Press, Ellmsford, NY.

Fan, D.F. and Maclachlan, G.A. (1967) Studies on the regulation of cellulase activity and growth in excised pea epicotyl sections. *Canadian Journal of Botany* **45**, 1837–1844.

Farkas, V. and Maclachlan, G. (1988) Stimulation of pea 1,4-β-glucanase activity by oligosaccharides derived from xyloglucan. *Carbohydrate Research* **184**, 213–220.

Farmer, E.E. and Ryan, C.A. (1990) Inter-plant communication: Airborne methyl jasmonate induces synthesis of proteinase inhibitors in plant leaves. *Proceedings of the National Academy of Sciences, USA* **87**, 7713–7716.

Farmer, E.E., Pearce, G. and Ryan, C.A. (1989) *In vitro* phosphorylation of plant plasma membrane proteins in response to the proteinase inhibitor inducing factor. *Proceedings of the National Academy of Sciences, USA* **86**, 1539–1542.

Farmer, E.E., Moloshok, T.D., Saxton, M.J. and Ryan, C.A. (1991) Oligosaccharide signalling in plants – specificity of oligouronide-enhanced plasma-membrane protein phosphorylation. *Journal of Biological Chemistry* **266**, 3140–3145.

Fath, A., Bethke, P.C. and Jones, R.L. (1999) Barley aleurone cell death is not apoptotic: Characterization of nuclease activities and DNA degradation. *Plant Journal* **20**, 305–315.

Feldman, K.A. and Marks, M.D. (1987) *Agrobacterium*-mediated transformation of germinating seeds of *Arabidopsis thaliana:* A non-tissue culture approach. *Molecular and General Genetics* **208**, 1–9.

Feldman, L.J. (1976) The *de novo* origin of the quiescent center in regenerating root apices of *Zea mays. Planta* **128**, 207–212.

Feldman, L.J. and Torrey, J.G. (1976) The isolation and culture *in vitro* of the quiescent centre of *Zea mays. American Journal of Botany* **63**, 345–355.

Fernandez, D.E., Heck, G.R., Perry, S.E., Patterson, S.E, Bleeker, A.B. and Fang, S.-C. (2000) The embryo MADS domain factor AGL 15 acts postembryonically: Inhibition of perianth senescence and abscission via constitutive expression. *Plant Cell* **12**, 183–198.

Feys, B.J., Benedetti, C.E., Penfold, C.N. and Turner, J.G. (1994) *Arabidopsis* mutants selected for resistance to the phytotoxin coronatine are male sterile, insensitive to methyl jasmonate, and resistant to a bacterial pathogen. *Plant Cell* **6**, 751–759.

Finkelstein, R.R., Gampala, S.S.L. and Rock, C.D. (2002) Abscisic acid signalling in seeds and seedlings. *Plant Cell*, **14**, Supplement, S15–S45.

Fletcher, J.C. 2002. Shoot and floral meristem maintenance in *Arabidopsis. Annual Review of Plant Physiology and Plant Molecular Biology* **53**, 45–66.

Fletcher, J.C., Brand, U., Running, M.P., Simon, R. and Meyerowitz, E.M. (1999) Signalling of cell fate decisions by CLAVATA3 in *Arabidopsis* shoot meristems. *Science* **283**, 1911–1914.

Forde, B.G. (2000) Nitrate transporters in plants: Structure, function and regulation. *Biochemica et Biophysica Acta* **1465**, 219–235.

Forde, B.G. (2002) The role of long-distance signalling in plant responses to nitrate and other nutrients. *Journal of Experimental Botany* **53**, 39–43.

Friedrichsen, D.M., Joazeiro, C.A.P., Li, J., Hunter, T. and Chory, J. (2000) Brassinosteroid-insensitive-1 is a ubiquitously expressed leucine-rich repeat receptor serine/threonine kinase. *Plant Physiology* **123**, 1247–1256.

Friml, J. and Palme, K. (2002) Polar auxin transport – old questions and new concepts? *Plant Molecular Biology* **49**, 273–284.

Friml, J., Wisniewska, J., Benkova, E., Mendgen, K. and Palme, K. (2002a) Lateral relocation of auxin efflux regulator PIN3 mediates tropism in *Arabidopsis. Nature* **415**, 806–809.

Friml, J., Benkova, E., Blilou, I., Wisniewska, J., Hamann, T., Ljung, K., Woody, S., Sandberg, G., Scheres, B., Jürgens, G. and Palme, K. (2002b) AtPIN4 mediates sink-driven auxin gradients and root patterning in *Arabidopsis. Cell* **108**, 661–673.

Friml, J., Vietin, A., Sauer, M., Weijers, D., Schwartz, H., Hamann, T., Offringa, R. and Jürgens, G. (2003) Efflux-dependent auxin gradients establish the apical-basal axis of *Arabidopsis. Nature* **426**, 147–153.

Fry, S.C., Aldington, S., Hetherington, P.R. and Aitken, J. (1993) Oligosaccharides as signals and substrates in the plant cell wall. *Plant Physiology* **103**, 1–5.

Fu, X. and Harberd, N. (2003) Auxin promotes *Arabidopsis* root growth by modulating gibberellin response. *Nature* **421**, 740–743.

Fu, X., Sudhakar, D., Peng, J., Richards, D.E., Christou, P. and Harberd, P. (2001) Expression of *Arabidopsis* GAI in transgenic rice represses multiple gibberellin responses. *Plant Cell* **13**, 1791–1802.

Fu, X., Richards, D.E., Ait-ali, T., Hynes, L.W., Ougham, H., Peng, J. and Harberd, N.P. (2002) Gibberellin-mediated proteasome-dependent degradation of the barley DELLA protein SLN1 repressor. *Plant Cell* **14**, 3191–3200.

Fujioka, S. and Yokota, T. (2003) Biosynthesis and metabolism of brassinosteroids. *Annual Review of Plant Biology* **54**, 137–164.

Fujita, H. and Syono, K. (1996) Genetic analysis of the effects of polar auxin transport inhibitors on root growth in *Arabidopsis thaliana*. *Plant and Cell Physiology* **37**, 1094–1101.

Fukaki, H., Wysocka-Diller, J., Kato, T., Fujisawa, H., Benfey, P.N. and Tasaka, M. (1998) Genetic evidence that the endodermis is essential for shoot gravitropism in *Arabidopsis thaliana*. *Plant Journal* **14**, 425–430.

Fukuda, H. (1994) Redifferentiation of single mesophyll cells into tracheary elements. *International Journal of Plant Science* **155**, 262–271.

Fukuda, H. (1996) Xylogenesis: Initiation, progression, and cell death. *Annual Review of Plant Physiology and Plant Molecular Biology* **47**, 299–325.

Fukuda, H. (1997) Tracheary element differentiation. *Plant Cell* **9**, 1147–1156.

Fukuda, H. and Komamine, A. (1980) Direct evidence for cytodifferentiation to tracheary elements without intervening mitosis in a culture of single cells isolated from the mesophyll of *Zinnia elegans*. *Plant Physiology* **65**, 61–64.

Gaff, D.F. and Ellis, R.P. (1974) Southern African grasses with foliage that revives after dehydration. *Bothalia* **11**, 305–308.

Gaffney, T., Freidrich, L., Vernooji, B., Negrotto, D., Nye, G., Uknes, S., Ward, E., Kessmann, H. and Ryals, J. (1993) Requirement of salicylic acid for the induction of systemic acquired resistance. *Science* **261**, 754–756.

Galway, M.E., Marucci, J.D., Lloyd, A.M., Walbot, V., Davis, R.W. and Schiefelbein, J.W. (1994) The *TTG* gene is required to specify epidermal cell fate and cell patterning in the *Arabidopsis* root. *Developmental Biology* **166**, 740–754.

Galweiler, L., Guan, C., Muller, A., Wisman, E., Mendgen, K., Yephremov, A. and Palme, K. (1998) Regulation of polar auxin transport by *AtPIN1* in *Arabidopsis* vascular tissue. *Science* **282**, 2226–2230.

Gamble, R., Coonfield, M. and Schaller, G.E. (1998) Histidine kinase activity of the ETR1 ethylene receptor from *Arabidopsis*. *Proceedings of the National Academy of Sciences, USA* **95**, 7825–7829.

Gamble, R.L., Qu, X. and Schaller, G.E. (2002) Mutational analysis of the ethylene receptor ETR1. Role of the histidine kinase domain in dominant ethylene insensitivity. *Plant Physiology* **128**, 1428–1438.

Gan, S. and Amasino, R.M. (1995) Inhibition of leaf senescence by autoregulated production of cytokinin. *Science* **270**, 1986–1988.

Gane, R. (1934) Production of ethylene by some ripening fruit. *Nature* **134**, 1008.

Gao, D., Knight, M.R., Trewavas, A.J., Sattelmacher, B. and Plieth, C. (2004) Self-reporting *Arabidopsis* expressing pH and $[Ca^{2+}]$ indicators unveil ion dynamics in the cytoplasm and in the apoplast under abiotic stress. *Plant Physiology* **134**, 898–908.

Gao, Z., Chen, Y.-F., Randlett, M.D., Zhao, X.-C., Findell, J.L., Kieber, J.J. and Schaller, G.E. (2003) Localization of the Raf-like kinase CTR1 to the endoplasmic reticulum of *Arabidopsis* through participation in ethylene receptor signalling complexes. *Journal of Biological Chemistry*, **278**, 34725–34732.

Geldner, M., Friml, J., York-Dieter, S., Jurgens, G. and Palme, K. (2001) Auxin transport inhibitors block PIN1 cycling and vesicle trafficking. *Nature* **413**, 425–428.

Ghoshroy, S., Lartey, R., Sheng, J. and Citousky, V. (1997) Transport of proteins and nucleic acids through plasmodesmata. *Annual Review of Plant Physiology and Plant Molecular Biology* **48**, 27–50.

Gilroy, S. and Jones, R.L. (1994) Perception of gibberellin and abscisic acid at the external face of plasma membrane of the barley (*Hordeum vulgare* L) aleurone protoplasts. *Plant Physiology* **104**, 1185–1192.

Girardin, J.P.L. (1864) Einfluss des Leuchtgases auf die Promenaden und Strassen Baume. *Jahresber Agrikulturchem, Versuchssta, Berlin,* **7**, 199–200.

Goffreda, J.C., Szymkowiak, E.J., Sussex, I.M. and Mutschler, M.A. (1990) Chimeric tomato plants show that aphid resistance and triacylglucose production are epidermal autonomous characters. *Plant Cell* **2**, 643–649.

Goldsmith, M.H.M. (1977) The polar transport of auxin. *Annual Reviews of Plant Physiology* **28**, 439–478.

Goldsworthy, A. and Mina, M.G. (1991) Electrical patterns of tobacco cells in media containing indole-3-acetic acid or 2,4-dichlorophenoxyacetic acid, their relation to organogenesis and herbicide action. *Planta* **183**, 368–373.

Gollin, D.J., Darvill, A.G. and Albersheim, P. (1984) Plant cell wall fragments inhibit flowering and promote vegetative growth in *Lemna minor. Biology of the Cell* **51**, 275–280.

Golub, S.J. and Wetmore, R.H. (1948) Studies of development in the vegetative shoot of *Equisetum arvense* L. I. The shoot apex. *American Journal of Botany* **35**, 755–762.

Gomez-Cadenas, A., Zentella, R., Walker-Simmons, M.K. and Ho, T.-H.D. (2001) Gibberellin/abscisic acid antagonism in barley aleurone cells: Site of action of the protein kinase PKABA1 in relation to gibberellin signaling molecules. *Plant Cell* **13**, 667–679.

Goring, D.R. and Rothstein, S.J. (1992) The *S*-locus receptor kinase gene in a self-incompatible *Brassica napus* line encodes a functional serine/threonine kinase. *Plant Cell* **4**, 1273–1281.

Gorst, J., Overall, R.L., and Wernicke, W. (1987) Ionic currents traversing cell clusters from carrot suspension cultures reveal perpetuation of morphogenetic potential as distinct from induction to embryogenesis. *Cell Differentiation* **21**, 101–110.

Goto, N., Starke, M. and Kranz, A.R. (1987) Effect of gibberellins on flower development of the *pin-formed* mutant of *Arabidopsis thaliana. Arabidopsis Information Services* **23**, 66–71.

Grabov, A. and Blatt, M.R. (1999) A steep dependence of inward-rectifying potassium channels on cytosolic free calcium concentration increase evoked by hyperpolarization in guard cells. *Plant Physiology* **119**, 277–287.

Granell, A., Cercos, M. and Carbonell, J. (1998) Plant cysteine proteinases in germination and senescence. In: *Handbook of Proteolytic Enzymes*, Barret A.J., Rawlings, N.D. and Woessner, J.F. (eds.). Academic Press, San Diego, London, pp. 578–583.

Gray, W.M., del Pozo, J.C., Walker, L., Hobbie, L., Risseeuw, E., Banks, T., Crosby, W.L., Yang, M., Hong, M. and Estelle, M. (1999) Identification of an SCF-ligase complex required for auxin response in *Arabidopsis thaliana. Genes and Development* **13**, 1678–1691.

Gray, W.M., Kepinski, S., Rouse, D., Leyser, O. and Estelle, M. (2001) Auxin regulates SCF[TIR1]-dependent degradation of AUX/IAA proteins. *Nature* **414**, 271–276.

Grbic, V. and Bleecker, A.B. (1995) Ethylene regulates the timing of leaf senescence in *Arabidopsis. Plant Journal* **8**, 595–602.

Green, P.B. (1999) Expression of pattern in plants: Combining molecules and calculus-based biophysical paradigms. *American Journal of Botany* **86**, 1059–1076.

Green, T.R. and Ryan, C.A. (1972) Wound-induced proteinase inhibitor in plant leaves: A possible defense mechanism against insects. *Science* **175**, 776–777.

Grove, M.D., Spencer, G.F., Rohwedder, W.K., Mandava, N., Worley, J.F., Warthen, J.D., Steffens, G.L., Flippen-Anderson, J.L. and Cook, J.C. (1979) Brassinolide, a plant growth-promoting steroid isolated from *Brassica napus* pollen. *Nature* **281**, 216–217.

Gubler, F., Falla, R., Roberts, J.K. and Jacobsen, J.V. (1995) Gibberellin-regulated expression of a myb gene in barley aleurone cells: Evidence for myb transactivation of a high-pl α-amylase gene promoter. *Plant Cell* **7**, 1879–1891.

Gubler, F., Chandler, P.M., White, R.G., Llewellyn, D.J. and Jacobsen, J.V. (2002) Gibberellin signaling in barley aleurone cells. Control of SLN1 and GAMYB expression. *Plant Physiology* **129**, 191–200.

Guilfoyle, T.J. and Hagan, G. (2001) Auxin response factors. *Journal of Plant Growth Regulation* **20**, 281–291.

Guinel, F.C. and Geil, R.D. (2002) A model for the development of the rhizobial and arbuscular mycorrhizal symbiosis in legumes and its use to understand the roles of ethylene in the establishment of these two symbiosis. *Canadian Journal of Botany* **80**, 695–720.

Gunawardena, A.H.L.A.N., Pearce, D.M., Jackson, M.B., Hawes, C.R. and Evans, D.E. (2001a) Characterisation of programmed cell death during aerenchyma formation induced by ethylene or hypoxia in roots of maize (*Zea mays* L.) *Planta* **212**, 205–214.

Gunawardena, A.H.L.A.N., Pearce, D.M.E., Jackson, M.B., Hawes, C.R. and Evans, D.E. (2001b) Rapid changes in cell wall pectic polysaccharides are closely associated with early stages of aerenchyma formation, a spatially localized form of programmed cell death in roots of maize (*Zea mays* L.) promoted by ethylene. *Plant, Cell and Environment* **24**, 1369–1375.

Guo, H. and Ecker, J.R. (2003) Plant responses to ethylene gas are mediated by SCF[EBP1/EBP2]-dependent proteolysis of EIN3 transcription factor. *Cell* **115**, 667–677.

Guzman, P. and Ecker, J.R. (1990) Exploring the triple response *Arabidopsis* to identify ethylene-related mutants. *Plant Cell* **2**, 513–523.

Haberer, G. and Kieber, J.J. (2002) Cytokinins: New insights into a classic phytohormone. *Plant Physiology* **128**, 354–362.

Hackett, R.M., Ho, C., Lin, Z., Foote, H.C.C., Fray, R.G. and Grierson, D. (2000) Antisense inhibition of the *Nr* gene restores normal ripening to the tomato *Never ripe* mutant, consistent with the ethylene receptor-inhibition model. *Plant Physiology* **124**, 1079–1086.

Hadfield, K.A. and Bennett, A.B. (1998) Polygalacturonases: Many genes in search of a function. *Plant Physiology* **117**, 337–343.

Hake, S. and Freeling, M. (1986) Analysis of genetic mosaics shows that extra epidermal cell divisions in *Knotted* mutant maize plants are induced by adjacent microphyll cells. *Nature* **320**, 621–623.

Hall, A.E., Findell, J.l., Schaller, G.E., Sisler, E.C. and Bleecker, A.B. (2000) Ethylene perception by the ERS1 protein in *Arabidopsis*. *Plant Physiology* **123**, 1449–1458.

Hamann, T., Benkova, E., Bäurle, I., Kientz, M. and Jürgens, G. (2002) The *Arabidopsis* BODENLOS gene encodes an auxin response protein inhibiting MONOPTEROS-mediated embryo patterning. *Genes and Development* **16**, 1610–1615.

Han, B., Berjak, P., Pammenter, N., Farrant, J., and Kermode, A.R. (1997) The recalcitrant plant species, *Castanospermum australe* and *Trichilia dregeana*, differ in their ability to produce dehydrin-related polypeptides during seed maturation and in response to ABA or water-deficit-related stresses. *Journal of Experimental Botany* **48**, 1717–1726.

Hanada, K., Nishiuchi, Y. and Hirano, H. (2003) Amino acid residues on the surface of the soybean 4-kDa peptide involved in the interaction with its binding protein. *European Journal of Biochemistry* **270**, 2583–2592.

Hangarter, R.P. and Good, N.E. (1981) Evidence that IAA conjugates are slow-release sources of free IAA in plant tissues. *Plant Physiology* **68**, 1424–1427.

Hardtke, C. and Berleth, T. (1998) The *Arabidopsis* gene MONOPTEROS encodes a transcription factor mediating embryo axis formation and vascular development. *EMBO Journal* **17**, 1405–1411.

Harper, J.R. and Balke, N.E. (1981) Characterization of the inhibition of K^+ absorption in oat roots by salicylic acid. *Plant Physiology* **68**, 1349–1353.

Hartung, W. (1983) The site of action of abscisic acid at the guard cell plasmalemma of *Valerianella locusta*. *Plant Cell and Environment* **6**, 427–428.

Hartung, W., Sauter, A. and Hose, E. (2002) Abscisic acid in the xylem: Where does it come from, where does it go? *Journal of Experimental Botany* **53**, 27–32.

Haughn, G.W. and Somerville, C.R. (1986) Sulfonylurea-resistant mutants of *Arabidopsis thaliana*. *Molecular and General Genetics* **204**, 430–434.

He, J.-X., Gendron, J.M., Yang, Y., Li, J. and Wang, Z.-Y. (2002) The GKS3-like kinase BIN2 phosphorylates and destabilizes BZR1, a positive regulator of the brassinosteroid signalling pathway in *Arabidopsis*. *Proceedings of the National Academy of Sciences, USA* **99**, 10185–10190.

He, Z., Wang, Z.-Y., Li, J., Zhu, Q., Lamb, C., Ronald, P. and Chory, J. (2000) Perception of brassinosteroidsby the extracellular domain of the receptor kinase BRI1. *Science* **288**, 2360–2363.

Hecht, K. (1912) Studien über den Vorgang der Plasmolyse. *Beitrage zür Biologie der Pflanzen* **11**, 133–189.

Hedden, P. and Phillips, A.L. (2000) Gibberellin metabolism: new insights revealed by the genes. *Trends in Plant Sciences* **5**, 523–530.

Heidstra, R., Yang, W.C., Yalcin, T., Peck, S., Emons, A.-M., van Kammen, A. and Biseling, T. (1997) Ethylene provides positional information on cortical cell division but is not involved in Nod factor-induced root hair tip growth in *Rhizobium*-legume interaction. *Development* **124**, 1781–1787.

Henderson, J., Bauly, J.M., Ashford, D.A., Oliver, S.C., Hawes, C.R., Lazarus, C.M. and Venis, M.A. (1997) Retention of maize auxin-binding protein in the endoplasmic reticulum: Quantifying escape and the role of auxin. *Planta* **202**, 313–323.

Henderson, J., Lyne, L. and Osborne, D.J. (2001a) Failed expression of an endo-β-1,4-glucan-hydrolase (cellulase) in a non-abscinding mutant of *Lupinus angustifolius* cv. Danja. *Phytochemistry* **58**, 1025–1034.

Henderson, J., Davies, H.A., Heyes, S.J. and Osborne, D.J. (2001b) The study of a monocotyledon abscission zone using microscopic, chemical, enzymatic and solid state [13]CCP/MAS NMR analyses. *Phytochemistry* **56**, 131–139.

Hensel, L.L., Grbic, V., Baumgerten, D.A. and Bleecker, A.B. (1993) Developmental and age-related processes that influence the longevity and senescence of photosynthetic tissues in *Arabidopsis*. *Plant Cell* **5**, 553–564.

Herschbach, C., van der Zalm, E., Schneider, A., Jouanin, L., De Kok, L.J. and Rennenberg, H. (2000) Regulation of sulfur nutrition in wild-type and transgenic poplar overexpressing 8-glutamylcysteine synthetase in the cytosol as affected by atmospheric H_2S. *Plant Physiology* **124**, 461–474.

Hertel, R., Thomson, K.-St. and Russo, V.E.A. (1972) *In vitro* auxin binding to particulate cell fractions from corn coleoptiles. *Planta* **107**, 325–340.

Hesse, T., Feldwisch, J., Balschusemann, D., Bauw, G., Puype, M., Vandekeckhove, J., Löbler, M., Klämbt, D., Schell, J. and Palme, K. (1989) Molecular cloning and structural anlysis of a gene from *Zea mays (L.)* coding for the plant hormone auxin. *EMBO Journal* **8**, 2453–2461.

Heuros, G., Varotto, S., Salamini, F. and Thompson, R.D. (1995) Molecular characterization of BET1, a gene expressed in the endosperm transfer cells of maize. *Plant Cell* **7**, 747–757.

Hey, S.J., Bacon, A., Burnett, E. and Neill, S.J. (1997) Abscisic acid signal transduction in epidermal cells of *Pisum sativum* L. Argenteum: Both dehydrin mRNA accumulation and stomatal response require protein phosphorylation and de-phosphorylation. *Planta* **202**, 85–92.

Himanen, K., Boucheron, E., Vanneste, S., Engler, J. de A., Inze, D. and Beeckman, T. (2002) Auxin-mediated cell cycle activation during early lateral root initiation. *Plant Cell* **14**, 2339–2351.

Holroyd, G.H., Hetherington, A.M. and Gray, J.E. (2002) A role for the cuticular waxes in the environmental control of stomatal development. *New Phytologist* **153**, 433–439.

Hooley, R., Beale, M.H. and Smith, S.J. (1991) Gibberellin perception at the plasma membrane of *Avena fatua* aleurone protoplasts. *Planta* **183**, 274–280.

Hooley, R., Beale, M.H., Smith, S.J., Walker, R.P., Rushton, P.J., Whitford, P.N. and Lazarus, C.M. (1992) Gibberellin perception and the *Avena fatua* aleurone: Do our molecular keys fit the correct locks. *Biochemical Society Transcations* **20**, 85–89.

Hooley, R., Smith, S.J., Beale, M.H. and Walker, R.P. (1993) *In vivo* photoaffinity labelling of gibberellin-binding proteins in *Avena fatua* aleurone. *Australian Journal of Plant Physiology* **20**, 573–584.

Hornberg, C. and Weiler, E.W. (1984) High affinity binding sites for abscisic acid on the plasmalemma of *Vicia faba* guard cells. *Nature* **310**, 321–324.

Horton, R.F. and Osborne, D.J. (1967) Senescence, abscission and cellulase activity in *Phaseolus vulgaris. Nature* **214**, 1086–1088.

Howe, G.A. and Ryan, C.A. (1999) Suppressors of systemin signalling identify genes in the tomato wound response pathway. *Genetics* **153**, 1411–1421.

Hua, J. and Meyerowitz, E.M. (1998) Ethylene responses are negatively regulated by a receptor gene family in *Arabidopsis thaliana. Cell* **94**, 261–271.

Hua, J., Chang, C., Sun, Q. and Meyerowitz, E.M. (1995) Ethylene insensitivity conferred *Arabidopsis* ERS gene. *Science* **269**, 1712–1714.

Hua, J., Sakai, S., Nourizadeh, S., Chen, Q.C., Bleecker, A.B., Ecker, J.R. and Meyerowitz, E.M. (1998) EIN4 and ERS2 are members of the putative ethylene receptor gene family in *Arabidopsis. Plant Cell* **10**, 1321–1332.

Huh, W.-K. I., Falvo, J.V., Gerke, L.C., Carroll, A.S., Howson, R.W., Weissman, J.S. and O'Shea, E.K. (2003) Global analysis of protein localization in budding yeast. *Nature* **425**, 686–691.

Hull, A.K., Vij, R. and Celenza, J.L. (2000) *Arabidopsis* cytochrome P450s that catalyze the first step of tryptophan-dependent indole-3-acetic acid biosynthesis. *Proceedings of the National Academy of Sciences USA* **97**, 2379–2384.

Hülskamp, M., Miséra, S. and Jürgens, G. (1994) Genetic dissection of trichome cell development in *Arabidopsis. Cell* **76**, 555–566.

Hwang, I. and Sheen, J. (2001) Two-component circuitry in *Arabidopsis* cytokinin signal transduction. *Nature* **413**, 383–389.

Hwang, I., Chen, H.-C. and Sheen, J. (2002) Two-component signal transduction pathways in *Arabidopsis. Plant Physiology* **129**, 500–515.

Ikeda, A., Ueguchi-Tanaka, M., Sonoda, Y., Kitano, H., Koshioka, M., Futsuhara, Y., Matsuoka, M. and Yamaguchi, J. (2001) *Slender* rice, a constitutive gibberellin response mutant is caused by a null mutation of the *SLR1* gene, an ortholog of the height-regulating gene *GAI/RGA/RHT/D8*. *Plant Cell* **13**, 999–1010.

Imamura, A., Hanaki, N., Umeda, H., Nakamura, A., Suzuki, T., Ueguchi, C. and Mizuno, T. (1998) Response regulators implicated in His-to-Asp phospho-transfer signalling in *Arabidopsis*. *Proceedings of the National Academy of Sciences, USA* **95**, 2691–2696.

Imamura, A., Hanaki, N., Nakamura, A., Suzuki, T., Taniguchi, M., Kiba, T., Ueguchi, C., Sugiyama, T. and Mizuno, T. (1999) Compilation and characterization of *Arabidopsis thaliana* response regulators implicated in His-Asp phospho-relay signal transduction. *Plant and Cell Physiology* **40**, 733–742.

Imaseki, H., Pjon, C.J. and Furuya, M. (1971) Phytochrome action in *Oryza sativa* L. *Plant Physiology* **48**, 241–244.

Imber, D. and Tal, M. (1970) Phenotypic reversion of *flacca*, a wilty mutant of tomato, by abscisic acid. *Science* **169**, 592–593.

Ingold, E., Sugiyama, M., and Komamine, A. (1990) L-α-aminoxy-β-phenylpropionic acid inhibits lignification but not the differentiation to tracheary elements of isolated mesophyll cells of *Zinnia elegans*. *Physiologia Plantarum* **78**, 67–74.

Inohara, N., Shimomura, S., Fukui, T. and Futai, M. (1989) Auxin-binding protein located in the endoplasmatic reticulum of maize shoots: Molecular cloning and complete primary structure. *Proceedings of the National Academy of Sciences, USA* **83**, 3654–3568.

Inoue, T., Higuchi, M., Hashimoto, Y., Seki, M., Kobayasjhi, M., Kato, T., Tabata, S., Shinozaki, K. and Katimoto, T. (2001) Identification of CRE1 as a cytokinin receptor from *Arabidopsis*. *Nature* **409**, 1060–1063.

Irvine, R.F. and Osborne, D.J. (1973) The effect of ethylene on 1-^{14}C glycerol incorporation into phospholipids of etiolated pea stems. *Biochemical Journal* **136**, 1133–1135.

Itoh, H., Ueguchi-Tanaka, M., Sato, Y., Ashikari, M. and Matsuoka, M. (2002) The gibberellin signalling pathway is regulated by the appearance and disappearance of SLENDER RICE1 in nuclei. *Plant Cell* **14**, 57–70.

Jacinto, T., McGurl, B., Franceschi, V., Delano-Freier, J. and Ryan, C.A. (1997) Tomato prosystemin promoter confers wound-inducible, vascular bundle-specific expression of the β-glucoronidase gene in transgenic tomato plants. *Planta* **203**, 406–412.

Jackson, D., Veit, B. and Hake, S. (1994) Expression of maize *KNOTTED1* relates homeobox genes in the shoot apical meristem predicts patterns of morphogenesis in the vegetative shoot. *Development* **120**, 405–413.

Jackson, M.B. (1993) Are plant hormones involved in root to shoot communication? *Advances in Botanical Research* **19**, 103–187.

Jackson, M.B. (2002) Long-distance signalling from roots to shoots assessed: The flooding story. *Journal of Experimental Botany* **53**, 175–181.

Jackson, M.B. and Armstrong, W. (1999) Formation of aerenchyma and the processes of plant ventilation in relation to soil flooding and submergence. *Plant Biology* **1**, 274–287.

Jackson, M.B., Summers, J.E. and Voesenek, L.A.C.J. (1997) *Potamogeton pectinatus*: A vascular plant that makes no ethylene. In: *Biology and Biotechnology of the Plant Hormone Ethylene*, Kanellis, A.K. et al. (eds.). Kluwer Academic Press, Dordrecht, pp. 229–237.

Jacobs, W.P., McCready, C.C. and Osborne, D.J. (1966) Transport of the auxin 2,4-dichlorophenoxyacetic acid through abscission zones, pulvini, and petioles of *Phaseolus vulgaris*. *Plant Physiology* **41**, 725–730.

Jaffe, L.F. (1958) Tropistic responses of zygotes of the Fucaceae to polarized light. *Experimental Cell Research* **15**, 282–299.

222 REFERENCES

Jaffe, L.F. (1966) Electrical currents through the developing *Fucus* egg. *Proceedings of the National Academy of Sciences, USA* **56**, 1102–1109.

Jaffe, M.J., Leopold, A.C. and Staples, R.C. (2002) Thigmo responses in plants and fungi. *American Journal of Botany* **89**, 375–382.

Jeong, S., Trotochaud, A.E. and Clark, S.E. (1999) The *Arabidopsis* CLAVATA2 gene encodes a receptor-like protein required for the stability of the CLAVATA1 receptor-like kinase. *Plant Cell* **11**, 1925–1934.

Jerie, P.H., Shaari, A.R. and Hall, M.A. (1979) The compartmentation of ethylene in developing cotyledons of *Phaseolus vulgaris* L. *Planta* **144**, 503–507.

John, I., Drake, R., Farrell, A., Cooper, W., Lee, P., Horton, P. and Grierson, D. (1995) Delayed leaf senescence in ethylene-deficient ACC oxidase anti-sense tomato plants: Molecular and physiological analysis. *Plant Journal* **7**, 483–490.

Johnson, M.A. and Preuss, D. (2003) On your mark, get set, grow! LePRK2-LAT52 interactions regulate pollen tube growth. *Trends in Plant Science* **8**, 97–99.

Jones, A.M. and Herman, E.M. (1993) KDEL-containing auxin-binding protein is secreted to the plasma membrane and cell wall. *Plant Physiology* **101**, 595–606.

Jones, A.M., Im, K.-H., Savka, M.A., Wu, M.-J., DeWitt, G., Shillito, R. and Binns, A.N. (1998a) Auxin-dependent cell expansion mediated by overexpressed auxin-binding protein 1. *Science* **282**, 1114–1117.

Jones, H.D., Smith, S.J., Desikan, R., Plakidou-Dymock, S., Lovegrove, A. and Hooley, R. (1998b) Heterotrimeric G proteins are implicated in gibberellin induction of α-amylase gene expression in wild oat aleurone. *Plant Cell* **10**, 245–253.

Kagan, M.L. and Sachs, T. (1991) Development of immature stomata: Evidence for epigenetic selection of a spacing pattern. *Developmental Biology* **146**, 100–105.

Kagan, M.L., Novoplansky, N. and Sachs, T. (1992) Variable cell lineages from the functional pea epidermis. *Annals of Botany* **69**, 303–312.

Kaihara, S., Watanabe, K. and Takimoto, A. (1981) Flower-inducing effect of benzoic and salicylic acids on various strains of *Lemna paucicostata* and *L. minor*. *Plant Cell Physiology* **22**, 819–825.

Kakimoto, T. (1996) CKI1, a histidine kinase homolog implicated in cytokinin signal transduction. *Science* **274**, 982–985.

Kakimoto, T. (1998) Cytokinin signalling. *Current Opinions in Plant Biology* **1**, 399–403.

Kakimoto, T. (2001) Identification of plant cytokinin biosynthetic enzymes as dimethylallyl diphosphate: ATP/ADP isopentenyltransferases. *Plant Cell Physiology* **42**, 677–685.

Kalla, R., Shimamoto, K., Potter, R., Nielsen, P.S., Linnestad, C. and Olsen, O.-A. (1994) The promoter of the barley aleurone-specific gene encoding a putative 7 kDa lipid transfer protein confers aleurone-specific gene expression in transgenic rice. *Plant Journal* **6**, 849–860.

Kaminek, M., Dobrev, P., Gaudinová, A., Motyka, V., Malbeck, J., Trávièkova, A. and Trčková, M. (2000) Potential physiological function of cytokinin binding proteins in seeds of cereals. *Plant Physiology and Biochemistry* **38**, Supplement, S79.

Kaminek, M., Trčková, M., Fox, J.E. and Gaudinová, A. (2003) Comparison of cytokinin-binding proteins from wheat and oat grains. *Physiologia Plantarum* **117**, 453–458.

Karlson, P. (1956) Biochemical studies on insect hormones. *Vitamins and Hormones* **14**, 227–266.

Kawai, M., Samarajeewa, P.K., Barrero, R.A., Nishiguchi, M. and Uchimiya, H. (1998) Cellular dissection of the degradation pattern of cortical cell death during aerenchyma formation of rice roots. *Planta* **204**, 277–287.

Kayes, J.M. and Clark, S.E. (1998) CLAVATA2, a regulator of meristem and organ development in *Arabidopsis*. *Development* **125**, 3843–3851.

Keefe, D., Hinz, U. and Meins, F. (1990) The effect of ethylene on the cell-type-specific and intracellular localization of β 1,3-glucanase and chitinase in tobacco leaves. *Planta* **182**, 43–51.

Kende, H. (1993) Ethylene biosynthesis. *Annual Review of Plant Physiology and Plant Molecular Biology* **44**, 283–307.

Kepenski, S. and Leyser, O. (2002) Ubiquitination and auxin signalling: A degrading story. *Plant Cell* **14**, Supplement, S81–95.

Kermode, A.R. (1997) Approaches to elucidate the basis of desiccation-tolerance in seeds. *Seed Science Research* **7**, 75–95.

Khurana, J.P. and Maheshwari, S.C. (1978) Induction of flowering in *Lemna paucicostata* by salicylic acid. *Plant Science Letters* **12**, 127–131.

Kiba, A., Sugimoto, M., Toyoda, K., Ichinose, Y., Yamada, T. and Shiraishi, T. (1998) Interaction between cell wall and plasma membrane via RGD motif is implicated in plant defense responses. *Plant and Cell Physiology* **39**, 1245–1249.

Kiba, T., Taniguchi, M., Imamura, A., Ueguchi, C., Mizuno, T. and Sugiyama, T. (1999) Differential expression of genes for response regulators in response to cytokinins and nitrate in *Arabidopsis thaliana*. *Plant and Cell Physiology* **40**, 767–771.

Kieber, J.J., Rothenberg, M., Roman, G., Feldmann, K.A. and Ecker, J.R. (1993) CTR1, a negative regulator of the ethylene response pathway in *Arabidopsis*, encodes a member of the Raf family of protein kinases. *Cell*, **72**, 427–441.

Kim, J., Harter, K. and Theologis, A. (1997) Protein-protein interactions among the Aux/IAA proteins. *Proceedings of the National Academy of Sciences, USA* **94**, 11786–11791.

Kim, Y.-S., Kim, D. and Jung, J. (2000) Two isoforms of soluble auxin receptor in rice (*Oryza sativa* L.) plants: Binding property for auxin and interaction with plasma membrane H^+-ATPase. *Plant Growth Regulation* **32**, 143–150.

Kim, Y.-S., Min, J.-K., Kim, D. and Jung, J. (2001) A soluble auxin-binding protein, ABP_{57}. *Journal of Biological Chemistry* **276**, 10730–10736.

Kiss, J.Z. and Sack, F.D. (1989) Reduced gravitropic sensitivity in roots of a starch-deficient mutant of *Nicotiana sylvestris*. *Planta* **180**, 123–130.

Kiss, J.Z., Guisinger, M.M., Miller, A.J. and Stackhouse, K.S. (1997) Reduced gravitropism in hypocotyls of a starch-deficient mutant of *Arabidopsis*. *Plant Cell Physiology* **38**, 518–525.

Klee, H.J. (2002) Control of ethylene-mediated processes in tomato at the level of receptors. *Journal of Experimental Botany* **53**, 2057–2063.

Klee, H.J. and Tieman, D. (2002) The tomato ethylene receptor gene family: Form and function. *Physiologia Plantarum* **115**, 336–341.

Klemsdal, S.S., Hughes, W., Lønneborg, A., Allen, R.B. and Olsen, O.-A. (1991) Primary structure of a novel barley gene differentially expressed in immature aleurone layers. *Molecular and General Genetics* **228**, 9–16.

Knox, J.P. (1995) Developmentally regulated proteoglycans and glycoproteins of the plant cell surface. *FASEB Journal* **9**, 1004–1012.

Knox, J.P. (1997) The use of antibodies to study the architecture and developmental regulation of plant cell walls. *International Review of Cytology* **171**, 79–120.

Knox, J.P., Day, S. and Roberts, K. (1989) A set of cell surface glycoproteins form an early marker of cell position, but not cell type in the root apical meristem of *Daucus carota* L. *Development* **106**, 47–56.

Knox, J.P., Linstead, P.J., King, J., Cooper, C. and Roberts, K. (1990) Pectin esterification is spatially regulated both within cell walls and between developing tissues of root apices. *Planta* **181**, 512–521.

224 REFERENCES

Knox, J.P., Linstead, P.J., Peart, J., Cooper, C. and Roberts, K. (1991) Developmentally regulated epitopes of cell surface arabinogalactan proteins and their relation to root tissue pattern formation. *Plant Journal* **1**, 317–326.

Koka, C.V., Cerny, R.E., Gardner, R.G., Noguchi, T., Fujioka, S., Takatsuto, S., Yoshida, S. and Clouse, S.D. (2000) A putative role for the tomato genes *DUMPY* and *CURL-3* in brassinosteroid biosynthesis and response. *Plant Physiology* **122**, 85–98.

Komalavilas, P., Zhu, J.-K. and Nothnagel, E.A. (1991) Arabinogalactan-proteins from the suspension culture medium and plasma membrane of rose cells. *Journal of Biological Chemistry* **266**, 15956–15965.

Koornneef, M. and van der Veen, J.H. (1980) Induction and analysis of gibberellin sensitive mutants in *Arabidopsis thaliana* (L.). *Heynh. Theoretical and Applied Genetics* **58**, 257–263.

Koornneef, M., van Eden, J., Hanhart, C.J., Stam, P., Braacksma, F.J. and Feenstra, W.J. (1983) Linkage map of *Arabidopsis thaliana*. *Journal of Heredity* **74**, 265–272.

Kowalczyk, M. and Sandberg, G. (2001) Quantitative analysis of indole-3-acetic acid metabolites in *Arabidopsis*. *Plant Physiology* **127**, 1845–1853.

Kreuger, M. and van Holst, G.-J. (1993) Arabinogalactan proteins are essential in somatic embryogenesis of *Daucus carota* L. *Planta* **189**, 243–248.

Kreuger, M. and van Holst, G.-J. (1995) Arabinogalactan-protein epitopes in somatic embryogenesis of *Daucus carota* L. *Planta* **197**, 135–141.

Kreuger, M. and van Holst, G.-J. (1996) Arabinogalactan proteins and plant differentiation. *Plant Molecular Biology* **30**, 1077–1086.

Ku, H.S., Suge, H., Rappaport, L. and Pratt, H.K. (1970) Stimulation of rice coleoptile growth by ethylene. *Planta* **90**, 333–339.

Kumar, A., Altabella, T., Taylor, M.A. and Tiburcio, A.F. (1997) Recent advances in polyamine research. *Trends in Plant Science* **2**, 124–130.

Kumar, D. and Klessig, D.F. (2003) High-affinity salicylic acid-binding protein 2 is required for plant innate immunity and has salicylic acid-stimulated lipase activity. *Proceedings of the National Academy of Sciences, USA* **100**, 16101–16106.

Kuo, A., Cappellutti, S., Cervantes-Cervantes, M., Rodriguez, M. and Bush, D.S. (1996) Okadaic acid, a protein phosphatase inhibitor, blocks calcium changes, gene expression and cell death induced by gibberellin in wheat aleurone cells. *Plant Cell* **8**, 259–269.

Kurosawa, K. (1926) Experimental studies on the secretion of *Fusarium heterosporum* on rice plants. *Transactions of the Natural History Society, Formosa* **16**, 213–227.

Kutschera, U. and Bette, A. (1998) In growing epidermal cells of rye coleoptiles microtubules are associated with the nuclei. *Journal Plant Physiology* **152**, 463–467.

Lang-Pauluzzi, I. and Gunning, B.E.S. (2000) A plasmolytic cycle: The fate of cytoskeletal elements. *Protoplasma* **212**, 174–185.

Lappartient, A.G., Vidmar, J.J., Leustek, T., Glass, A.D.M. and Touraine, B. (1999) Inter-organ signalling in plants: Regulation of ATP sulfurylase and sulfate transporter gene expression in roots mediated by phloem-translocated compound. *Plant Journal* **18**, 89–95.

Lashbrook, C.C., Tieman, D.M. and Klee, H.J. (1998) Differential regulation of the tomato ETR gene family throughout plant development. *Plant Journal* **15**, 243–252.

Laval, V., Chabannes, M., Carrière, M., Canut, H., Barre, A., Rougé, P., Pont-Lezica, R. and Galaud, J., 1999. A family of *Arabidopsis* plasma membrane receptors presenting animal β-integrin domains. *Biochimica and Biophysica Acta* **1435**, 61–70.

Leblanc, N., Perrot-Reichenmann, C. and Barbier-Brygoo, H. (1999a) The auxin-binding protein Nt-Erabp1 alone activates an auxin-like transduction pathway. *FEBS Letters* **449**, 57–60.

Leblanc, N., David, K., Grosclaude, J., Pradier, J.-M., Barbier-Brygoo, H., Labiau, S. and Perrot-Rechenmann, C. (1999b) A novel immunological approach establishes that the auxin-binding protein, Nt-abp1, is an element involved on auxin signalling at the plasma membrane. *Journal of Biological Chemistry* **274**, 28314–28320.

LeClere, S., Tellez, R., Rampey, R.A., Matsuda, S.P.T. and Bartel, B. (2002) Characterization of a family of IAA-amino acid conjugate hydrolases from *Arabidopsis Journal of Biological Chemistry* **277**, 20446–20452.

Lee, G.I. and Howe, G.A. (2003) The tomato mutant spr1 is defective in systemin perception and the production of a systemic wound signal for defense gene expression. *Plant Journal* **33**, 567–576.

Lee, M.M. and Schiefelbein, J. (2001) Developmentally distinct MYB genes encode functionally equivalent proteins in *Arabidopsis*. *Development* **128**, 1539–1546.

Leiberman, M., Kanisti, A.T., Mapson, L.W. and Wardale, A. (1966) Stimulation of ethylene production in apple tissue slices by methionine. *Plant Physiology* **41**, 376–382.

Lembi, C.A., Morré, D.J., St-Thompson, K. and Hertel, R. (1971) *N*-1-naphthylphthalamic acid-binding of a plasma membrane-rich fraction from maize coleoptiles. *Planta* **99**, 37–45.

Lenhard, M. and Laux, T. (2003) Stem cell homeostasis in the *Arabidopsis* shoot meristem is regulated by intercellular movement of CLAVATA3 and its sequestration by CLAVATA1. *Development* **130**, 3163–3173.

Leon, J., Lawton, M.A. and Raskin, I. (1995) Hydrogen peroxide stimulates salicylic acid biosynthesis in tobacco. *Plant Physiology* **108**, 1673–1678.

Leon, P. and Sheen, J. (2003) Sugar and hormone connections. *Trends in Plant Science* **8**, 110–116.

Leopold, A.C. and Guernsey, F.S. (1953) Auxin polarity in the *Coleus* plant. *Botanical Gazette* **115**, 147–154.

Leshem, Y.Y. and Haramaty, E. (1996) The characterisation and contrasting effects of the nitric oxide free radical in vegetative stress and senescence of *Pisum sativum* Linn. foliage. *Journal of Plant Physiology* **148**, 258–263.

Leshem, Y.Y. and Pinchasov, Y. (2000) Non-invasive photoacoustic spectroscopic determination of relative endogenous nitric oxide and ethylene content stoichiometry during ripening of strawberries, *Fragaria anannasa* (Duch.) and avocado *Persea americana* (Mill.). *Journal of Experimental Botany* **51**, 1471–1473.

Leslie, C.A. and Romani, R.J. (1986) Salicylic acid: A new inhibitor of ethylene biosynthesis. *Plant Cell Reports* **5**, 144–146.

Leslie, C.A. and Romani, R.J. (1988) Inhibition of ethylene biosynthesis by salicylic acid. *Plant Physiology* **88**, 833–837.

Letham, D.S. (1963) Zeatin, a factor inducing cell division from *Zea mays*. *Life Sciences* **8**, 569–573.

Lewis, N.G. and Yamamoto, E. (1990) Lignin: Occurrence, biogenesis and biodegradation. *Annual Review of Plant Physiology and Plant Molecular Biology* **41**, 455–496.

Leyser, H.M.O. (2002) Molecular genetics of auxin signalling. *Annual Reviews of Plant Biology* **53**, 377–398.

Leyser, H.M.O. and Furner, I.J. (1992) Characterisation of three shoot apical meristem mutants of *Arabidopsis thaliana*. *Development* **116**, 397–403.

Leyser, H.M.O., Lincoln, C., Timpte, C., Lammer, D., Turner, J. and Estelle, M. (1993) The auxin-resistance gene AXR1 of *Arabidopsis* encodes a protein related to ubiquitin-activase enzyme E1. *Nature* **304**, 161–164.

Lhernould, S., Karamanos, Y., Priem, B. and Morvan, H. (1994) Carbon starvation increases endoglycosidase activities and production of "unconjugated N-glycans" in *Silene alba* cells. *Plant Physiology* **106**, 776–784.

Li, J. and Chory, J., (1997) A putative leucine-rich repeat receptor kinase involved in brassinosteroid signal transduction. *Cell* **90**, 929–938.

Li, J. and Chory, J. (1999) Brassinosteroid action in plants. *Journal of Experimental Botany* **50**, 275–282.

Li, J. and Nam, K.H. (2002) Regulation of brassinosteroid signalling by a GSK3/SHAGGY-like kinase. *Science* **295**, 1299–1301.

Li, J., Nagpal, P., Vitart, V., McMorris, T.C. and Chory, J. (1996) A role for brassinosteroids in light-dependent development of *Arabidopsis. Science* **272**, 398–401.

Li, J., Biswas, M.G., Chao, A., Russel, D.W. and Chory, J. (1997) Conservation of function between mammalian and plant steroid 5α-reductases. *Proceedings of the National Academy of Sciences, USA* **94**, 3554–3559.

Li, J., Nam, K.H., Vafeados, D. and Chory, J. (2001) BIN2, A new brassinosteroid-insensitive locus in *Arabidopsis. Plant Physiology* **127**, 14–22.

Li, N., Parsons, B.L., Liu, D. and Mattoo, A.K. (1992) Accumulation of wound-inducible ACC synthase transcripts in tomato fruit is inhibited by salicylic acid and polyamines. *Plant Molecular Biology* **18**, 477–487.

Lincoln, J.E., Cordes, S., Read, E. and Fischer, R.L. (1987) Regulation of expression by ethylene during *Lycopersicon esculentum* (tomato) fruit development. *Proceedings of the National Academy of Sciences, USA* **84**, 2793–2797.

Lindsey, K., Casson, S. and Chilley, P. (2002) Peptides: New signalling molecules in plants. *Trends in Plant Science,* **7**, 78–83.

Liu, D.H., Post-Beiltenmiller, D. (1995) Discovery of an epidermal stearoyl-acyl carrier protein thio esterase: Its potential role in wax biosynthesis. *Journal of Biological Chemistry,* **270**, 16962–16969.

Ljung, K., Bhalerao, R.P. and Sandberg, G. (2001) Sites and homeostatic control of auxin biosynthesis in *Arabidopsis* during vegetative growth. *Plant Journal.* **28**, 465–474.

Lobler, M. and Klambt, D. (1985) Auxin-binding protein from coleoptile membranes of corn (*Zea mays* L.) I. Purification by immunological methods and characterization. *Journal of Biological Chemistry* **260**, 9848–9853.

Lopes, M.A. and Larkins, B.A. (1993) Endosperm origin, development, and function. *Plant Cell* **5**, 1383–1399.

López-Serrano, M., Fernández, M.D., Pomar, F., Pedreño, M.A. and Barceló, A.R. (2004) *Zinnia elegans* uses the same peroxidase isoenzyme complement for cell wall lignification in both single-cell tracheary elements and xylem vessels. *Journal of Experimental Botany* **55**, 423–431.

Lovegrove, A. and Hooley, R. (2000) Gibberellin and abscisic acid signalling in aleurone. *Trends in Plant Science* **5**, 102–110.

Lovegrove, A., Barratt, D.H.P., Beale, M.H. and Hooley, R. (1998) Gibberellin-photoaffinity labelling of two polypeptides in plant plasma membranes. *Plant Journal* **15**, 311–320.

Lu, C. and Fedoroff, N. (2000) A mutation in the *Arabidopsis* HYL1 gene encoding a dsRNA binding protein affects responses to abscisic acid, auxin and cytokinin. *Plant Cell* **12**, 2351–2366.

Lucas, W.J., Bouche-Pillon, S., Jackson, D.P., Nguyen, L., Baker, L., Ding, B. and Hake, S. (1995) Selective trafficking of KNOTTED1 homeodomain protein and its mRNA through plasmodesmata. *Science* **270**, 1980–1983.

Ludwig-Müller, J., Epstein, E. and Hilgenberg, W. (1996) Auxin-conjugate hydrolysis in Chinese cabbage: Characterization of an amidohydrolase and its role during infection with clubroot disease. *Physiologia Plantarum* **97**, 627–634.

Macdonald, H., Henderson, J., Napier, R.M., Venis, M.A., Hawes, C. and Lazarus, C.M. (1994) Authentic processing and targeting of active maize auxin-binding protein in the baculovirus expression system. *Plant Physiology* **105**, 1049–1057.

Macdonald, M.M. (1984) Dormancy, growth and differentiation of tuber buds of *Solanum tuberosum*. D. Phil Thesis, Oxford University, U.K., 171 pp.

MacMillan, J. (1997) Biosynthesis of the gibberellin plant hormones. *Natural Product Reports* **14**, 221–244.

MacRobbie, E.A.C. (2000) ABA activates multiple Ca^{2+} fluxes in stomatal guard cells, triggering vacuolar K^+ (Rb^+) release. *Proceedings of the National Academy of Sciences, USA* **97**, 12361–12368.

Mahonen, A.P., Bonke, M., Kauppinen, L., Riikone, M., Benfey, P.N. and Helariutta, Y. (2000) A novel two-component hybrid molecule regulates vascular morphogenesis of the *Arabidopsis* root. *Genes and Development* **14**, 2938–2943.

Majewska-Sawka, A. and Nothnagel, E.A. (2000) The multiple roles of arabinogalactan proteins in plant development. *Plant Physiology* **122**, 3–10.

Malone, M. (1993) Hydraulic signals. *Philosophical Transactions of the Royal Society (London)* **B341**, 33–39.

Malone, M. (1994) Wound-induced hydraulic signals and stimulus transmission in *Mimosa pudica* L. *New Phytologist* **128**, 49–56.

Mandava, N.B. (1988) Plant growth-promoting brassinosteroids. *Annual Review of Plant Physiology and Plant Molecular Biology* **39**, 23–52.

Marfa, V., Gollin, D.J., Eberhard, S., Mohnen, D., Darvill, A. and Albersheim, P. (1991) Oligogalacturonides are able to induce flowers to form on tobacco explants. *Plant Journal* **1**, 217–225.

Martin, A.C., del Pozo, J.C., Iglesias, J., Rubio, V., Solano, R., de la Pena, A., Leyva, A. and Paz-Ares, J. (2000) Influence of cytokinins on the expression of phosphate starvation responsive genes in *Arabidopsis*. *Plant Journal* **24**, 559–567.

Martin, C. and Thimann, K.V. (1972) The role of protein synthesis in the senescence of leaves. *Plant Physiology* **49**, 64–71.

Martin, J.P. and Juniper, B.E. (1970) *The Cuticles of Plants*. Edward Arnold Ltd., Sevenoaks, U.K.

Martinez, P.G., Gomez, R.L. and Gomez-Lim, L.A. (2001) Identification of an ETR1-homologue from mango fruit expressing during fruit ripening and wounding. *Journal of Plant Physiology* **158**, 101–108.

Masuda, Y. and Yamamoto, R. (1972) Control of auxin-induced stem elongation by the epidermis. *Physiologia Plantarum* **27**, 109–115.

Mathesius, U., Charon, C., Rolfe, B.G., Kondorosoi, A. and Crespi, M. (2000) Temporal and spatial order of events during the induction of cortical cell divisions in white clover by *Rhizobium leguminosarum* bv. *trifolii* inoculation or localized cytokinin addition. *Molecular Plant-Microbe Interactions* **13**, 617–628.

Mathieu, Y., Kurkdjian, A., Xia, H., Guern, J., Koller, A., Spiro, M.D., O'Neill, M.A., Albersheim, P.A. and Darvill, A. (1991) Membrane responses induced by oligogalacturonides in suspension-cultured tobacco cells. *Plant Journal* **1**, 333–343.

Matsubayashi, Y. and Sakagami, Y. (1996) Phytosulfokine, sulfated peptides that induce the proliferation of single mesophyll cells of *Asparagus officinalis* L. *Proceedings of the National Academy of Sciences, USA* **93**, 7623–7627.

Matsubayashi, Y. and Sakagami, Y. (2000) 120- and 160-kDa receptors from endogenous mitogenic peptide, phytosulfokine-α in rice plasma membranes. *Journal of Biological Chemistry* **275**, 15520–15525.

Matsubayashi, Y., Omura, N., Morita, A. and Sakagami, Y. (2002) An LRR receptor kinase involved in perception of a peptide plant hormone, phytosulfokine. *Science* **296**, 1470–1472.

Matsubayashi, Y., Takagi, L., Omura, N., Morita, A. and Sakagami, Y. (1999) The endogenous sulfated pentapeptide phytosulfokine-α stimulates tracheary element differentiation of isolated mesophyll cells of *Zinnia*. *Plant Physiology* **120**, 1043–1048.

Mauro, M.L., De Lorenzo, G., Costantino, P. and Bellincampi, D. (2002) Oligogalacturonides inhibit the induction of late but not early auxin-responsive genes in tobacco. *Planta* **215**, 494–501.

McAinsh, M.R., Brownlee, C. and Hetherington, A.M. (1997) Calcium ions as second messengers in guard cell signal transduction. *Physiologia Plantarum* **100**, 16–29.

McCabe, P.F., Valentine, T.A., Forsberg, L.S. and Pennell, R.I. (1997) Soluble signals from cells identified at the cell wall establish a developmental pathway in carrot. *Plant Cell* **9**, 2225–2241.

McDougall, G.J. and Fry, S.C. (1989) Structure-activity relationships for xyloglucan oligosaccharides with antiauxin activity. *Plant Physiology* **89**, 883–887.

McDougall, G.J. and Fry, S.C. (1990) Xyloglucan oligosaccharides promote growth and activate cellulase: Evidence for a role of cellulase in cell growth. *Plant Physiology* **93**, 1042–1048.

McDougall, G.J. and Fry, S.C. (1991) Xyloglucan nonasaccharide, a naturally-occurring oligosaccharin, arises *in vivo* by polysaccharide breakdown. *Journal of Plant Physiology* **137**, 332–336.

McGaw, B.A. and Burch, C.A. (1995) Cytokinin biosynthesis and metabolism. In: *Plant Hormones: Physiology, Biochemistry and Molecular Biology*, Davies, P.J. (ed.), 2nd Edition. Kluwer Academic Publishers, Dordrecht, pp. 98–117.

McGurl, B., Pearce, G., Orozco-Cardenas, M. and Ryan, C.A. (1992) Structure, expression and anti-sense inhibition of the systemin precursor gene. *Science* **255**, 1570–1573.

McManus, M.T. (1983) Identification studies of the ethylene responsive target cells in leaf abscission zones. D.Phil Thesis, University of Oxford, U.K., 185 pp.

McManus, M.T. (1994) Peroxidases in the separation zone during ethylene-induced bean leaf abscission. *Phytochemistry* **35**, 567–572.

McManus, M.T. and Osborne, D.J. (1989) Identification and characterisation of a specific class of target cells for ethylene. In: *Cell Separation in Plants*, NATO ASF Series, Vol. H35. Springer Verlag, Berlin, Heidelberg, pp. 201–210.

McManus, M.T and Osborne, D.J. (1990a) Evidence for the preferential expression of particular polypeptides in leaf abscission zones of the bean, *Phaseolus vulgaris* L. *Journal of Plant Physiology* **136**, 391–397.

McManus, M.T. and Osborne, D.J. (1990b) Identification of polypeptides specific to rachis abscission zone cells of *Sambucus nigra*. *Physiologia Plantarum* **79**, 471–478.

McManus, M.T. and Osborne, D.J. (1991) Identification and characterisation of ionically-bound cell wall glycoprotein expressed preferentially in the leaf rachis abscission zone of *Sambucus nigra* L. *Journal of Plant Physiology* **137**, 251–255.

McManus, M.T., Thompson, D.S., Merriman, C., Lyne, L. and Osborne, D.J. (1998) Transdifferentiation of mature cortical cells to functional abscission cells in bean. *Plant Physiology* **116**, 891–899.

McManus, M.T., McKeating, J., Secher, D.S., Osborne, D.J., Ashford, D.A., Dwek, R.A. and Rademacher, T.W. (1988) Identification of a monoclonal antibody to abscission tissue that recognises xylose/fucose-containing N-linked oligosaccharides from higher plants. *Planta* **175**, 506–512.

Mergemann, H. and Sauter, M. (2000) Ethylene induces epidermal cell death at the site of adventitious root emergence in rice. *Plant Physiology* **124**, 609–614.

Milborrow, B.V. (2001) The pathway of biosynthesis of abscisic acid in vascular plants: A review of the present state of knowledge of ABA biosynthesis. *Journal of Experimental Botany* **52**, 1145–1164.

Mita, S., Kawamura, S. and Asai, T. (2002) Regulation of the expression of a putative ethylene receptor, PePRS2, during the development of passion fruit (*Passiflora edulis*). *Physiologia Plantarum* **114**, 271–280.

Miyawaki, K., Matsumoto-Kitano, M. and Kakimoto, T. (2004) Expression of cytokinin biosynthetic isopentenyltransferase genes in *Arabidopsis*: Tissue specificity and regulation by auxin, cytokinin and nitrate. *Plant Journal* **37**, 128–138.

Miyazawa, Y., Nakajima, N., Abe, T., Sakai, A., Fujioka, S., Kawano, S., Kuroiwa, T. and Yoshida, S. (2003) Activation of cell proliferation by brassinolide application in tobacco BY-2 cells: Effects of brassinolide on cell multiplication, cell-cycle-related gene expression, and organellar DNA contents. *Journal of Experimental Botany* **54**, 2669–2678.

Mockaitis, K. and Howell, S.H. (2000) Auxin induces mitogenic activated protein kinase (MAPK) activation in roots of *Arabidopsis* seedlings. *Plant Journal* **24**, 785–796.

Mohnen, D., Eberhard, S., Marfa, V., Doubrava, N., Toubart, P., Gollin, D.J., Gruber, T.A., Nuri, W., Albersheim, P. and Darvill, A. (1990) The control of root, vegetative shoot and flower morphogenesis in tobacco thin cell-layer explants (TLCs). *Development* **108**, 191–201.

Molisch, H. (1938) *The Longevity of Plants*, Fullington, H. (transl.). Science Press, Lancaster, PA.

Mollet, J.-C., Park, S.-Y., Nothnagel, E.A. and Lord, E.M. (2000) A lily stylar pectin is necessary for pollen tube adhesion to an *in vitro* stylar matrix. *Plant Cell* **12**, 1737–1750.

Monteiro, A.M., Crozier, A. and Sandberg, G. (1988) The biosynthesis and conjugation of indole-3-acetic acid in germinating seed and seedlings of *Dalbergia dolichopetala*. *Planta* **174**, 561–568.

Montoya, T., Nomura, T., Farrar, K., Kaneta, T., Yokota, T. and Bishop, G.J. (2002) Cloning of the tomato Curl3 gene highlights the putative dual role of the leucine-rich receptor kinase tBRI1/SR160 in plant steroid hormone and peptide signalling. *Plant Cell* **14**, 3163–3176.

Moore, D., Hock, B., Greening, J.P., Kern, V.D., Novak Frazer, L. and Monzer, J. (1996) Gravimorphogenesis in agarics. *Mycological Research* **100**, 257–273.

Moore, R. (1986) Calcium movement, graviresponsiveness and the structure of columella cells in primary roots of Amylomaize mutants of *Zea mays*. *American Journal of Botany* **73**, 417–426.

Morris, K., Mackerness, A.S.-H., Page, T., John, C.F., Murphy, A.M., Carr, J.P. and Buchanan-Wollaston, V. (2000) Salicylic acid has a role in regulating gene expression during leaf senescence. *Plant Journal* **23**, 677–685.

Moshkov, I.E., Mur, L.A.J., Novikova, G.V., Smith, A.R. and Hall, M.A. (2003) Ethylene regulates monomeric GTP-binding protein gene expression and activity in *Arabidopsis*. *Plant Physiology* **131**, 1705–1717.

Mott, K.A. and Buckley, T.N. (2000) Patchy stomatal conductance: Emergent collective behaviour of stomata. *Trends in Plant Science* **5**, 258–262.

Muday, G.K. and Murphy, A.S. (2002) An emerging model of auxin transport regulation. *Plant Cell* **14**, 293–299.

Muller, A. and Weiler, E.W. (2000) Indolic constituents and indole-3-acetic acid biosynthesis in the wild-type and a tryptophan auxotroph mutant of *Arabidopsis*. *Planta* 211, 855–863.

Muller, A., Guan, C., Galweiler, L., Tanzler, P., Huijser, P., Marchant, A., Parry, G., Bennett, M., Wisman, E. and Palme, K. (1998) AtPIN2 defines a locus of *Arabidopsis* for root gravitropism control. *EMBO Journal* 17, 6903–6911.

Muller, J.F., Goujaud, J. and Caboche, M. (1985) Isolation *in vitro* of naphthaleneacetic acid-tolerant mutants of *Nicotiana tabacum*, which are impaired in root morphogenesis. *Molecular and General Genetics* 199, 194–200.

Mundree, S.G., Whittaker, A., Thomson, J.A. and Farrant, J.M. (2000) An aldose reductase homolog from the resurrection plant *Xerophyta viscosa*, Baker. *Planta* 211, 693–670.

Musgrave, A., Jackson, M.B. and Ling, E. (1972) *Callitriche* stem elongation is controlled by ethylene and gibberellin. *Nature* 238, 93–96.

Nadeau, J.A. and Sack, F.D. (2002) Control of stomatal distribution on the *Arabidopsis* leaf surface. *Science* 296, 1697–1700.

Nadeau, J.A., Zhang, X.S., Nair, H. and O'Neill, S.D. (1993) Temporal and spatial regulation of 1-aminocyclopropane-1-carboxylate oxidase in the pollination-induced senescence of orchid flowers. *Plant Physiology* 103, 31–39.

Nagahashi, G. and Douds, D.D. (1997) Appressorium formation by arbuscular mycorrhiza fungi on isolated cell walls. *New Phytologist* 136, 299–304.

Nakamura, A., Higuchi, K., Goda, H., Fujiwara, M.T., Sawa, S., Koshiba, T., Shimada, Y. and Yoshida, S. (2003) Brassinolide induces *IAA5, IAA19*, and DR5, a synthetic auxin response element in *Arabidopsis*, implying a cross-talk point of brassinosteroid and auxin signaling. *Plant Physiology* 133, 1843–1853.

Nakatsuka, A., Murachi, S., Okunishi, H., Shiomi, S., Nakano, R., Kubo, Y. and Inaba, A. (1998) Differential expression and internal feedback regulation of 1-aminocyclopropane-1-carboxylate synthase, 1-amino cyclopropane-1-carboxylate oxidase and ethylene receptor genes in tomato fruit during development and ripening. *Plant Physiology* 118, 1295–1305.

Napier, R.M. (2001) Models of auxin binding. *Journal of Plant Growth Regulation* 20, 244–254.

Napier, R.M. and Venis, M.A. (1990) Monoclonal antibodies detect an auxin-induced conformational change in the maize auxin-binding protein. *Planta* 182, 313–318.

Napier, R.M., Venis, M.A., Bolton, M.A., Richardson, L.I. and Butcher, G.W. (1988) Preparation and characterization of monoclonal and polyclonal antibodies to maize membrane auxin-binding protein. *Planta* 176, 519–526.

Napier, R.M., Fowke, L.C., Hawes, C., Lewis, M. and Pelham, H.R.B. (1992) Immunological evidence that plants use both HDEL and KDEL for target proteins to the endoplasmic reticulum. *Journal of Cell Science* 102, 261–271.

Narváez-Vásquez, J. and Ryan, C.A. (2004) The cellular localisation of prosystemin: A functional role for phloem parenchyma in systemic wound signalling. *Planta* 218, 360–369.

Narváez-Vásquez, J., Pearce, G., Orozco-Cardenas, M.L., Franceschi, V.R. and Ryan, C.A. (1995) Autoradiographic and biochemical evidence for the systemic translocation of systemin in tomato plants. *Planta* 195, 593–600.

Neill, S.J., Desikan, R. and Hancock, J.T. (2003) Nitric oxide signalling in plants. *New Phytologist* 159, 11–35.

Neljubov, D.N. (1901) Uber die horizontale nutation der Stengel von *Pisum sativum* und eineger anderen Pflanzen. *Beihefte zum Botanischen Zentralblatt* 10, 128–138.

Nick, P. (1999) Signals, motors, morphogenesis – the cytoskeleton in plant development. *Plant Biology* **1**, 169–179.

Noel, A.R.A. and Van Staden, J. (1975) Phyllomorph senescence in *Streptocarpus molweniensis*. *Annals of Botany* **39**, 921–929.

Nomura, T., Nakayama, M., Reid, J.B., Takeuchi, Y. and Yokota, T. (1997) Blockage of brassinosteroid biosynthesis and sensitivity causes dwarfism in garden pea. *Plant Physiology* **113**, 31–37.

Noodén, L.D. and Leopold, A.C. (eds.) (1988) *Senescence and Ageing in Plants.* Academic Press, San Diego, 526 pp.

Normanly, J. and Bartel, B. (1999) Redundancy as a way of life – IAA metabolism. *Current Opinion in Plant Biology* **2**, 207–213.

Normanly, J., Slovin, J.P. and Cohen, J.D. (1995) Rethinking auxin biosynthesis and metabolism. *Plant Physiology* **107**, 323–329.

Obara, K., Kuriyama, H. and Fukuda, H. (2001) Direct evidence of active and rapid nuclear degradation triggered by vacuole rupture during programmed cell death in *Zinnia*. *Plant Physiology* **125**, 615–626.

Obendorf, R.L. (1997) Oligosaccharides and galactosyl cyclitols in seed desiccation tolerance. *Seed Science Research* **7**, 63–74.

O'Donnell, P.J., Calvert, C., Atzorn, R., Wasternek, C., Leyser, H.M.O. and Bowles, D.J. (1996) Ethylene as a signal mediating the wound response to tomato plants. *Science* **274**, 1914–1917.

Oeller, P.W., Keller, J.A., Parks, J.A., Silbert, J.E. and Theologis, A. (1993) Structural characterization of the early indoleacetic acid-inducible genes, PS-IAA4/5, and PS-IAA6, of pea (*Pisum sativum L.*) *Journal of Molecular Biology* **233**, 789–798.

Oh, M.-H., Romanow, W.G., Smith, R.C., Zamski, E., Sasse, J. and Clouse, S.D. (1998) Soybean BRU1 encodes a functional xyloglucan endotransglycosylase that is highly expressed in inner epicotyl tissues during brassinosteroid-promoted elongation. *Plant and Cell Physiology* **39**, 124–130.

Oh, M.-H., Ray, W.K., Huber, S.C., Asara, J.M., Gage, D.A. and Clouse, S.D. (2000) Recombinant brassinosteroid insensitive 1 receptor-like kinase autophosphorylates on serine and threonine residues and phosphorylates a conserved peptide motif *in vitro*. *Plant Physiology* **124**, 751–766.

Ohkuma, K., Lyon, J.L., Addicott, F.T. and Smith, O.E. (1963) Abscisin II, an abscission-accelerating substance from young cotton fruit. *Science* **142**, 1592–1593.

Oka, M., Miyamoto, K., Okada, K. and Ueda, J. (1999) Auxin polar transport and flower formation in *Arabidopsis thaliana* transformed with indoleacetamide hydrolase (*iaaH*) gene. *Plant and Cell Physiology* **40**, 231–237.

Okada, K., Ueda, J., Komaki, M.K., Bell, C.J. and Shimura, Y. (1991) Requirement of the auxin polar transport system in early stages of *Arabidopsis* floral bud formation. *Plant Cell* **3**, 677–684.

Oliver, A.E., Crowe, L.M. and Crowe, J.H. (1998) Methods for dehydration-tolerance: Depression of the phase transition temperature in dry membranes and carbohydrate vitrification. *Seed Science Research* **8**, 211–221.

Olsen, O.-A. (2001) Endosperm development: Cellularization and cell fate specification. *Annual Review of Plant Physiology and Plant Molecular Biology* **52**, 233–267.

Olsen, O.-A., Lemmon, B.E. and Brown, R.C. (1998) A model for aleurone cell development. *Trends in Plant Science* **3**, 168–169.

Olszewski, N., Sun, T.-P. and Gubler, F. (2002) Gibberellin signalling: Biosynthesis, catabolism, and response pathways. *Plant Cell* **14**, Supplement, S61–S80.

O'Neill, S.D., Nadeau, J.A., Zhang, X.S., Bui, A.Q. and Halevy, A.H. (1993) Inter-organ regulation of ethylene biosynthetic genes by pollination. *Plant Cell* **5**, 419–432.

Oparka, K.J. and Santa Cruz, S. (2000) The great escape: Phloem transport and unloading of macromolecules. *Annual Review of Plant Physiology and Plant Molecular Biology* **51**, 323–347.

Osborne, D.J. (1976) Control of cell shape and cell size by the dual regulation of auxin and ethylene. In: *Perspectives in Experimental Biology*, Vol. 2, 'Botany', Sunderland, N. (ed.). Pergamon Press, Oxford, pp. 89–102.

Osborne, D.J. (1977a) Ethylene and target cells in the growth of plants. *Science Progress* (Oxford) **64**, 51–63.

Osborne, D.J. (1977b) Auxin and ethylene and the control of cell growth. The identification of three classes of target cells. In: *Plant Growth Regulation*, Pilet, P.E. (ed.). Springer, Heidelberg, pp. 161–171.

Osborne, D.J. (1979) Target cells – new concepts for plant regulation in horticulture. *Scientific Horticulture* **30**, 1–13.

Osborne, D.J. (1984) Ethylene and plants of aquatic and semi-aquatic environments: A review. *Plant Growth Regulation* **2**, 167–185.

Osborne, D.J. (1989) Abscission. *CRC Critical Reviews in Plant Sciences* **8**, 103–129.

Osborne, D.J. (1990) Ethylene formation, cell types and differentiation. In: *Polyamines and Ethylene: Biochemistry, Physiology, and Interactions*. Flores, H.E., Arteca, R.N. and Shannon, J.C. (eds.). American Society of Plant Physiologists, pp. 203–215.

Osborne, D.J. and Boubriak, I. (2002) Telomeres and their relevance to the life and death of seeds. *Critical Reviews in Plant Sciences* **21**, 127–141.

Osborne, D.J. and Cheah, K.S.E. (1982) Hormones and foliar senescence. In: *Growth Regulators in Plant Senescence*, Jackson, M.B., Grout, B. and Mackenzie, I.A. (eds.), British Plant Growth Regulator Group Monograph 8, pp. 57–83.

Osborne, D.J. and Hallaway, M. (1964) The auxin, 2,4-dichlorophenoxyacetic acid as a regulator of protein synthesis and senescence in detached leaves of *Prunus*. *New Phytologist* **63**, 334–347.

Osborne, D.J. and Sargent, J.A. (1976) The positional differentiation of abscission zones during the development of leaves of *Sambucus nigra* and the response of the cells to auxin and ethylene. *Planta* **132**, 197–204.

Osborne, D.J., McManus, M.T. and Webb, J. (1985) Target cells for ethylene action. In: *Ethylene and Plant Development*, Roberts, J.A. and Tucker, G.A. (eds.). Butterworths, London. pp. 197–212.

Osborne, D.J., Walters, J., Milborrow, B.V., Norville, A. and Stange, L.M.C. (1996) Evidence for a non-ACC ethylene biosynthesis pathway in lower plants. *Phytochemistry* **42**, 51–60.

Ottenschlager, I., Wolff, P., Wolverton, C., Bhalerao, R.P., Sandberg, G., Ishikawa, H., Evans, M. and Palme, K. (2002) Gravity-regulated differential auxin transport from columella to lateral root cap cells. *Proceedings of the National Academy of Sciences, USA* **100**, 2987–2991.

Ouaked, F., Rozhon, W., Lecourieux, D. and Hirt, H. (2003) A MAPK pathway mediates ethylene signalling in plants. *EMBO Journal* **22**, 1282–1288.

Ouellet, F., Overoorde, P.J. and Theologis, A. (2001) IAA17/AXR3: Biochemical insight into an auxin phenotype. *Plant Cell* **13**, 829–842.

Paleg, L.G. (1960) Physiological effects of gibberellic acid: I. On carbohydrate metabolism and amylase activity of barley endosperm. *Plant Physiology* **35**, 293–299.

Palme, K., Hesse, T., Campos, N., Garbers, C., Yanofsky, M.F. and Schell, J. (1992) Molecular analysis of an auxin binding protein gene located on chromosome 4 of *Arabidopsis*. *Plant Cell* **4**, 193–201.

Palmgren. M.G. (2001) Plant plasma membrane H$^+$ ATPases: Powerhouses for nutrient uptake. *Annual Review of Plant Physiology and Plant Molecular Biology* **52**, 817–845.

Parry, A.D., Neill, S.J. and Horgan, R. (1988) Xanthoxin levels and metabolism in the wild-type and wilty mutants of tomato. *Planta* **173**, 397–404.

Patterson, S.E. and Bleecker, A.B. (2004) Ethylene-dependent and -independent processes associated with floral organ abscission in *Arabidopsis*. *Plant Physiology* **134**, 194–203.

Payton, S., Fray, R., Brown, S. and Grierson, D. (1996) Ethylene receptor expression is regulated during fruit ripening, flower senescence and abscission. *Plant Molecular Biology* **31**, 1227–1231.

Pearce, G., Strydom, D., Johnson, S. and Ryan, C.A. (1991) A polypeptide from tomato leaves induces wound-inducible proteinase inhibitor proteins. *Science* **253**, 895–898.

Pearce, G., Moura, D.S., Stratmann, J. and Ryan, C.A. (2001a) Production of multiple plant hormones from a single polyprotein precursor. *Nature* **411**, 817–820.

Pearce, G., Moura, D.S., Stratmann, J. and Ryan, C.A. (2001b) RALF, a 5-kDa ubiquitous polypeptide in plants, arrests root growth and development. *Proceedings of the National Academy of Sciences, USA* **99**, 12843–12847.

Peng, J. and Harberd, N.P. (1993) Derivative alleles of the *Arabidopsis* gibberellin-insensitive (gai) mutation confers a wild-type phenotype. *Plant Cell* **5**, 351–360.

Peng, J., Carol, P., Richards, D.E., King, K.E., Cowling, R.J., Murphy, G.P. and Harberd, N.P. (1997) The *Arabidopsis* GAI gene defines a signalling pathway that negatively regulates gibberellin responses. *Genes and Development* **11**, 3194–3205.

Pennell, R.I. and Roberts, K. (1990) Sexual development in the pea is presaged by altered expression of arabinogalactan protein. *Nature* **344**, 547–549.

Penninckx, I.A.M.A., Thomma, B.P.H.J., Buchala, A., Metraux, J.-P. and Broekaert, W.F. (1998) Concomitant activation of jasmonate and ethylene response pathways is required for induction of a plant defensin gene in *Arabidopsis*. *Plant Cell* **10**, 2103–2113.

Perbal, G. and Driss-Ecole, D. (2003) Mechanotransduction in gravisensing cells. *Trends in Plant Science* **8**, 498–504.

Philippar, K., Fuchs, I., Luthen, H., Hoth, S., Bauer, C.S., Haga, K., Thiel, G., Ljung, K., Sandberg, G., Bottger, M., Becker, D. and Hedrich, R. (1999) Auxin-induced K$^+$ channel expression represents an essential step in coleoptile growth and gravitropism. *Proceedings of the National Academy of Sciences, USA* **96**, 12186–12191.

Phinney, B.O. (1956) Biochemical mutants in maize: Dwarfism and its reversal with gibberellins. *Plant Physiology* **31**, Supplement, 20.

Piquemal, J., Larierre, C., Myton, K., O'Connell, A., Schuch, W., Grima-Pettenati, J. and Boudet, A.-M. (1998) Down-regulation of cinnamoyl-CoA reductase induces significant changes in lignin profiles in tobacco plants. *Plant Journal* **13**, 71–83.

Poethig, R.S. (1987) Clonal analysis of cell lineage patterns in plant development. *American Journal of Botany* **74**, 581–594.

Poethig, R.S. (1989) Genetic mosaics and cell lineage analysis in plants. *Trends in Genetics* **5**, 273–277.

Poli, D.B., Jacobs, M. and Cooke, T.J. (2003) Auxin regulation of axial growth in bryophyte sporophytes: Its potential significance for the evolution of early land plants. *American Journal of Botany* **90**, 1405–1415.

Priem, B. and Gross, K.C. (1992) Mannosyl- and xylosyl-containing glycans promote tomato (*Lycopersicon esculentum* Mill.) fruit ripening. *Plant Physiology* **98**, 399–401.

Priem, B., Morvan, H., Monin, A., Hafez, A. and Morvan, C. (1990a) Influence of a plant glycan of the oligomannoside type on the growth of flax plantlets. *Comptus Rendus Academic Press, Paris* **311**, 411–416.

Priem, B., Solo-Kwan, J., Wieruszeski, J.M., Strecker, G., Nazih, H. and Morvan, H. (1990b) Isolation and characterization of free glycans from the oligomannoside type from the extracellular medium of a plant cell suspension. *Glycoconjugate J.* **7**, 121–132.

Priem, B., Morvan, H. and Gross, K.C. (1994) Unconjugated *N*-glycans as a new class of plant oligosaccharins. *Biochemical Society Transactions* **22**, 398–402.

Quatrano, R.S. (1978) Development of cell polarity. *Annual Review of Plant Physiology* **29**, 487–510.

Racusen, R.H. and Schiavone, F. (1990) Positional cues and differential gene expression in somatic embryos of higher plants. *Cell Differentiation and Development* **30**, 159–169.

Ramos, J., Zenser, N., Leyser, O. and Callis, J. (2001) Rapid degradation of auxin/indoleacetic acid proteins requires conserved amino acids of domain II and is proteasome dependent. *Plant Cell* **13**, 2349–2360.

Rashotte, A.M., Brady, S.R., Reed, R.C., Ante, S.J. and Muday, G.K. (2000) Basipetal auxin transport is required for gravitropism in roots of *Arabidopsis*. *Plant Physiology* **122**, 481–490.

Raskin, I. (1992) Role of salicylic acid in plants. *Annual Review of Plant Physiology and Plant Molecular Biology* **43**, 439–463.

Raskin, I. (1995) Salicylic acid. In: *Plant Hormones: Physiology, Biochemistry and Molecular Biology*, Davies, P.J. (ed.), 2nd Edition. Kluwer Academic Publishers, Dordrecht, pp. 188–205.

Raskin, I., Ehmann, A., Melander, W.R. and Meeuse, B.J.D. (1987) Salicylic acid – a natural inducer of heat production in *Arum* lilies. *Science* **237**, 1601–1602.

Rasmussen, J.B., Hammerschmidt, R. and Zook, M.N. (1991) Systemic induction of salicylic acid accumulation in cucumber after inoculation with *Pseudomonas syringae* pv. *syringae*. *Plant Physiology* **97**, 1342–1347.

Rasori, A., Ruperti, B., Bonghi, C., Tonutti, P. and Ramina, A. (2002) Characterization of two putative ethylene receptor genes expressed during peach fruit development and abscission. *Journal of Experimental Botany* **53**, 2333–2339.

Ray, P. M. (1977) Auxin-binding sites of maize coleoptiles are localized on membranes of the endoplasmic reticulum. *Plant Physiology* **59**, 594–599.

Ray, P. M., Dohrmann, U. and Hertel, R. (1977) Characterization of napthaleneacetic acid binding to receptor sites on cellular membranes of maize coleoptiles tissue. *Plant Physiology* **59**, 357–364.

Reymond, P., Grunberger, S., Paul, K., Muller, M. and Farmer, E.E. (1995) Oligogalacturonide defense signals in plants: Large fragments interact with the plasma membrane *in vitro*. *Proceedings of the National Academy of Sciences, USA* **92**, 4145–4149.

Richmond, A. and Lang, A. (1957) Effect of kinetin on protein content and survival of detached *Xanthium* leaves. *Science* **125**, 650–651.

Ridge, I. (1992) Sensitivity in a wider context: Ethylene and petiole growth in *Nymphoides peltata*. In: *Progress in Plant Growth Regulation*. Karssen, C.M., van Loon, L.C. and Vreugdenhil, D. (eds.). Kluwer Academic Publishers, Dordrecht, pp. 254–263.

Ridge, I. and Osborne, D.J. (1969) Cell growth and cellulases: Regulation by ethylene and indole-3-acetic acid in shoots of *Pisum sativum*. *Nature* **223**, 318–319.

Ridge, I. and Osborne, D.J. (1989) Wall extensibility, wall pH and tissue osmalality: significance for auxin and ethylene-enhanced petiole growth in semi-aquatic plants. *Plant, Cell and Environment* **12**, 383–393.

Ridge, I., Omer, J., Osborne, D.J. and Walters, J. (1991) Cell expansion and wall pH in the fern *Regnellidium diphyllum*, a plant lacking acid-induced growth. *Journal of Experimental Botany* **42**, 1171–1179.

Ridge, I., Omer, J. and Osborne, D.J. (1998) Different effects of vanadate on net proton secretion in the fern *Regnellidium diphyllum* and the dicotyledon *Nymphoides peltata*: Relevance to cell growth. *Journal of Plant Physiology* **153**, 430–436.

Rinne, P., Tuominen, H. and Junttila, L. (1992) Arrested leaf abscission in the non-abscising variety of pubescent birch: Developmental morphological and hormonal aspects. *Journal of Experimental Botany* **43**, 975–982.

Ritchie, S., McCubbin, A., Ambrose, G., Kao, T.-H. and Gilroy, S. (1999) The sensitivity of barley aleurone tissue to gibberellin is heterogeneous and may be spatially determined. *Plant Physiology* **120**, 361–370.

Rober-Kleber, N., Albrechtová, J.T.P., Fleig, S., Huck, N., Michalke, W., Wagner, E., Speth, V., Neuhaus, G. and Fischer-Iglesias, C. (2003) Plasma membrane H⁺-ATPase is involved in auxin-mediated cell elongation during wheat embryo development. *Plant Physiology* **131**, 1302–1312.

Roberts, I.N., Murray, P.F., Caputo, C.P., Passeron, S. and Barneix, A.J. (2003) Purification and characterization of a subtilisin-like serine protease induced during the senescence of wheat leaves. *Physiologia Plantarum* **118**, 483–492.

Roberts, J.A. (1984) Tropic responses of hypocotyls from normal tomato plants and the gravitropic mutant *Lazy-1*. *Plant Cell and Environment* **7**, 515–520.

Robinson, P.M., Wareing, P.F. and Thomas, T.H. (1963) Dormancy regulators in woody plants. Isolation of the inhibitor varying with photoperiod in *Acer pseudoplatanus*. *Nature* **199**, 875–876.

Rogg, L., Lasswell, J. and Bartel, B. (2001) A gain-of-function mutation in IAA28 suppresses lateral root development. *Plant Cell* **13**, 465–480.

Rojo, E., Sharma, V.K., Kovaleva, V., Raikhel, N.V. and Fletcher, J.C. (2002) CLV3 is localised to the extracellular space, where it activates the *Arabidopsis* CLAVATA stem cell signalling pathway. *Plant Cell* **14**, 969–977.

Roman, G., Lubarsky, B., Kieber, J.J., Rothenberg, M. and Ecker, J.R. (1995) Genetic analysis of ethylene signal transduction in *Arabidopsis thaliana;* Five novel mutant loci integrated into stress-response pathway. *Genetics* **139**, 1393–1409.

Rouse, D., Mackay, P., Stirnberg, P., Estelle, M. and Leyser, O. (1998) Changes in auxin response from mutations in an AUX/IAA gene. *Science* **279**, 1371–1373.

Ruegger, M., Dewey, E., Gray, W.M., Hobbie.L., Turner, J. and Estelle, M. (1998) The TIR1 protein of *Arabidopsis* functions in auxin response and is related to human SKP2 and yeast Grr1p. *Genes and Development* **12**, 198–207.

Ruel, K., Chabannes, M., Bondet, A.-M., Legrand, M. and Joseleau, J.-P. (2001) Reassessment of qualitative changes in lignification of transgenic tobacco plants and their impact on cell wall assembly. *Phytochemistry* **57**, 875–882.

Ryals, J., Lawton, K.A., Delaney, T.P., Friedrich, L., Kessmann, H., Neuenschwander, U., Uknes, S., Vernooij, B. and Weymann, K. (1995) Signal transduction in systemic acquired resistance. *Proceedings of the National Academy of Sciences, USA* **92**, 4202–4205.

Ryals, J.A., Neuenschwander, U.H., Willits, M.G., Molina, A., Steiner, H.Y. and Hunt, M.D. (1996) Systemic acquired resistance. *Plant Cell* **8**, 1809–1819.

Ryan, C.A. (1974) Assay and biochemical properties of the proteinase inhibitor-inducing factor, a wound hormone. *Plant Physiology* **54**, 328–332.

Ryan, C.A. and Moura, D.S. (2002) Systemic wound signalling in plants: A new perception. *Proceedings of the National Academy of Sciences, USA* **99**, 6519–6520.

Ryan, C.A., Pearce, G., Scheer, J. and Moura, D.S. (2002) Polypeptide hormones. *Plant Cell* **14**, Supplement, S251–S264.

Saibo, N.J.M., Vriezen, W.H., Beemster, G.T.S. and Van Der Straeten, D. (2003) Growth and stomata formation of *Arabidopsis* hypocotyls is controlled by gibberellins and modulated by ethylene and auxins. *Plant Journal* **33**, 989–1000.

Sachs, T. (1991) Cell polarity and tissue patterning in plants. *Development*, Supplement 1, 83–93.

Sachs, T. (2000) Integrating cellular and organismic aspects of vascular differentiation. *Plant and Cell Physiology* **41**, 649–656.

Sakai, H., Hua, J., Chen, G.Q., Chang, C., Medrano, L.J., Bleecker, A.B. and Meyerowitz, E.M. (1998) ETR2 is an ETR1-like gene involved in ethylene signal transduction in *Arabidopsis.Proceedings of the National Academy of Sciences, USA* **95**, 5812–5817.

Sakai, H., Honma, T., Aoyama, T., Sato, S., Kato, T., Tabata, S. and Oka, A. (2001) ARR1, a transcription factor for genes immediately responsive to cytokinins. *Science* **294**, 1519–1521.

Sakakibara, H. and Takei, K. (2002) Identification of cytokinin biosynthesis genes in *Arabidopsis*: A breakthrough for understanding the metabolic pathway and the regulation in higher plants. *Journal of Plant Growth Regulation* **21**, 17–23.

Sakakibara, H., Suzuki, M., Takei, K., Deji, A., Taniguchi, M. and Sugiyama, T. (1998) A response-regulator homologue possibly involved in nitrogen signal transduction mediated by cytokinin in maize. *Plant Journal* **14**, 337–344.

Sakakibara, H., Taniguchi, M. and Sugiyama, T. (2000) His-Asp phospho-relay signalling: A communication avenue between plants and the environment. *Plant Molecular Biology* **42**, 273–278.

Salisbury, F.B. (1963) *Flowering Process.* Pergamon Press, Oxford, London, New York and Paris.

Samejima, M. and Sibaoka, T. (1983) Identification of the excitable cells in the petiole of *Mimosa pudica* by intracellular injection of procion yellow. *Plant and Cell Physiology* **24**, 33–39.

Samuel, G. (1927) On the shot-hole disease caused by *Clasterosporium carpophilum* and on the "shot-hole" effect. *Annals of Botany* **41**, 375–404.

Sanders, P.M., Lee, P.Y., Biesgen, C., Boone, J.D., Beals, T.P., Weiler, E.W. and Goldberg, R.B. (2000) The *Arabidopsis DELAYED DEHISCENCE 1* gene encodes an enzyme in the jasmonic acid synthesis pathway. *Plant Cell* **12**, 1041–1061.

Sasaki, A., Itoh, H., Gomi, K., Ueguchi-Tanaka, M., Ishiyama, K., Kobayashi, M., Jeong, D.-H., An, G., Kitano, H., Ashikari, M. and Matsuoka, M. (2003) Accumulation of phosphorylated repressor for gibberellin signaling in an F-box mutant. *Science* **299**, 1896–1998.

Satina, S., Blakeslee, A.F. and Avery, A.G. (1940) Demonstration of the three germ layers in the shoot apex of *Datura* by means of induced polyploidy in periclinal chimeras. *American Journal of Botany* **27**, 895–905.

Sato-Nara, K., Yuhashi, K.-I., Higashi, K., Hosoya, K., Kubota, M. and Ezura, H. (1999) Stage- and tissue-specific expression of ethylene receptor homolog genes during fruit development in muskmelon. *Plant Physiology* **120**, 321–330.

Saunders, M.J. and Hepler, P.K. (1983) Calcium antagonists and calmodulin inhibitors block cytokinin-induced bud formation in *Funaria. Developmental Biology* **99**, 41–49.

Savill, J., Gregory, C. and Haslett, C. (2003) Eat me or die. *Science* **302**, 1516–1517.

Schaller, G.E. and Bleecker, A.B. (1995) Ethylene-binding sites generated in yeast expressing the *Arabidopsis* ETR1 gene. *Science* **270**, 1809–1811.

Scheer, J.M. and Ryan, C.A. (1999) A 160-kD systemin receptor on the surface of *Lycopersicon peruvianum* suspension-cultured cells. *Plant Cell* **11**, 1525–1536.

Scheer, J.M. and Ryan, C.A. (2002) The systemin receptor SR160 from *Lycopersicon peruvianum* is a member of the LRR receptor kinase family. *Proceedings of the National Academy of Sciences, USA* **99**, 9585–9590.

Scheres, B., Di Laurenzo, L., Willemsen, V., Hauser, M.T., Janmaat, K., Weisbeek, P. and Benfey, P.N. (1995) Mutations affecting the radial organization of the *Arabidopsis* root display specific defects throughout the embryonic axis. *Development* **121**, 53–62.

Schlagnhaufer, C.D. and Arteca, R.N. (1991) The uptake and metabolism of brassinosteroid by tomato *Lycopersicon esculentum* plants. *Journal of Plant Physiology.* **138**, 191–194.

Schopfer, P. (1990) Cytochemical identification of arabinogalactan protein in the outer epidermal wall of maize coleoptiles. *Planta* **183**, 139–142.

Schroeder, J.I. and Hagiwara, S. (1990) Repetitive increases in cytosolic Ca^{2+} of guard cells by abscisic acid activation of nonselective Ca^{2+} permeable channels. *Proceedings of the National Academy of Sciences, USA* **87**, 9305–9309.

Schroeder, J.I., Allen, G.J., Hugouvieux, V., Kwak, J.M. and Waner, D. (2001) Guard cell signal transduction. *Annual Review of Plant Physiology and Plant Molecular Biology* **52**, 627–658.

Schumaker, K.S. and Gizinski, M.J. (1993) Cytokinin stimulates dihyropyridine-sensitive calcium uptake in moss protoplasts. *Proceedings of the National Academy of Sciences, USA* **90**, 10937–10941.

Segovia, M., Haramaty, L., Berges, J.A. and Falkowski, P.G. (2003) Cell death in the unicellular chlorophyte *Dunaliella tertiolecta.* A hypothesis on the evolution of apoptosis in higher plants and metazoans. *Plant Physiology* **132**, 99–105.

Seo, H.S., Song, J.T., Cheong, J.-J., Lee, Y.-H., Lee, Y.-W., Hwang, I., Lee, J.S. and Choi, Y.D. (2001) Jasmonic acid carboxyl methyltransferase: A key enzyme for jasmonate-regulated plant responses. *Proceedings of the National Academy of Sciences, USA* **98**, 4788–4793.

Setlow, P. (1992) DNA in dormant spores of *Bacillus* species is in an A-like conformation. *Molecular Microbiology* **6**, 563–567.

Setlow, P. (1994) Mechanisms which contribute to the long-term survival of spores of *Bacillus* species. *Journal of Applied Bacteriology* **76**, 49S–60S.

Shantz, E.M. and Steward, F.C. (1952) Coconut milk factor: The growth-promoting substances in coconut milk. *Journal of the American Chemical Society* **74**, 6133–6135.

Sharma, Y.K., Leon, J., Raskin, I. and Davis, K.R. 1996. Ozone-induced responses in *Arabidopsis thaliana*: The role of salicylic acid in the accumulation of defense-related transcripts and induced resistance. *Proceedings of the National Academy of Sciences, USA* **93**, 5099–5104.

Shimada, Y., Goda, H., Nakamura, A., Takasuto, S., Fujioka, S. and Yoshida, S. (2003) Organ-specific expression of brassinosteroid-biosynthetic genes and distribution of endogenous brassinosteroids in *Arabidopsis*. *Plant Physiology* **131**, 287–297.

Shimomura, S., Sotobayashi, T., Futai, M. and Fuhui, T. (1986) Purification and properties of an auxin-binding protein from maize shoot membranes. *Journal of Biochemistry* **99**, 1513–1524.

Shiu, O.Y., Oetiker, J.H., Yip, W.K. and Yang, S.F. (1998) The promoter of *LE-ACS7*, an early flooding-induced 1-aminocyclopropane-1-carboxylate synthase gene of the tomato, is tagged by a *Sol3* transposon. *Proceedings of the National Academy of Sciences, USA* **95**, 10334–10339.

Shulaev, V., Silverman, P. and Raskin, I. (1997) Airborne signalling by methyl salicylate in plant pathogen resistance. *Nature* **385**, 718–721.

Siegel, B.A. and Verbeke, J.A. (1989) Diffusable factors essential for epidermal cell redifferentiation in *Catharanthus roseus. Science* **244**, 580–582.

Sievers, A. and Schmitz, M. (1982) Röntgen-Mikroanalyse von Barium, Schwefel und Strontium in Statolithen-Kompartimenten von Chara-Rhizoiden. *Berichte der Deutschen Botanischen Gessellschaft* **95**, 353–360.

Sievers, A., Braun, M. and Monshausen, G.B. (2002) The root cap: Structure and function. In: *Plant Roots: The Hidden Half*, Waisel, Y., Eshel, A. and Kafkefi, U. (eds.), 3rd Edition. Marcel Dekker, New York, Basel, pp. 33–47.

Sievers, A.F. and True, R.H. (1912) *U.S. Department of Agricultural Bureau Plant Industry Bulletin* 232.

Silverstone, A.L., Mak, P.Y.A., Martinez, E.C. and Sun, T.-P. (1997) The new RGA locus encodes a negative regulator of gibberellin response in *Arabidopsis thaliana*. *Genetics* **146**, 1087–1099.

Silverstone, A.L., Clampaglio, C.N. and Sun, T.-P. (1998) The *Arabidopsis RGA* gene encodes a transcriptional regulator repressing the gibberellin signal transduction pathway. *Plant Cell* **10**, 155–170.

Silverstone, A.L., Jung, H.-S., Dill, A., Kawaide, H., Kamiya, Y. and Sun, T.-P. (2001) Repressing a repressor: Gibberellin-induced rapid reduction of the RGA protein in *Arabidopsis*. *Plant Cell* **13**, 1555–1566.

Simpson, R.J., Lambers, H. and Dalling, M.J. (1982) Kinetin application to roots and its effect on uptake, translocation and distribution of nitrogen in wheat (*Triticum aestivum*) grown with a split root system. *Physiologia Plantarum* **56**, 430–435.

Simpson, S.D., Ashford, D.A., Harvey, D.J. and Bowles, D.J. (1998) Short chain oligogalacturonides induce ethylene production and expression of the gene encoding aminocyclopropane-1-carboxylic acid oxidase in tomato plants. *Glycobiology* **8**, 579–583.

Sisler, E.C. (1979) Measurement of ethylene binding in plant tissue. *Plant Physiology* **64**, 538–542.

Sisler, E.C. (1980) Partial purification of an ethylene-binding component for plant tissue. *Plant Physiology* **66**, 404–406.

Skoog, F. and Miller, C.O. (1957) Chemical regulation of growth and organ formation in plant tissues cultured *in vitro*. *Symposium of the Society of Experimental Biology* **XX**, 118–131.

Smallwood, M., Beven, A., Donovan, N., Neill, S.J., Peart, J., Roberts, K. and Knox, J.P. (1994) Localization of cell wall proteins in relation to the developmental anatomy of the carrot root apex. *Plant Journal* **5**, 237–246.

Smertenko, A.P., Bozhkov, P.V., Filonova, L.H., von Arnold, S. and Hussey, P.J. (2003) Re-organisation of the cytoskeleton during developmental programmed cell death in *Picea abies* embryos. *Plant Journal* **33**, 813–824.

Smigocki, A.C. and Owens, L.D. (1988) Cytokinin gene fused with a strong promoter enhances shoot organogenesis and zeatin levels in transformed plant cells. *Proceedings of the National Academy of Sciences, USA* **85**, 5131–5135.

Spiro, M.D., Bowers, J.F. and Cosgrove, D.J. (2002) A comparison of oligogalacturonide- and auxin-induced extracellular alkalinization and growth responses in roots of intact cucumber seedlings. *Plant Physiology* **130**, 895–903.

Sponsel, V.M. (1995) The biosynthesis and metabolism of gibberellins in higher plants. In: *Plant Hormones*, Davies, P.J. (ed.), 2nd Edition. Kluwer Academic Publishers, Dordecht, pp. 66–97.

Stacey, N.J., Roberts, K. and Knox, J.P. (1990) Patterns of expression of the JIM4 arabinogalactan protein epitope in cell cultures and during somatic embryogenesis in *Daucus carota* L. *Planta* **180**, 285–292.

Stange, L. and Osborne, D.J. (1988) Cell specificity in auxin- and ethylene-induced 'super growth' in *Riella helicophylla*. *Planta* **175**, 341–347.

Staswick, P.E., Su, W. and Howell, S.H. (1992) Methyl jasmonate inhibition of root growth and induction of a leaf protein are decreased in an *Arabidopsis thaliana* mutant. *Proceedings of the National Academy of Sciences, USA* **89**, 6837–6840.

Staswick, P.E., Yuen, G.Y. and Lehman, C.C. (1998) Jasmonate signalling mutants of *Arabidopsis* are susceptible to the soil fungus, *Pythium irregulare*. *Plant Journal* **15**, 747–754.

Staswick, P.E., Tiryaki, I. and Rowe, M.L. (2002) Jasmonate response locus *JAR1* and several related *Arabidopsis* genes encode enzymes of the firefly luciferase superfamily that show activity on jasmonic, salicylic, and indole-3-acetic acids in an assay for adenylation. *Plant Cell* **14**, 1405–1415.

Steeves, T.A. and Sussex I.M. (1989) *Patterns in Plant Development*, 2nd edition. Cambridge University Press, Cambridge, U.K.

Steffens, B., Feckler, C., Palme, K., Christian, M., Bötter, M. and Lüthen, H. (2001) The auxin signal for protoplast swelling is perceived by extracellular ABP1. *Plant Journal* **27**, 591–599.

Stein, J.C., Howlett, B., Boyes, D.C., Nasrallah, M.E. and Nasrallah, J.B. (1991) Molecular cloning of a putative receptor protein kinase gene encoded at the self-incompatability locus of *Brassica oleracea*. *Proceedings of the National Academy of Sciences, USA* **88**, 8816–8820.

Steinmann, T., Geldner, N., Grebe, M., Mangold, S., Jackson, C.L., Paris, S., Galweiler, L., Palme, K. and Jurgens, G. (1999) Coordinated polar localization of auxin efflux carrier PIN1 by GNOM ARF GEF. *Science* **286**, 316–318.

Stewart, R.N., Meyer, F.G. and Desmene, H. (1972) Camellia + 'Daisy Eggleson' a graft chimera of *Camellia sasangua* and *C. japonica*. *American Journal of Botany* **59**, 515–524.

Stintzi, A. and Browse, J. (2000) The *Arabidopsis* male-sterile mutant *opr3* lacks the 12-oxophytodienoic acid reductase required for jasmonate biosynthesis. *Proceedings of the National Academy of Sciences, USA* **97**, 10625–10630.

Stintzi, A., Weber, H., Reymond, P., Browse, J. and Farmer, E.E (2001) Plant defense in the absence of jasmonic acid: The role of cyclopentenones. *Proceedings of the National Academy of Sciences, USA* **98**, 12837–12842.

Stratmann, J.W. (2003) Long distance run in the wound response – jasmonic acid is pulling ahead. *Trends in Plant Science* **8**, 247–250.

Stratmann, J.W. and Ryan, C.A. (1997) Myelin basic protein kinase activity in tomato leaves is induced systemically by wounding and increases in response to systemin and oligosaccharide elicitors. *Proceedings of the National Academy of Sciences, USA* **94**, 11085–11089.

Su, W. and Howell, S.H. (1992) A single genetic locus, *Ckr1*, defines *Arabidopsis* mutants in which root growth is resistant to low concentrations of cytokinin. *Plant Physiology* **99**, 1569.

Sun, T.-P. (2000) Gibberellin signal transduction. *Current Opinion in Plant Biology* **3**, 374–380.

Sun, T.-P. and Kamiya, Y. (1994) The *Arabidopsis* GA1 locus encodes the cyclase ent-kaurene synthetase A of gibberellin biosynthesis. *Plant Cell.* **6**, 1509–1518.

Surplus, S.L., Jordan, B.R., Murphy, A.M., Carr, J.P., Thomas, B. and Mackerness, A.-H. (1998) Ultraviolet-B induced responses in *Arabidopsis thaliana*: Role of salicylic acid and reactive oxygen species in the regulation of transcripts encoding photosynthetic and acidic pathogenesis-related proteins. *Plant, Cell and Environment* **21**, 685–694.

Suzuki, T., Sakurai, K., Ueguchi, C. and Mizuno, T. (2001a) Two types of putative nuclear factors that physically interact with histidine-containing phototransfer (Hpt) domains, signalling mediators in histo-Asp phosphorelay, in *Arabidopsis thaliana. Plant Cell Physiology* **42**, 37–45.

Suzuki, T., Miwa, K., Ishikawa, K., Yamada, H., Aiba, H. and Mizuno, T. (2001b) The *Arabidopsis* sensor His-kinase, AHK4, can respond to cytokinins. *Plant Cell Physiology* **42**, 107–113.

Suzuki, Y., Kitagawa, M., Knox, J.P. and Yamaguchi, I. (2002) A role for arabinogalactan proteins in gibberellin-induced α-amylase production in barley aleurone cells. *Plant Journal* **29**, 733–741.

Swarup, R., Friml, J., Marchant, A., Ljung, K., Sandberg, G., Palme, K. and Bennett, M. (2001) Localisation of the auxin permease AUX1 suggests two functionally distinct hormone transport pathways operate in the *Arabidopsis* root apex. *Genes and Development* **15**, 2648–2653.

Sweere, U., Eichenberg, K., Lohrmann, J., Mira-Rodado, V., Baurle, I., Kudla, J., Nagy, F., Schafer, E. and Harter, K. (2001) Interaction of the response regulator ARR4 with phytochrome B in modulating red light signalling. *Science* **294**, 1108–1111.

Szekeres, M., Németh, K., Koncz-Kálmán, Z., Mathur, J., Kauschmann, A., Altmann, T., Rédei, G.P., Nagy, F., Schell, J. and Koncz, C. (1996) Brassinosteroids rescue the deficiency of CYP90, a cytochrome P450, controlling cell elongation and de-etiolation in *Arabidopsis. Cell* **85**, 171–182.

Sztein, A.E., Ilic, N., Cohen, J.D. and Cooke, T.J. (2002) Indole-3-acectic acid biosynthesis in isolated axes from germinating bean seeds: The effect of wounding on the biosynthetic pathway. *Plant Growth Regulation.* **36**, 201–207.

Szymanski, D.B. and Marks, M.D. (1998) GLABROUS1 over expression and TRIPTYCHON alter the cell cycle and trichome fate in *Arabidopsis. Plant Cell* **10**, 2047–2062.

Szymkowiak, E.J. and Irish, E.E. (1999) Interactions between *jointless* and wild-type tomato tissues during development of the pedicel abscission zone and the inflorescence meristem. *Plant Cell* **11**, 159–176.

Szymkowiak, E.J. and Sussex, I.M. (1992) The internal meristem layer (L3) determines floral meristem size and carpel number in tomato periclinal chimeras. *Plant Cell* **4**, 1089–1100.

Tajima, Y., Imamura, A., Kiba, T., Amano, Y., Yamashino, T. and Mizuno T. (2004) Comparative studies on the type-B response regulators revealing their distinctive properties in the His-to-Asp phospho-relay signal transduction of *Arabidopsis thaliana. Plant Cell Physiology* **45**, 28–39.

Takahashi, H., Saito, T. and Suge, H. (1983) Separation of the effects of photoperiod and hormones on sex expression in cucumber. *Plant and Cell Physiology* **24**, 147–154.

Takahashi, H., Kobayashi, T., Sato-Nara, K., Tomita, K.-O. and Ezura, H. (2002) Detection of ethylene receptor protein Cm-ERS1 during fruit development in melon (*Cucumis melo L.*). *Journal of Experimental Botany* **53**, S415–422.

Takasaki, T., Hatakeyama, K., Suzuki, G., Watanabe, M., Isogai, A. and Hinata, K. (2000) The *S* receptor kinase determines self-incompatability in *Brassica* stigma. *Nature* **403**, 913–916.

Takayama, S., Shimosato, H., Shiba, H., Funato, M., Che, F.S., Watanabe, M., Iwano, M. and Isogai, A. (2001) Direct ligand-receptor complex interaction controls *Brassica* self-incompatability. *Nature* **413**, 534–538.

Takeda, T., Furuta, Y., Awano, T., Mizuno, K., Mitsuishi, Y. and Hayashi, T. (2002) Suppression and acceleration of cell elongation by integration of xyloglucans in pea stem segments. *Proceedings of the National Academy of Sciences, USA* **99**, 9055–9060.

Takei, K., Sakakibara, H. and Sugiyama, T. (2001) Identification of genes encoding adenylate isopentenyltransferase, a cytokinin biosynthesis enzyme in *Arabidopsis thaliana*. *Journal of Biological Chemistry* **276**, 26405–26410.

Takei, K., Takahashi, T., Sugiyama, T., Yamaya, T. and Sakakibara, H. (2002) Multiple routes communicating nitrogen availability from roots to shoots: A signal transduction pathway mediated by cytokinin. *Journal of Experimental Botany* **53**, 971–977.

Tal, M. and Nevo, Y. (1973) Abnormal stomatal behaviour and root resistance, and hormonal imbalance in three wilty mutants of tomato. *Biochemical Genetics* **8**, 291–300.

Tang, W., Ezcurra, I., Muschietti, J. and McCormick, S. (2002) A cysteine-rich extracellular protein, LAT52, interacts with the extracellular domain of the pollen receptor kinase LePRK2. *Plant Cell* **14**, 2277–2287.

Taylor, J.G., Owen Jr., T.P., Koonce, L.T. and Haigler, C.H. (1992) Dispersed lignin in tracheary elements treated with cellulose synthesis inhibitors provides evidence that molecules of the secondary cell wall mediate wall patterning. *Plant Journal* **2**, 959–970.

The *Arabidopsis* Genome Initiative (2000) Analysis of the genome sequence of the flowering plant *Arabidopsis thaliana*. *Nature* **408**, 796–815.

Tieman, D. and Klee, H. (1999) Differential expression of two novel members of the tomato ethylene-receptor family. *Plant Physiology* **120**, 165–172.

Tieman, D.V., Taylor, M.G., Ciardi, J.A. and Klee, H.J. (2000) The tomato ethylene receptors NR and LeETR4 are negative regulators of ethylene response and exhibit functional compensation within a multigene family. *Proceedings of the National Academy of Sciences, USA* **97**, 5663–5668.

Thelen, M.P. and Northcote, D.H. (1989) Identification and purification of a nuclease from *Zinnia elegans* L.: A potential molecular marker for xylogenesis. *Planta* **179**, 181–195.

Theologis, A., Huynh, T.V. and Davis, R.W. (1985) Rapid induction of specific mRNAs by auxin in pea epicotyl tissue. *Journal of Molecular Biology* **183**, 53–68.

Thiel, G., Blatt, M.R., Fricker, M.D., White, I.R. and Millner, P. (1993) Modulation of K+ channels in *Vicia* stomatal guard cells by peptide homologs to the auxin binding protein C-terminus. *Proceedings of the National Academy of Sciencse, USA* **90**, 11493–11497.

Thimann, K.V. (1980) *Senescence in Plants*. CRC Press Inc., Boca Raton, FL, 276 pp.

Thomas, H., Ougham, H.J., Wagstaff, C. and Stead, A.D. (2003) Defining senescence and death. *Journal of Experimental Botany* **54**, 1127–1132.

Thompson, D.S. and Osborne, D.J. (1994) A role for the stele in intertissue signaling in the initiation of abscission in bean leaves (*Phaseolus vulgaris* L.). *Plant Physiology* **105**, 341–347.

Thompson, D.S, Davies, W.J. and Ho, L.C. (1998) Regulation of tomato fruit growth by epidermal cell wall enzymes. *Plant Cell and Environment* **21**, 589–599.

Tillmann, U., Viola, G., Kayser, B., Seimeister, G., Hesse, T., Palme, K., Löbler, M. and Klämbt, D. (1989) cDNA clones of the auxin binding protein from corn coleoptiles (*Zea mays* L.): Isolation and characterization by immunological methods. *EMBO Journal* **8**, 2463–2467.

Timpte, C. (2001) Auxin binding protein: Curiouser and curiouser. *Trends in Plant Science* **6**, 586–590.

Tiryaki, I. and Staswick, P. (2002) An *Arabidopsis* mutant defective jasmonate response is allelic to the auxin-signaling mutant *axr1*. *Plant Physiology* **130**, 887–894.

Tiwari, S.B., Wang, X.-J., Hagen, G. and Guilfoyle, T.J. (2001) Auxin/IAA proteins are active repressors, and their stability and activity are modulated by auxin. *Plant Cell* **13**, 2809–2822.

Tran Thanh Van, K., Toubert, P., Cousson, A., Darvill, A.G., Gollin, D.J., Chelf, P. and Albersheim, P. (1985) Manipulation of the morphogenetic pathway of tobacco explants by oligosaccharins. *Nature* **314**, 615–617.

Traw, M.B., Bergelson, J. (2003) Interactive effects of jasmonic acid, salicylic acid and gibberellin on induction of trichomes in *Arabidopsis*. *Plant Physiology* **133**, 1367–1375.

Turner, S.R. and Somerville, C.R. (1997) Collapsed xylem phenotype of *Arabidopsis* identifies mutants deficient in cellulose deposition in the secondary cell wall. *Plant Cell* **9**, 689–701.

Turner, J.G., Ellis, C. and Devoto, A. (2002) The jasmonate signal pathway. *Plant Cell*, **14**, Supplement, S153–S164.

Ueda, J. and Kato, J. (1980) Isolation and identification of a senescence-promoting substance from wormwood (*Artemisia absinthium* L.). *Plant Physiology* **66**, 246–249.

Ulmasov, T., Hagen, G. and Guilfoyle, T.J. (1997a) ARF1, a transcription factor that binds to auxin response elements. *Science* **276**, 1865–1868.

Ulmasov, T., Murfett, J., Hagen, G. and Guilfoyle, T. J. (1997b) Aux/IAA proteins repress expression of reporter genes containing natural and highly active synthetic auxin reponse elements. *Plant Cell* **9**, 1963–1971.

Ulmasov, T., Hagen, G. and Guilfoyle, T.J. (1999a) Activation and repression of transcription by auxin-response factors. *Proceedings of the National Academy of Sciences, USA* **96**, 5844–5849.

Ulmasov, T., Hagen, G. and Guilfoyle, T.J. (1999b) Dimerization and DNA binding of auxin response factors. *Plant Journal* **19**, 309–319.

Urao, T., Yakubov, B., Yamaguchi-Shinozaki, K. and Shinozaki, K. (1998) Stress-responsive expression of genes for two-component response regulator-like proteins in *Arabidopsis thaliana*. *FEBS Letters* **427**, 175–178.

Urao, T., Miyata, S., Yamaguchi-Shinozaki, K. and Shinozaki, K. (2000) Possible His-to-Asp phospho-relay signalling in an *Arabidopsis* two-component system. *FEBS Letters* **478**, 227–232.

Vahatalo, M. and Virtanen, A. (1957) A new cyclic α-aminocarboxylic acid in berries of cowberry. *Acta Chemica Scandinavica* **11**, 741–756.

van de Sande, K., Pawlowski, K., Czaja, I., Wieneke, U., Schell, J., Schmidt, J., Walden, R., Matvienko, M., Wellink, J., van Kammen, A., Franssen, H. and Bisseling, T. (1996) Modification of phytohormone response by a peptide encoded by ENOD40 of legumes and a nonlegume. *Science* **273**, 370–373.

van den Berg, C., Willemsen, V., Hage, W., Weisbeck, P. and Scheres, B. (1995) Cell fate in the *Arabidopsis* root meristem determined by directional signalling. *Nature* **378**, 62–65.

Van Der Schoot, C., Dietrich, M.A., Storms, M., Verbeke, J.A. and Lucas, W.J. (1995) Establishment of cell-to-cell communication pathway between separate carpels during gynoecium development. *Planta* **195**, 450–455.

Van Doorn, W. and Stead, A. (1997) Abscission of flowers and floral parts. *Journal of Experimental Botany* **48**, 821–837.

Van Hengel, A.J., Tadesse, Z., Immerzeel, P., Schols, H., van Kammen, A. and de Vries, S.C. (2001) *N*-acetylglucosamine and glucosamine-containing arabinogalactan proteins control somatic embryogenesis. *Plant Physiology* **125**, 1880–1890.

Van Huystee, R.B. and McManus, M.T. (1998) Glycans of higher plant peroxidases: Recent observations and future speculations. *Glycoconjugate Journal* **15**, 101–106.

Van Overbeek, J. and Went, F.W. (1937) Mechanism and quantitative application of the pea test. *Botanical Gazette* **99**, 22–41.

Van Overbeek, J., Conklin, M.E. and Blakeslee, A.F. (1941) Factors in coconut milk essential for growth and development of *Datura* embryos. *Science* **94**, 350–351.

Veit, B., Briggs, S.P., Schmidt, R.J., Yanofsky, M.F. and Hake, S. (1998) Regulation of leaf initiation by the terminal ear 1 gene of maize. *Nature* **393**, 166–168.

Venis, M.A. and Napier, R.M. (1995) Auxin receptors and auxin binding proteins. *Critical Reviews in Plant Sciences* **14**, 27–47.

Venis, M.A., Napier, R., Barbier-Brygoo, H., Maurel, C., Perrit-Rechenmann, C. and Guern, J. (1992) Antibodies to a peptide from the maize auxin-binding protein have auxin agonist activity. *Proceedings of the National Academy of Sciences, USA* **89**, 7208–7212.

Verbeke, J.A. (1992) Fusion events during floral morphogenesis. *Annual Review of Plant Physiology and Plant Molecular Biology* **43**, 583–598.

Verbeke, J.A. and Walker, D.B. (1985) Rate of induced cellular dedifferentiation in *Catharanthus roseus*. *American Journal of Botany* **72**, 1314–1317.

Vernooij, B., Friedrich, L., Morse, A., Reist, R., Kolditz-Jawhar, R., Ward, E., Uknes, S., Kessmann, H. and Ryals, J. (1994) Salicylic acid is not the translocated signal responsible for inducing systemic acquired resistance but is required in signal transduction. *Plant Cell* **6**, 959–965.

Verpy, E., Leibovici, M. and Petot, C. (1999) Characterization of otoconin-95, the major protein of murine otoconia, provides insights into the formation of these inner ear biomaterials. *Proceedings of the National Academy of Sciences, USA* **96**, 529–534.

Vick, B.A. and Zimmerman, D.C. (1984) Biosynthesis of jasmonic acid by several plant species. *Plant Physiology* **75**, 458–461.

Voesenek, L.A.C.J., Banga, M., Their, R.H., Mudde, C.M., Harren, F.M., Barendse, G.W.M. and Blom, C.W.P.M. (1993) Submergence-induced ethylene synthesis, entrapment, and growth in two plant species with contrasting flooding resistance. *Plant Physiology* **103**, 783–791.

Vogel, J.P., Woeste, K.E., Theologis, A. and Kieber, J.J. (1998) Recessive and dominant mutations in the ethylene biosynthetic gene *ACS5* of *Arabidopsis* confer cytokinin insensitivity and ethylene over-production, respectively. *Proceedings of the National Academy of Sciences, USA* **95**, 4766–4771.

von Groll, U. and Altmann, T. (2001) Stomatal cell biology. *Current Opinion in Cell Biology* **4**, 555–560.

von Groll, U., Berger, D. and Altmann, T. (2002) The subtilisin-like serine protease SDD1 mediates cell-to-cell signalling during *Arabidopsis* stomatal development. *Plant Cell* **14**, 1527–1539.

Voznesenskaya, E.V., Edwards, G.E., Kiirats, O., Artyusheva, E.G. and Franceschi, V.R. (2003) Development of biochemical specialization and organelle partitioning in the single-cell C4 system in leaves of *Borszczowia aralocaspica* L. (Chenopodiaceae). *American Journal of Botany* **90**, 1669–1680.

Vriezen, W.H., van Rijn, C.P.E., Voesenek, A.C.J. and Mariani, C. (1997) A homolog of the *Arabidopsis thaliana ERS* gene is actively regulated in *Rumex palustris* upon flooding. *Plant Journal* **11**, 1265–1271.

Vriezen, W.H., de Graaf, B., Mariani, C. and Voesenek, L.A.C.J. (2000) Submergence induces expansin gene expression in flooding-tolerant *Rumex palustris* and not in flooding-intolerant *R. acetosa*. *Planta* **210**, 956–963.

Wada, T, Tachibana, T., Shimura, Y. and Okada, K. (1997) Epidermal cell differentiation in *Arabidopsis* determined by a myb homolog, CPC. *Science* **277**, 1113–1116.

Walker, J.C. and Key, J.L. (1982) Isolation of cloned cDNAs to auxin-responsive poly(A) + RNAs of elongating soybean hypocotyl. *Proceedings of the National Academy of Sciences, USA* **79**, 7185–7189.

Walker, L.M. and Sack, F.D. (1990) Amyloplasts as possible statoliths in gravitropic protonemata of the moss *Ceratodon purpureus*. *Planta* **181**, 71–77.

Wang, M., Oppedijk, B., Lu, X., Van Duijn, B. and Schilperoort, R.A. (1996) Apoptosis in barley aleurone during germination and its inhibition by abscisic acid. *Plant Molecular Biology* **32**, 1125–1134.

Wang, M., Oppedijk, B.J., Caspers, M.P.M., Lamers, G.E.M., Boot, M.J., Geerlings, D.N.G., Bakhuizen, B., Meijer, A.J. and Van Duijin, B. (1998) Spatial and temporal regulation of DNA fragmentation in aleurone of germinating barley. *Journal of Experimental Botany* **49**, 1293–1301.

Wang, W., Hall, A.E., O'Malley, R. and Bleecker, A.B. (2003) Canonical histidine kinase activity of the transmitter domain of the ETR1 ethylene receptor from *Arabidopsis* is not required for signal transmission. *Proceedings of the National Academy of Sciences, USA* **100**, 352–357.

Wang, Z.-Y. and He, J.-X. (2004) Brassinosteroid signal transduction – choices of signals and receptors. *Trends in Plant Science* **9**, 91–96.

Wang, Z.-Y., Seto, H., Fujioka, S., Yoshida, S. and Chory, J. (2001) BRI-1 is a critical component of a plasma-membrane receptor for plant steroids. *Nature* **410**, 380–383.

Wang, Z.-Y., Nakano, T., Gendron, J., He, J., Chen, M., Vafeados, D., Yang, Y., Fujioka, S., Yoshida, S., Asami, T. and Chory, J. (2002) Nuclear-localized BZR1 mediates brassinosteroid-induced growth and feedback suppression of brassinosteroid biosynthesis. *Developmental Cell* **2**, 505–513.

Warneck, H. and Seitz, H.U. (1993) Inhibition of gibberellic acid-induced elongation-growth of pea epicotyls by xyloglucan oligosaccharides. *Journal of Experimental Botany* **44**, 1105–1109.

Warren-Wilson, J., Roberts, L.W., Warren-Wilson, P.M. and Gresshoff, P.M. (1994) Stimulatory and inhibiting effects of sucrose concentration in xylogenesis in lettuce pith explants: Possible mediation by ethylene biosynthesis. *Annals of Botany* **73**, 65–73.

Warwicker, J. (2001) Modelling of auxin-binding protein 1 suggests that its C-terminus and auxin could compete for a binding site that incorporates a metal ion and tryptophan residue 44. *Planta* **212**, 343–347.

Wasternack, C. and Parthier, B. (1997) Jasmonate-signalled plant gene expression. *Trends in Plant Sciences* **2**, 302–307.

Watanabe, A. and Imaseki, H. (1982) Changes in translatable mRNA in senescing wheat leaves. *Plant and Cell Physiology* **23**, 489–497.

Wayne, R., Staves, M.P. and Leopold, A.C. (1992) The contribution of the extracellular matrix to gravisensing in *Characean* cells. *Journal of Cell Science* **101**, 611–623.

Webster, B.D. and Leopold, A.C. (1972) Stem abscission in *Phaseolus vulgaris* explants. *Botanical Gazette* **133**, 292.

Wei, N., Kwok, S.F., von Arnim, A.G., Lee, A., McNellis, T.W., Piekas, B. and Deng, X.W. (1994) *Arabidopsis* COP8, COP10 and COP11 genes are involved in repression of photomorphogenic development in darkness. *Plant Cell* **6**, 629–643.

Wen, C.-K. and Chang, C. (2002) *Arabidopsis* RGL1 encodes a negative regulator of gibberellin responses. *Plant Cell* **14**, 87–100.

Wendehemme, D., Pugin, A., Klessig, D.F. and Durner, J. (2001) Nitric oxide: comparative synthesis and signaling in animal and plant cells. *Trends in Plant Science* **6**, 177–183.

Went, F.W. (1928) Wuchsstoff and Wachstum. *Recueil des Travaux Botaniques Neerlandais* **25**, 1–116.

Went, F.W. (1936) Allgemaine betrachtungen über das auxin-problem. *Biologishes Zentralblatt* **56**, 449–463.

Wenzel, C.L., Chandler, P.M., Cunningham, R.B. and Passioura, J.B. (1997) Characterization of the leaf epidermis of barley (*Hordeum vulgare* L. Himalaya). *Annals of Botany* **79**, 41–46.

Weterings, K., Apuya, N.R., Bi, Y., Fisher, R.L., Harada, J.J. and Goldberg, R.B. (2001) Regional localization of suspensor mRNAs during early embryo development. *Plant Cell* **13**, 2409–2425.

Whitelaw, C.A., Lyssenko, N.N., Chen, L., Zhou, D., Mattoo, A.K. and Tucker, M.L. (2002) Delayed abscission and shorter internodes correlate with a reduction in the ethylene receptor *LeETR1* transcript in transgenic tomato. *Plant Physiology* **128**, 978–987.

Whiting, P. and Goring, D.A.I. (1983) The composition of the carbohydrates in the middle lamella and secondary wall of tracheids from black spruce wood. *Canadian Journal of Chemistry* **61**, 506–508.

Wildon, D.C., Thain, J.F., Minchin, P.E.H., Gubb, I.R., Reilly, A.J., Skipper, Y.D., Doherty, H.M., O'Donnell, P.J. and Bowles, D.J. (1992) Electrical signalling and systemic proteinase inhibitor induction in the wounded plant. *Nature* **360**, 62–65.

Wilkinson, J.Q., Lanahan, M.B., Yen, H.-C., Giovannoni, J.J. and Klee, H.J. (1995) An ethylene-inducible component of signal transduction encoded by Never-ripe. *Science* **270**, 1807–1809.

Wilkinson, S. and Davies, W.J. (2002) ABA-based chemical signalling: The co-ordination of responses to stress in plants. *Plant, Cell and Environment* **25**, 195–210.

Willats, W.G.T. and Knox, J.P. (1996) A role for arabinogalactan-proteins in plant cell expansion: Evidence from studies on the interaction of β-glycosyl Yariv reagent with seedlings of *Arabidopsis thaliana. Plant Journal* **9**, 919–925.

Willats, W.G.T., McCartney, L. and Knox, J.P. (2001a) *In situ* analysis of pectic polysaccharides in seed mucilage and at the root surface of *Arabidopsis thaliana. Planta* **213**, 37–44.

Willats, W.G.T., Orfila, C., Limberg, G., Buchholt, H.C., van Alebeek, G-J. W.M., Voragen, G.J., Marcus, S.E., Christensen, T.M.I.E., Mikkelson, J.D., Murray, B.S. and Knox, J.P. (2001b) Modulation of the degree and pattern of methyl-esterification of pectic homogalacturanan in plant cell walls: Implications for pectin methylesterase action, matrix properties, and cell adhesion. *Journal of Biological Chemistry* **276**, 19404–19413.

Williams, R.W., Wilson, J.M. and Meyerowitz, E.M. (1997) A possible role for kinase-associated protein phosphatase in the *Arabidopsis* CLAVATA1 signalling pathway. *Proceedings of the National Academy of Sciences, USA* **94**, 10467–10472.

Williamson, R.E. (1991) Orientation of cortical microtubules in interphase plant cells. *International Review of Cytology* **129**, 135–206.

Wilson, M.A., Sawyer, J., Hatcher, P.G. and Lerch, H.E. (1989) 1,3,5-hydroxybenzene structures in mosses. *Phytochemistry* **28**, 1395–1400.

Wilson, M.P.K. and Bruck, D.K. (1999) Lack of influence of the epidermis on underlying cell development in leaflets of *Pisum sativum* var. argenteum (Fabaceae). *Annals of Botany* **83**, 1–10.

Wisman, E., Cardon, G.H., Fransz, P. and Saedler, H. (1998) The behaviour of the autonomous maize transposable element *En/Spm* in *Arabidopsis thaliana* allows efficient mutagenesis. *Plant Molecular Biology* **37**, 989–999.

Wolbang, C.M., Chandler, P.M., Smith, J.J. and Ross, J.J. (2004) Auxin and the developing inflorescence is required for the biosynthesis of active gibberellins in barley stems. *Plant Physiology* **134**, 769–776.

Wong, C.H. and Osborne, D.J. (1978) The ethylene-induced enlargement of target cells in flower buds of *Ecballium elaterium* (L.) A. Rich. and their identification by the content of endo-reduplicated DNA. *Planta* **139**, 103–111.

Wong, L.M., Abel, S., Shen, N., de la Foata, M., Mall, Y. and Theologis, A. (1996) Differential activation of the primary auxin response genes, *PS-IAA4/5* and *PS-IAA6*, during early plant development. *Plant Journal* **9**, 587–600.

Woodward, F.I. and Kelly, C.K. (1995) The influence of CO_2 concentration on stomatal density. *New Phytologist* **131**, 311–327.

Worley, C.K., Zenser, N., Ramos, J., Rouse, D., Leyser, O., Theologis, A. and Callis, J. (2000) Degradion of aux/IAA proteins is essential for normal auxin signalling. *Plant Journal* **21**, 553–562.

Wright, A.D., Sampson, M.B., Neuffer, M.G., Michalczuk, L., Slovin, J.P. and Cohen, J.D. (1991) Indole-3-acetic acid biosynthesis in the mutant maize *orange pericarp*, tryptophan auxotroph. *Science* **254**, 998–1000.

Wright, M. (1982) The polarity of movement of endogenously produced IAA in relation to a gravity perception mechanism. *Journal of Experimental Botany* **33**, 929–934.

Wright, M. (1986) The acquisition of gravisensitivity during the development of nodes of *Avena fatua*. *Journal of Plant Growth Regulation* **5**, 37–47.

Wright, M. and Osborne, D.J. (1974) Abscission in *Phaseolus vulgaris*: The positional differentiation and ethylene-induced expansion growth of specialised cells. *Planta* **120**, 163–170.

Wright, M., Mousdale, D.M.A. and Osborne, D.J. (1978) Evidence for a gravity-regulated level of endogenous auxin controlling cell elongation and ethylene production during geotropic bending in grass nodes. *Biochemistry, Physiology Pflanzen* **172**, 581–596.

Wright, S.T.C. and Hiron, R.W.P. (1969) (+)-Abscisic acid, the growth inhibitor induced in detached wheat leaves by a period of wilting. *Nature* **224**, 719–720.

Xie, D.-X., Fey, B.F., James, S., Nieto-Rostro, M. and Turner, J.G. (2003) *COI1*: An *Arabidopsis* gene required for jasmonate-regulated defense and fertility. *Science* **280**, 1091–1094.

Xu, D.P., Duan, X., Wang, B., Hong, B., Ho, T.H.D. and Wu, R. (1996) Expression of a late embryogenesis abundant protein gene HVA1 from barley confers tolerance to water deficit and salt stress in transgenic rice. *Plant Physiology* **110**, 249–257.

Xu, W., Purugganan, M.M., Polisensky, D.H., Antosiewicz, D.M., Fry, S.C. and Braam, J. (1995) *Arabidopsis TCH4*, regulated by hormones and the environment, encodes a xyloglucan endotransglycosylase. *Plant Cell* **7**, 1555–1567.

Yabata, T. and Sumiki, Y. (1938) Biochemical studies on "Bakanae" fungus. Crystals with plant growth promoting activity. *Journal of the Agricultural Chemistry Society, Japan* **14**, 1526.

Yamagami, M., Haga, K., Napier, R.M. and Iino, M. (2004) Two distinct signalling pathways participate in auxin-induced swelling of pea epidermal protoplasts. *Plant Physiology* **134**, 735–747.

Yamasaki, S., Fujii, N. and Takahashi, H. (2000) The ethylene-regulated expression of *CS-ETR2* and *CS-ERS* genes in cucumber plants and their possible involvement with sex expression in flowers. *Plant Cell Physiology* **41**, 608–616.

Yamazaki, T., Takaoka, M., Katoh, E., Hanada, K., Sakita, M., Sakata, K., Nishiuchi, Y. and Hirano, H. (2003) A possible physiological function and the tertiary structure of a 4-kDa peptide in legumes. *European Journal of Biochemistry* **270**, 1269–1276.

Yang, H., Matsubayashi, Y., Nakamura, K. and Sakagami, Y. (1999) *Oryza sativa PSK* gene encodes a precursor of phytosulfokine-α, a sulfated peptide growth factor found in plants. *Proceedings of the National Academy of Sciences, USA* **96**, 13560–13565.

Yang, H., Matsubayashi, Y., Nakamura, K. and Sakagami, Y. (2001) Diversity of *Arabidopsis* genes encoding precursors for phytosulfokine, a peptide growth factor. *Plant Physiology* **127**, 842–851.

Yang, S.F. and Hoffman, N.E. (1984) Ethylene biosynthesis and its regulation in higher plants. *Annnual Review of Plant Physiology* **35**, 155–189.

Yao, C., Conway, W.S. and Sams, C.E. (1995) Purification and characterization of a polygalacturonase-inhibiting protein from apple fruit. *Phytopathology* **85**, 1373–1377.

Ye, Z.-H. (2002) Vascular tissue differentiation and pattern formation in plants. *Annual Review of Plant Biology* **53**, 183–202.

Ye, Z.-H., Zhong, R., Morrison, W.H. and Himmelsbank, D.S. (2001) Caffeoyl coenzyme A O-methyltransferase and lignin biosynthesis. *Phytochemistry* **57**, 1177–1185.

Yin, Y., Wang, Z.-Y., Mora-Garcia, S., Li, J., Yoshida, S., Asami, T. and Chory, J. (2002) BES1 accumulates in the nucleus in response to brassinosteroids to regulate gene expression and promote stem elongation. *Cell* **109**, 181–191.

Yokota, T. (1997) The structure, biosynthesis and function of brassinosteroids. *Trends in Plant Sciences* **2**, 137–143.

York, W.S., Darvill, A.G. and Albersheim, P. (1984) Inhibition of a 2,4-dichlorophenoxyacetic acid-stimulated elongation of pea stem segments by a xyloglucan oligosaccharide. *Plant Physiology* **75**, 295–297.

Youl, J.J., Bacic, A. and Oxley, D. (1998) Arabinogalactan-proteins from *Nicotiana alata* and *Pyrus communis* contain glycosylphosphatidylinositol membrane anchors. *Proceedings of the National Academy of Sciences, USA* **95**, 7921–7926.

Young, T.E. and Gallie, D.R. (1999) Analysis of programmed cell death in wheat endosperm reveals differences in endosperm development between cereals. *Plant Molecular Biology* **39**, 915–926.

Young, T.E. and Gallie, D.R. (2000) Regulation of programmed cell death in maize endosperm by abscisic acid. *Plant Molecular Biology* **42**, 397–414.

Young, T.E., Gallie, D.R. and DeMason, D.A. (1997) Ethylene mediated programmed cell death during maize endosperm development of wild type and Shrunken2 genotypes. *Plant Physiology* **115**, 737–751.

Yuan, M., Warn, R.M., Shaw, P.J. and Lloyd, C.W. (1992) Dynamic microtubules under the radial and outer tangential walls of microinjected pea epidermal cells observed by computer reconstruction. *Plant Journal* **7**, 17–23.

Zablackis, E., York, W.S., Pauly, M., Hantus, S., Rieter, W.-D., Chapple, C.C.S., Albersheim, P. and Darvill, A. (1996) Substitution of L-fucose by L-galactose in cell walls of *Arabidopsis mur1*. *Science* **272**, 1808–1810.

Zeevaart, J.A.D. (1976) Physiology of flower formation. *Annual Review of Plant Physiology* **27**, 321–348.

Zenser, N., Ellsmore, A., Leasure, C. and Callis, J. (2001) Auxin modulates the degradation rate of aux/IAA proteins. *Proceedings of the National Academy of Sciences, USA* **98**, 11795–11800.

Zhang, D.-P., Wu, Z.-Y., Li, X.-Y. and Zhao, Z.-X. (2002) Purification and identification of a 42-kilodalton abscisic acid-specific-binding protein from epidermis of broad bean leaves. *Plant Physiology* **128**, 714–725.

Zhao, J., Peng, P., Schmitz, R.J., Decker, A.D., Tax, F.E. and Li, J. (2002a) Two putative BIN2 substrates are nuclear components of brassinosteroid signaling. *Plant Physiology* **130**, 1221–1229.

Zhao, Y., Christensen, S.K., Fankhauser, C., Cashman, J.R., Cohen, J.D., Weigel, D. and Chory, J. (2001) A role for flavin monooxygenase-like enzymes in auxin biosynthesis. *Science* **291**, 306–309.

Zhao, Y., Hull, A.K., Gupta, N.R., Goss, K.A., Alonso, J., Ecker, J.R., Normanly, J., Chory, J. and Celenza, J.L. (2002b) Trp-dependent auxin biosynthesis in *Arabidopsis*: Involvement of cytochrome P450s CYP79B2 and CYP79B3. *Genes and Development* **16**, 3100–3112.

Zhong, R. and Ye, Z.-H. (2001) Alteration of auxin polar transport in *Arabidopsis ifl1* mutants. *Plant Physiology* **126**, 549–563.

Zhou, D., Kalaitzis, P., Mattoo, A. and Tucker, M. (1996) The mRNA for an ETR1 homologue in tomato is constitutively expressed in the vegetative and reproductive tissues. *Plant Molecular Biology* **30**, 1331–1338.

Zureck, D.M. and Clouse, S.D. (1994) Molecular cloning and characterisation of a brassinosteroid-regulated gene from elongating soybean (*Glycine max* L.) epicotyls. *Plant Physiology* **104**, 161–170.

Zurfluh, L.L. and Guilfoyle, T.J. (1982) Auxin-induced changes in the population of translatable messenger RNA in elongating sections of soybean hypocotyl. *Plant Physiology* **69**, 332–337.

Index

tpl, 48
transdifferentiation, 43, 48, 84, 99, 114, 115, 116, 119, 127, 128, 130
trichomes, 78, 91, 95, 99, 100, 101, 104, 138, 142, 143, 158, 200
ttg, 104
TTG, 101, 104
TUNEL, 137
tunica, 48, 78
two-component signalling, 164, 168, 181, 182

ubiquitin-mediated degradation, 160, 161
 ASK, 161
 Cdc53p, 161
 EBF1, 170
 EBF2, 170
 ECR1, 161, 200
 RUB1, 161
 SCF, 161
 Skp1p, 161
 tir1, 160, 161
 TIR1, 161

Valerianella locusta, 16
Vicia faba, 66, 140
vp1, 134
vp9, 134
VSP, 27

WER, 143
wol, 183
WOL, 84, 183, 184, 185
wooden leg, 183
WUSCHEL (WUS), 50

Xanthium, 93
Xenopus, 146
XET (xyloglucan endotransglycosylase), 20, 58, 106, 112
xylogenesis, 20, 43, 115, 145
xyloglucan, 20, 30, 58, 59, 60, 106, 112, 231, 246
xyloglucans, 30, 58
 XXFG, 30, 58
 XXXG, 30, 58, 59

ypd1, 183
YUCCA, 8

Zea mays, 81, 122
ZEN1, 144
Zinnia, 20, 38, 43, 84, 99, 115, 116, 144, 145, 228
Zinnia elegans, 144
ZMK1, 154
ZMK2, 154
Zosterophyllum, 78